PRINCIPLES OF
AEROSOL TECHNOLOGY

PRINCIPLES
OF AEROSOL
TECHNOLOGY

PAUL A. SANDERS

*Research Associate, Freon Products
Laboratory, E. I. du Pont de Nemours
& Company, Wilmington, Delaware*

VAN NOSTRAND REINHOLD COMPANY

NEW YORK / CINCINNATI / TORONTO / LONDON / MELBOURNE

Van Nostrand Reinhold Company Regional Offices:
New York   Cincinnati   Chicago   Millbrae   Dallas

Van Nostrand Reinhold Company International Offices:
London   Toronto   Melbourne

Manufactured in the United States of America.

Published by Van Nostrand Reinhold Company
450 West 33rd Street, New York, N.Y. 10001

Published simultaneously in Canada by
Van Nostrand Reinhold Ltd.

15   14   13   12   11   10   9   8   7   6   5   4   3   2

TO ALICE, ALAN AND RICK

PREFACE

The "Freon"* Products Division of the Du Pont Company has provided a course in aerosol technology for over ten years. The course, designed to present a comprehensive picture of the technical areas in the aerosol field, consists of lectures combined with laboratory experiments. The lectures have been given to a cross section of aerosol technical personnel, ranging from those who needed basic instruction to experienced aerosol chemists who wished to discuss specific areas in detail. The course has been well received by the aerosol industry; therefore it seemed desirable to make the lecture material more readily available to the academic and industrial scientific community in the form of a textbook. The present volume on aerosol technology is a modified and expanded version of the lectures.

Particular emphasis in the book has been placed upon the fundamental principles that govern the characteristics and behavior of aerosols. Only through an understanding of these principles will many of the problems in the aerosol field be solved. For this reason, the basic properties of aerosols, such as spray characteristics, vapor pressure, solubility, and flammability, have been treated as individual subjects. Aerosol emulsion and foam technology, particularly in the fields of cosmetic and pharmaceutical products, has become increasingly important in recent years. Many of the fundamental concepts of surface chemistry have been reviewed in some detail in the chapters of Part Two.

In any field, there is a considerable amount of information useful to laboratory chemists which is difficult to find in the literature. Much of this type of information has been included in the present volume. Examples in the aerosol field which fall into this category are methods for determining

* "Freon" is Du Pont's registered trademark for its fluorocarbons.

the compositions of propellant blends with a specified density or vapor pressure, and the pitfalls that may be encountered in the measurement of properties such as solubility, flammability, and vapor pressure.

The class of fluorinated hydrocarbon propellants is generally referred to in this text by the Freon® nomenclature. This seemed appropriate because most of the data listed in the text for the fluorinated hydrocarbon propellants were obtained with the Freon® propellants. The Freon® compounds were the first fluorinated hydrocarbon propellants to appear on the market and were the only fluorinated hydrocarbon propellants available for a considerable number of years. At the present time, these same chemical compounds are also manufactured by a number of other companies in the world who use the same numerical nomenclature prefixed by their respective trade names.

The preparation of a book is a difficult undertaking under the best of conditions, but the task is made much easier by the assistance of others. The author has been fortunate in the help extended by his associates in the Du Pont Company. It is impossible to list all who have contributed to the book. However, the author would like to express his particular appreciation to the following four associates: Mr. N. W. Kent, Manager, Special Services and Training, for his assistance in preparing the original manuscript; Dr. R. L. McCarthy, Director of the "Freon" Products Laboratory, for his continued encouragement during the preparation of the manuscript; Dr. R. P. Ayer, Aerosol Project Leader; and the late Dr. F. T. Reed, Division Head, Aerosol Section, "Freon" Products Laboratory, for their careful and patient reading of the manuscript. In addition, the comments and suggestions of the following "Freon" Products Division personnel were helpful during the preparation of the manuscript: Mr. C. E. Kimble, Manager, Aerosol Customer Services; Dr. D. E. Kvalnes, Manager, Technical Section; Mr. A. H. Lawrence, Jr., Marketing Manager, Aerosol Propellants; and Mr. D. C. Miller, Assistant Director of Sales. Finally, the author would like to extend his thanks to the three contributors for their chapters: Mr. T. D. Armstrong, Jr., Project Leader, "Freon" Products Laboratory; Dr. J. Wesley Clayton, Jr., Hazleton Laboratories, Inc., formerly Assistant Director of the Haskell Laboratory for Toxicology and Industrial Medicine; and Dr. J. H. Fassnacht, Technical Associate in the Technical Section.

PAUL A. SANDERS

Wilmington, Delaware
December, 1969

CONTENTS

III MISCELLANEOUS

HOMOGENEOUS SYSTEMS AND THEIR PROPERTIES

1

INTRODUCTION

DEFINITION OF TERMS

The word *aerosol* is a familiar term in the United States. It encompasses practically all products that have been packaged under pressure and can be dispensed from the container as a result of pressure, including such diversified products as shaving lathers, insecticides, hair sprays, whipped cream, room deodorants, paints, and powder sprays. In the scientific realm, the word aerosol was first employed in the field of colloid chemistry in an effort to describe a suspension of small particles in air or gas in which the radius of the particles was less than 50 μ.[1] The usual range of particle size in an aerosol was considered to be from 0.01–10 μ. [A micron μ is .001 mm or .0000 39 in.]

The suspended particles could be either liquid or solid. Examples of aerosols that fall into this classification are dust, smoke, and fog. In dusts (suspensions of solids in air), the particles may have diameters as small as 0.1 μ or less. These suspensions produce haze. In fogs (suspensions of water droplets in air), the particle sizes are larger and range from about 4–50 μ in diameter. Smoke is an aerosol of solid particles, such as carbon. Carbon smoke consists of small particles with a radius of about 0.01 μ, and these will coagulate into long, irregular filaments that may reach several microns in length.[1]

In the aerosol industry, the term aerosol initially had the same meaning as it did in the field of colloid chemistry. The first major aerosol product on the market was the aerosol insecticide. The Department of Agriculture classified an aerosol insecticide as a produce in which all of the particles in the spray had a diameter less than 50 μ, and 80% of the particles had a diameter less than 30 μ.

A tentative glossary of terms for the aerosol industry was published by

3

the Chemical Specialties Manufacturers Association (CSMA) in 1955.[2] An aerosol was defined in the glossary as a suspension of fine solid or liquid particles in air or gas, as smoke, fog, or mist. This definition was similar to the scientific definition in colloid chemistry. In this glossary, a considerable distinction was made between the terms *aerosol* and *aerosol product*. The latter was defined as a "self-contained sprayable product in which the propellant force was supplied by a liquefied gas." This classification included space sprays, residual sprays, surface coatings, foams, and various other products. It did not include gas pressurized products, such as whipped cream.

In recent years, the difference between aerosol, aerosol product, or *pressurized product,* as far as general use of the terms is concerned, has disappeared. According to the most recent definition approved by the Aerosol Scientific Committee of the Chemical Specialties Manufacturers Association, an aerosol or pressurized product may be a liquid, solid, gas, or mixture thereof discharged by a propellant force of liquefied and/or nonliquefied compressed gas, usually from a disposable type of dispenser through a valve.[3] The terms *aerosol foods, aerosol dispensing, aerosol packaging,* and *aerosols* are now commonly used.

With respect to meaning, other terms have also changed in meaning as the aerosol industry has matured. *Concentrate* was originally defined in the glossary of terms as a basic ingredient or mixture of ingredients to which other ingredients, active or inactive, were added. At the present time, it is more commonly used to indicate all of the components of an aerosol product except the propellant. A *propellant* was originally defined as a liquefied gas with a vapor pressure greater than atmospheric pressure (14.7 psia) at a temperature of 105°F. The term *compressed gas propellant* is now used throughout the aerosol industry in addition to the term *liquefied gas propellant.*

COMMON AEROSOL SYSTEMS

Most aerosol products contain three major components: propellants, solvents, and active ingredients. Depending upon the manner in which these components are combined in the product, aerosols can be divided into two broad classes, homogeneous or heterogeneous. Products are homogeneous or have been formulated as homogeneous systems when all the components are mutually soluble. Homogeneous products are two-phase systems consisting of a single liquid phase in equilibrium with a vapor phase. Examples of common products that are homogeneous aerosols are hair sprays, personal deodorants, and many insecticides. Products formu-

lated as homogeneous systems do not have to be shaken before use.
Products in which all the components are not mutually soluble are classified as heterogeneous systems. There are several types of heterogeneous aerosols. Powder sprays consist of suspensions of an insoluble powder in a liquefied gas propellant. Aerosol bath powders and various foot sprays are examples of typical powder sprays.

Other heterogeneous systems contain several immiscible liquids that can be combined as an emulsion (a dispersion of one liquid in another). Products formulated as emulsion systems are normally shaken before use in order to obtain a uniform dispersion. Many space deodorants and insecticides are formulated as emulsion systems. Most aerosol foams are emulsion systems in the container and do not become foams until they are discharged.

All heterogeneous systems contain at least three phases. The powder sprays consist of a solid phase and a liquid phase in equilibrium with a vapor phase. The emulsion systems contain two liquid phases in equilibrium with a vapor phase.

FUNCTIONS OF THE AEROSOL COMPONENTS

The three main components of most aerosol products are the propellants, solvents, and active ingredients. In some products, a component may have only a single use while in others it may serve a number of purposes. Of the three components, the liquefied gas propellants and the solvents generally have the most varied functions. The propellants provide the pressure that forces the product out of the container when the valve is opened. They also influence the form in which the product is discharged—foam, stream, or spray. Variations in the type and concentration of a propellant can change a coarse, wet spray to a fine, dry spray. Similar propellant variations can also change a wet, sloppy foam to a dry, coherent foam.

Since the liquefied gas propellants are liquids, they are a part of the solvent system in all homogeneous products. Therefore, they influence the solubility properties of the liquid phase and the solvent characteristics of the propellants in the liquid state must be taken into consideration in formulating products. In some products, the propellants are also the active ingredients.

Typical solvents, such as ethyl alcohol, isopropyl alcohol, methylene chloride, methyl chloroform, or odorless mineral spirits, are usually present in homogeneous aerosols. Like the propellants, solvents generally perform a number of functions. One of the major uses of solvents is to bring the active ingredients into solution with the propellants. Most propellants have poor solubility characteristics; in many cases, the active ingredients are not soluble in the propellants. In order to obtain a homogeneous mixture, it is necessary to add a solvent with the necessary solubility properties. It is sometimes desirable to have another liquid, which is not miscible with the propellant, such as water or propylene glycol, present. In these cases, a co-solvent such as ethyl alcohol is added to obtain a homogeneous system.

Another function of solvents is to help produce a spray with a particle size that is most effective for the particular application involved. Aerosol insecticides, for example, are most efficient when the particle size is neither too large nor too small. If no solvents were present, the propellants would evaporate completely in air shortly after discharge, leaving extremely small particles. High boiling solvents evaporate only slowly in the air, once the propellant has vaporized; therefore, droplets of a specified size can be obtained by proper formulation.

Still another use of solvents in some products is to reduce the vapor pressure of the propellant. There are various pressure regulations governing the packaging of aerosol products. Some of the propellants, such as "Freon" 12, have too high a pressure to be used alone. In such a case, a vapor pressure depressant is added to reduce the pressure to an acceptable level. The vapor pressure depressant may be a solvent such as ethyl alcohol or odorless mineral spirits, or it may be a high boiling propellant like "Freon" 11, which can be considered both a solvent and a propellant.

The active ingredients are the materials (perfumes, pharmaceuticals, insecticides, deodorants, hair spray resins, etc.) essential for the specific application for which the aerosol was designed. It is the job of the aerosol chemist to combine the active ingredients, solvents, and propellants so that an efficient, attractive, and acceptable product is obtained.

In addition to the three major components (propellants, solvents, and active ingredients), aerosols usually contain a number of other materials that have been added to improve the product. Some of these, such as perfumes, enhance the aesthetic appeal while others increase the performance or efficiency of the product. Typical additives in this group include moisturizers, emollients, plasticizers, stabilizers, surfactants, and skin penetrants. Often, the formulation of a successful aerosol product is quite difficult and may require the services and knowledge of many different suppliers and companies. The perfuming of aerosols, for example, is an art that requires extensive experience and skill, and for this reason a reputable fragrance house should be consulted when questions regarding perfuming of aerosol products arise.

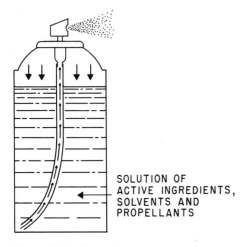

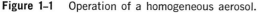

SOLUTION OF
ACTIVE INGREDIENTS,
SOLVENTS AND
PROPELLANTS

Figure 1-1 Operation of a homogeneous aerosol.

OPERATION OF AN AEROSOL

Basically, an aerosol is a product consisting of a mixture of a propellant and a concentrate that is packaged under pressure in a suitable container equipped with a valve for discharging the product when it is needed. The source of pressure in the aerosol is the propellant that is present. The liquefied gas propellants, such as the "Freon" propellants, are gases under normal conditions at room temperature and pressure. In aerosol products, however, these propellants are present as liquids and, therefore, are confined at a temperature above their boiling point; thus, the pressure in the container is produced. When the valve is opened by pushing down on the actuator, the pressure in the container forces the product up the dip tube, through the valve and out the actuator. When the aerosol leaves the actuator and enters the atmosphere, the propellants flash from a liquid to a gas. In doing so, they break the concentrate up into fine droplets. This action, illustrated in Figure 1-1, is the process that results in the formation of a spray from an aerosol.

REFERENCES

1. D. Sinclair, "Handbook on Aerosols," p. 64. Washington, D.C., 1950.
2. "Glossary of Terms Used in the Aerosol Industry," Chemical Specialties Manufacturers Association, August 25, 1955
3. Bulletin No. 211-66, Chemical Specialties Manufacturers Association, December 27, 1966.

2

HISTORICAL BACKGROUND

The present aerosol industry is generally considered to have received its stimulus from the development of the insecticide aerosols used in World War II. Although there were few aerosol products on the market before World War II, a considerable amount of information was already known about aerosols and aerosol systems. The patents issued in this field date as far back as 1862. During the years from 1862 to the 1940's, the patents that were granted to various workers show that they recognized many of the fundamental aspects of aerosols, such as the convenience of application provided by aerosol packaging, the fundamentals of aerosol formulation, the advantages of liquefied gas propellants over compressed gases, the use of three-phase systems for discharging water, the value of an expansion chamber in a valve in promoting finer particle size, and the conception of a mechanical breakup valve for decreasing the particle size.

In 1862, Lynde[1] was granted a patent for a valve with a dip tube for discharging an aerated liquid from a bottle. After this early work, there appeared to be little activity in the field of aerosols until 1899. In that year, Helbing and Pertsch[2] described a method for producing a fine jet or spray from a solution of materials, such as gums, resins, nitrocellulose, etc., in methyl or ethyl chloride. The heat of the hand caused the methyl or ethyl chloride inside the container to generate sufficient pressure so that the solution was ejected through the orifice in a fine jet or spray. The main object of the invention was to apply a thin, uniform coating of a product, such as collodion, to the skin.

Gebauer recognized the value of an expansion chamber in producing a finer spray in 1901.[3] In that year, he obtained a patent on an improved receptacle for containing and discharging volatile liquids. When the valve on the receptacle was opened, the pressure inside the container forced the liquid through the valve into a capillary tube and then into an expansion chamber. The expansion chamber ended in a nozzle that opened to the atmosphere. The diameter of the capillary tube was smaller than that of the outer nozzle.

Gebauer determined that it was the partial vaporization of the liquid in the expansion chamber that promoted a finer spray, and he also stated that the design of the expansion chamber affected the spray characteristics. The section of his unit that contained the capillary tube, expansion chamber, and outer nozzle could be removed for cleaning when the capillary tube became clogged.

Gebauer improved the design of his receptacle in 1902.[4] The capillary tube leading into the expansion chamber was eliminated and the valve between the expansion chamber and the main body of the receptacle was used as the inlet orifice. When the valve was barely opened, the diameter of the valve orifice was smaller than that of the orifice in the outer nozzle. Under these conditions, the liquid discharged as a spray. At valve openings where the valve orifice was larger in diameter than that of the outer nozzle, the product discharged as a stream. The discharge characteristics of the products were controlled by the ratios of the orifice diameters of the inner and outer orifices and this, in turn, was controlled by the degree of valve opening.

In 1903, Moore[5] obtained a patent on an atomizer for perfumes in which carbon dioxide was employed as the propellant. The atomizer was constructed so that carbon dioxide had two functions. The pressure of the carbon dioxide forced the solution of perfume up the dip tube and through a nozzle. Carbon dioxide was also discharged from the vapor phase as a gas through a separate nozzle. The carbon dioxide nozzle was arranged so that the gas passed over the tip of the perfume nozzle and created a partial vacuum, producing a fine spray or stream, depending upon the opening of the discharge valve of the perfume solution.

The propellant properties of carbon dioxide were also recognized by Mobley in 1921[6] who obtained a patent on a method for dispensing liquid antiseptics using carbon dioxide as the propellant. One of the most interesting features of Mobley's apparatus was the discharging device on his receptacle which contained the equivalent of four orifices. After the carbonated antiseptic liquid passed through the first metering orifice, it entered a chamber equipped with what Mobley described as a *block-partition* containing two small orifices. The construction of the block-partition was such that the two orifices imparted a spirally rotating motion to the jet stream before it was discharged through the outer orifice.

A patent on the use of carbon dioxide for atomizing perfumes was also granted to Lemoine[7] in 1926.

Some of the most significant advances in the field of aerosols were achieved by Rotheim[8] in 1931. Rotheim reported a method for spraying coating compositions which involved dissolving materials, such as lacquers, soaps, resins, cosmetic products, etc., in liquefied dimethyl ether in a closed container and discharging the solutions through a valve. Rotheim observed that the spray characteristics of the product could be varied from coarse to fine by changing the percentage of dimethyl ether in the formulation. He also pointed out that the volume of gas resulting when dimethyl ether was vaporized was about 350 times the volume of the initial liquefied dimethyl ether and that this was considerably larger than could be obtained with a compressed gas, such as carbon dioxide.

In a subsequent patent,[9] Rotheim reported that the use of an expansion chamber in a valve gave a considerably finer spray from a liquefied gas aerosol than a valve with a single orifice. He also stated that it was necessary to construct the valve so that the orifice leading into the expansion chamber had a smaller diameter than the orifice that opened to the atmosphere. These principles of valve construction were recognized much earlier by Gebauer. In addition, Rotheim disclosed other compounds that could be used as liquefied gas propellants, such as methyl chloride, isobutane, methyl nitrite, vinyl chloride, and ethylene chloride.

In the early 1930's, the fluorinated hydrocarbons were developed as replacements with low toxicity for sulfur dioxide and ammonia in refrigeration systems by Midgley, Henne, and McNary. Subsequently, the fluorinated hydrocarbons were recognized as effective aerosol propellants, and in 1933, a patent was granted to Midgley et al.,[10] on the use of these compounds in fire extinguishers in which the function of the propellant was to create sufficient pressure to expel itself and a flame arrester from a container. Dichlorodifluoromethane was specifically mentioned.

The use of the fluorinated hydrocarbon propellants for fire extinguishers was extended by Bichowsky in 1935,[11] who disclosed the use of compounds, such as dichlorodifluoromethane, dichloromonofluoromethane, dichlorotetrafluoroethane, etc., as propellants. These propellants were used in combination with fire extinguishing agents, such as trichloromonofluoromethane, carbon tetrachloride, sodium bicarbonate, and water. Bichowsky may have been the first to disclose that combinations, such as dichlorodifluoromethane and carbon tetrachloride, had a constant pressure during the discharge of the unit. He also disclosed systems containing mixtures of propellant and water in which the water was layered over the propellant and the standpipe extended only to the bottom of the aqueous phase. The function of the propellant was to provide sufficient pressure to discharge the water from the container.

Rotheim continued to work on aerosols and in 1938[12] was granted a patent covering insecticidal compositions in which propellants boiling below $-20°F$ were used. Rotheim claimed that the use of such compounds provided a more efficient atomization of the insecticidal aerosols. Examples of propellant mixtures that Rotheim disclosed were combinations of butane and ethane, propane and methyl chloride, ethane and dimethyl ether, propane and methane, etc.

One of the most important developments in the aerosol field occurred during World War II. During an investigation to find a way of combating the insects which caused disease among overseas troops, Goodhue and Sullivan, of the Department of Agriculture, developed a portable aerosol dispenser which used "Freon" 12 fluorocarbon as the propellant.[13]

The aerosol insecticides used by the military during World War II were dispensed from heavy steel containers which operated at about 70 psig at 70°F. These containers were made of two shells, drawn of 0.044 steel, and welded together. A brass, screw-type valve was brazed to the top center, and a blowoff release was set at 300 psig. It had a capacity of 16 oz, and was filled with 90% "Freon" 12 and 10% concentrate made of pyrethrin and sesame oil. Attached to the valve, inside the container, was a 4-in. metal dip tube, 0.017-in. inside diameter. From July 1942, when the requirement was 10,000 containers per day, until the end of the war, one company alone, Westinghouse Corporation, supplied over 30,-000,000 aerosols to the Armed Forces. Other companies engaged in supplying aerosols for the military included Regal Chemical Company, Brooklyn, New York; Airosol, Inc., Neodesha, Kansas; and Bridgeport Brass Company, Bridgeport, Connecticut. Immediately after the war, Westinghouse Corporation dropped out of the aerosol field, but the others continued to produce aerosol products.

The first aerosols for consumer use appeared on the market in 1947 and production that year was estimated at 4.3 million units, almost entirely insecticides. The containers were still the heavy, awkward military type. They were expensive and had no "sales appeal." Through the efforts of the can companies, a low-pressure container was developed, and through research at the Du Pont Company, a low-pressure propellant mixture consisting of "Freon" 12–"Freon" 11 (50/50) was found. The combination of the low-cost container and the low-pressure propellant made possible the production of aerosol products at a price that was within reach by a large section of the American public. The first aerosols were mainly insecticides, but the list of products soon began to expand as the advantages of aerosol packaging became more and more apparent.

REFERENCES

1. J. D. Lynde, U. S. Patent 34,894 (1862).
2. H. Helbing and G. Pertsch, U. S. Patent 628,463 (1899).
3. C. L. Gebauer, U. S. Patent 668,815 (1901).

4. C. L. Gebauer, U. S. Patent 711,045 (1902).
5. R. W. Moore, U. S. Patent 746,866 (1903).
6. L. K. Mobley, U. S. Patent 1,378,481 (1921).
7. R. M. L. Lemoine, D. R. Patent 532,194 (1926).
8. E. Rotheim, U. S. Patent 1,800,156 (1931).
9. E. Rotheim, U. S. Patent 1,892,750 (1933).
10. T. Midgley, Jr., A. L. Henne, and R. R. McNary, U. S. Patent 1,926,396 (1933).
11. R. R. Bichowsky, U. S. Patent 2,021,981 (1935).
12. E. Rotheim, U. S. Patent 2,128,433 (1938).
13. L. D. Goodhue, and W. N. Sullivan, U. S. Patent 2,321,023 (1943).

Reprinted from "Freon" Aerosol Report, A-69, "Historical Background of the Aerosol Industry," with permission of the copyright owner, E. I. du Pont de Nemours and Company.

3

THE
STRUCTURE
OF LIQUIDS
AND GASES

The behavior of propellants in aerosol products is governed by the same fundamental laws that apply to all other gases and liquids. A knowledge of these basic laws is necessary in order to understand how the propellants act in their various capacities in aerosol products. Such phenomena as the liquefaction of gases, the formation of sprays, the effect of solvents upon vapor pressure of the propellants, etc., become much easier to understand when considered on the basis of the structures of gases and liquids. Therefore, these concepts are reviewed briefly at this point.

The structures and properties of gases are explained by the kinetic theory. According to this theory, gases are considered to consist of a number of small particles called molecules. Under normal conditions of room temperature and atmospheric pressure, the average distance between the molecules is large compared to the size of the molecules themselves. The molecules are in constant motion. As a result of this motion, gas molecules collide with one another and also strike the walls of any vessel in which they are confined. It has been estimated that a gas molecule collides and is halted about 60,000 times in travelling a distance of 1 cm.[1] The average distance a molecule travels between collisions is referred to as the mean-free-path of the molecule.[2] As a result of these collisions, the speeds of the molecules range from zero to very high values. The molecules of hydrogen, for example, have been reported to move at an average speed at 0°C of over one mile per second.[3] A consequence of the motion of the molecules is the property of gases called pressure. The pressure of a gas results from the impact of the gas molecules on the object against which the pressure is being exerted. This was first suggested by Bernoulli in 1740.

The properties of gases have been investigated for many years. For example, in 1662, Boyle discovered that the volume of a gas is inversely proportional to the pressure if the temperature is constant. This is known as Boyle's Law and can be illustrated as follows, where V = volume, P = pressure, and T = temperature. $\propto$ is a proportionality symbol:

$$V \propto \frac{1}{P} \quad (T \text{ is constant}).$$

Later, Charles and Gay-Lussac found that the volume of a gas is directly proportional to its temperature if the pressure remains constant.

$$V \propto T \quad (P \text{ is constant}).$$

These discoveries lead to the derivation of an expression that relates the pressure, volume, and temperature. This is called the ideal gas law and is illustrated mathematically as follows:

$$PV = nRT. \tag{3-1}$$

In this equation, P is the pressure exerted by the gas, V is the volume occupied by the gas, n is the number of mols of gas (which indicates its concentration), T is the absolute temperature, and R is a number which is called a gas constant.

Initially, it was assumed that Equation 3-1 could be used to calculate the exact properties of any gas if a sufficient number of the other variables were already known. For example, if $V, n, R,$ and T were known, then P could be calculated. However, careful experiments by a number of investigators showed that the properties of the gases that were calculated using Equation 3-1 did not agree exactly with the experimentally determined values. It was also noted that the differences between the calculated and experimental values were the least under conditions of high temperature and low pressure. The fact that gases do not follow the ideal gas law (Equation 3-1) is referred to as the deviation of the gas from the gas laws.

Budde[4] was the first to suggest that one reason for the deviation of actual gases from the ideal gas law was that the volume V in Equation 3-1 did not take into account the volume occupied by the gas molecules themselves. If a correction for the volume of the gas molecules, designated by the letter b is applied then the preceding equation becomes:

$$P (V - b) = nRT. \tag{3-2}$$

Equation 3-2 was now found to be satisfactory for hydrogen but not for any other gas. In 1879, van der Waals pointed out that the ideal gas law also did not take into account the intermolecular forces of attraction that exist between all molecules. This attractive force is considered to be electrical in nature and tends to bring the molecules closer together.[5]

The intermolecular forces of attraction are designated by the term a/V^2 where a is the constant of intermolecular attraction and V is the volume occupied by the gas. When the term a/V^2 is added to Equation 3–2, the following equation, known as van der Waals equation, is obtained.[4,5]

$$(P + a/V^2)\ (V - b) = nRT. \tag{3-3}$$

Van der Waals equation not only applies to gases but also to many liquids.

The intermolecular forces of attraction between the molecules only operate over a very short distance of several molecular diameters.[2] When the gas is at a comparatively high temperature, the velocity of the gas molecules is sufficient to overcome the attractive forces between the molecules. Also, at low pressures, there are so few molecules present that the volume occupied by the gas molecules themselves is small compared to the total volume occupied by the gas. Therefore, under conditions of low pressure and high temperature, the corrections b and a/V^2 become negligible. The gas then approaches ideal behavior.

The liquefaction of gases can be explained from the preceding discussion. It is the attractive force between molecules, designated by a in van der Waals equation, that is responsible for the condensation of gases to liquids.[4,5] When pressure is applied to a gas, the molecules are brought closer together and the attractive forces between the molecules then become increasingly stronger. Likewise, if the temperature of the gas is lowered, the velocity of the gas molecules is decreased and this also increases the molecular attraction. If the pressure is high enough and the temperature is low enough, the attractive forces between the molecules become so predominant that the molecules begin to cling to each other.[1] These clumps of molecules grow larger and ultimately settle to the bottom and the gas then changes to a liquid.

According to Richards, the molecules of a liquid are in actual contact.[1] This is why liquids are so difficult to compress compared to gases. However, the molecules in a liquid still possess a limited amount of motion and, therefore, a liquid is capable of flowing and taking the shape of the vessel the liquid is in. The strength of the attractive forces between molecules depends upon the structure of the molecules. The main difference between gases and liquids is the different order of magnitude of the intermolecular forces of attraction in the two cases.[2] In compounds, such as water and ethyl alcohol, which are liquids under normal conditions of room temperature and atmospheric pressure, the attractive forces between the molecules are very strong.

In all liquids, there is a continuous flight of molecules from the surface of the liquid into the vapor space above the liquid. In order for the molecules to escape from the surface of the liquid and enter the vapor phase, the molecules must overcome the intermolecular forces of attraction in the liquid. Only the molecules with the highest speeds, i.e., the highest kinetic energy, are capable of doing this. If the liquid is open to the air, these molecules are lost to the atmosphere and the liquid evaporates. Since these molecules have the highest kinetic energy, their loss results in a decrease in the total energy in the liquid. When the kinetic energy decreases the liquid cools. This is why a liquid cools during evaporation, unless heat is supplied by an external source.

If the liquid is confined in a closed vessel, there is also a continuous flight of molecules from the surface of the liquid into the vapor space. However, at the same time, a reverse process of condensation of vapor molecules at the surface of the liquid takes place. In time, a condition of equilibrium will be established in the container when the rates of vaporization and condensation are equal. The vapor is said to be saturated at this point. The pressure exerted by the molecules of the vapor in equilibrium with the liquid is known as the vapor pressure of the liquid. The concentration of molecules in the vapor phase at equilibrium is constant for any liquid at a given temperature and, therefore, the vapor pressure remains constant. When the temperature is increased, both the number of molecules in the vapor phase and their velocities are increased. Therefore, the pressure increases when the temperature is increased.

Aerosol propellants have boiling points below room temperature and, therefore, are gases under conditions of room temperature and atmospheric pressure. However, when the gaseous propellants are cooled or compressed sufficiently, they are converted into liquids because of the intermolecular forces of attraction, as previously discussed. If the propellants are cooled below their boiling points, they can be poured from one container to another as any normal liquid. However, at room temperature they must be confined in a closed container in order to keep them from changing into vapor.

These concepts can be applied to aerosol products to explain the breakup of a liquid concentrate into a spray of small droplets by the propellant as follows: In a homogeneous aerosol product, there is an intimate mixture of the liquefied propellant, the solvents, and the active ingredients. The addition of the propellant to the concentrate dilutes the solvents and thus separates the solvent molecules. The strong intermolecular forces of attraction between the solvent molecules are reduced or broken because the average distance between the solvent molecules is increased. Instead of each solvent molecule being surrounded by other solvent molecules, the solvent molecules are surrounded by both solvent and liquefied propellant molecules. The intermolecular forces of attraction between the propellant molecules or between propellant–solvent molecules are weaker than those between the solvent molecules. It is also probable that the relatively high numbers of propellant molecules that continue to vaporize and condense in the container tend to keep the molecules in the liquid phase in more of an agitated condition than if the propellant were not present. This would reduce the effect of the intermolecular forces.

The vapor pressure of the aerosol is greater than atmospheric pressure, and when the valve is opened the pressure forces the aerosol up the dip tube and out the valve. As soon as the aerosol leaves the container and is exposed to the atmosphere, the propellant vaporizes and the propellant molecules fly off into the atmosphere. In doing so, they undoubtedly collide with many of the solvent molecules and cause collisions among the solvent molecules themselves, disrupting the attractive forces between them. Also, as a result of the dilution effect of the liquefied propellant, many of the solvent molecules are already too far apart for the intermolecular forces to operate. As the propellant vaporizes in the air, the solvent molecules are left almost stranded, so to speak, in relatively small groups or particles. The liquefied propellant, therefore, provides the pressure to force the aerosol out of the container when the valve is opened and it also serves to break up the concentrate into small droplets or particles once the aerosol has reached the atmosphere.

REFERENCES

1. M. C. Sneed and J. L. Maynard, "General Inorganic Chemistry," 2nd ed., Chap. 8, D. Van Nostrand and Company, Inc., Princeton, N.J., 1942.
2. F. A. Saunders, "A Survey of Physics for College Students," Chap. 10, Henry Holt & Company, New York, N.Y., 1930.
3. F. H. MacDougall, "Physical Chemistry," Chap. 3, The Macmillan Company, New York, N.Y., 1936.
4. F. H. Getman and F. Daniels, "Outlines of Theoretical Chemistry," 5th ed., Chap. 2, John Wiley & Sons, Inc., New York, N.Y., 1931.
5. F. Daniels and R. A. Alberty, "Physical Chemistry," Chap. 2, John Wiley & Sons, Inc., New York, N.Y., 1955.

4

PROPELLANTS

"FREON" PROPELLANTS AND OTHER FLUORINATED PROPELLANTS

The Numbering System

The various "Freon" propellants are usually distinguished only by a number after the trademark, "Freon," such as "Freon" 12, "Freon" 114, "Freon" 11, etc. At first glance, the numbers assigned to the propellants may seem to have been selected arbitrarily without any reason or logic. However, the numbering system used for the different fluorinated compounds was developed by the Du Pont Company years ago, so that the chemical formula of a fluorinated hydrocarbon could be determined from the number of the compound alone. Thus, it is possible to write both the empirical and structural formulas for "Freon" 12, "Freon" 114, "Freon" 11, etc., from the numbers of these products. The numbering system also applies to hydrocarbons and chlorinated hydrocarbons. The system was given to the refrigeration industry several years ago[1] in order to establish uniformity in numbering the various refrigerants and it is now used by all the U. S. manufacturers of the fluorinated hydrocarbon compounds. It applies to aerosol propellants as well as the refrigerants.

The formulas of the aerosol propellants may be determined from the propellant numbers by the following seven rules.

RULE 1. The first digit on the right is the number of fluorine (F) atoms in the compound.

RULE 2. The second digit from the right is one more than the number of hydrogen (H) atoms in the compound.

RULE 3. The third digit from the right is one less than the number of carbon (C) atoms in the compound. When this digit is zero, it is omitted from the number.

18

RULE 4. The number of chloride (Cl) atoms in the compound is found by subtracting the sum of the fluorine and hydrogen atoms from the total number of atoms that can be connected to the carbon atoms. When only one carbon atom is involved, the total number of attached atoms is four. When two carbon atoms are present, the total number of attached atoms is six.

An application of the first four rules in determining the formula for "Freon" 12 propellant is as follows:

Derivation of the Formula for "Freon" 12 Propellant

(a) The first digit on the right in "Freon" 12 is 2. Therefore, the compound contains two fluorine atoms in the molecule. (Rule 1)

(b) The second digit from the right is 1. Therefore, there are no hydrogen atoms in the compound since 1 is one more than zero. (Rule 2)

(c) The third digit from the right has been omitted from the number and therefore is zero. This shows that the compound contains one carbon atom. (Rule 3)

(d) Since the compound contains one carbon atom, there are four other atoms attached to the carbon atom. The number of chlorine atoms is found by subtracting the sum of the fluorine atoms and the hydrogen atoms from four. There are two fluorine atoms and no hydrogen atoms. Therefore, there are two chlorine atoms in the compound. (Rule 4)

Based upon the above information, it is known that "Freon" 12 contains one carbon atom, two fluorine atoms and two chlorine atoms. Therefore, the chemical formula for "Freon" 12 is CCl_2F_2.

Additional rules and their application are as follows:

RULE 5. In the case where isomers exist, each has the same number, but the most symmetrical one is indicated by the number alone. As the isomers become more and more unsymmetrical, the letters a, b, c, etc., are appended. Symmetry is determined by dividing the molecule in two and adding the atomic weights of the groups attached to each carbon atom. The closer the total weights are to each other, the more symmetrical the product. This is illustrated by deriving the structure of "Freon" 114 as follows:

Derivation of the Formula for "Freon" 114 Propellant

(a) The first digit on the right is 4. Therefore, the compound contains four fluorine (F) atoms. (Rule 1)

(b) The second digit from the right is 1. Therefore, there are no hydrogen atoms in the compound. (Rule 2)

(c) The third digit from the right is 1. Therefore, the compound contains two carbon atoms. (Rule 3)

(d) Since the compound contains two carbon atoms, there are six other atoms attached to the carbon atoms. Since there are four fluorine atoms and no hydrogen atoms, the compound contains two chlorine atoms. (Rule 4) The empirical formula for "Freon" 114 is, therefore, $C_2Cl_2F_4$. However, two isomeric forms of this compound exist, $CClF_2CClF_2$ and CCl_2FCF_3. The most symmetrical, $CClF_2CClF_2$, is designated "Freon" 114 and has the following structure:

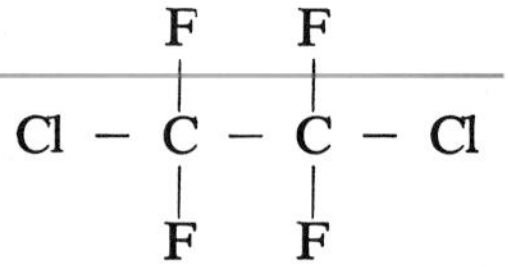

$$
\begin{array}{ccc}
& F & F & \\
& | & | & \\
Cl - & C - & C - & Cl \qquad \text{"Freon" 114} \\
& | & | & \\
& F & F &
\end{array}
$$

The unsymmetrical form, CCl_2FCF_3, is designated "Freon" 114a and has the following structure:

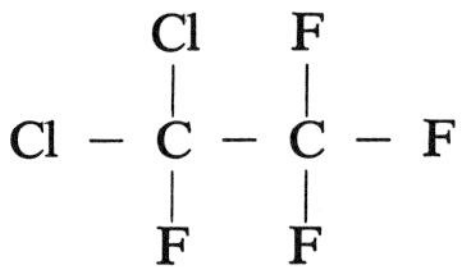

$$
\begin{array}{ccc}
& Cl & F & \\
& | & | & \\
Cl - & C - & C - & F \qquad \text{"Freon" 114a} \\
& | & | & \\
& F & F &
\end{array}
$$

RULE 6. For cyclic derivatives, the letter C is used before the identifying number. Thus, by applying the first six rules, the formula for "Freon" C-318 is found to be:

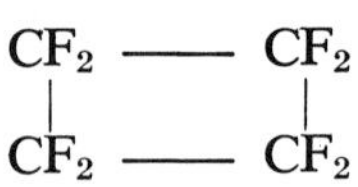

$$
\begin{array}{ccc}
CF_2 & —— & CF_2 \\
| & & | \qquad \text{"Freon" C-318} \\
CF_2 & —— & CF_2
\end{array}
$$

RULE 7. In case the compound is unsaturated, the above rules apply except that the number one (1) is used as the fourth digit from the right to indicate an unsaturated double bond. The number for vinyl chloride on this basis is 1140 and for tetrafluoroethylene is 1114.

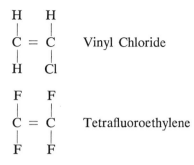

Properties of the Fluorinated Propellants

Most of the fluorinated hydrocarbon propellants are nonflammable, i.e., do not form an explosive mixture with air at any concentration. The trademark, "Freon" propellants, is reserved exclusively for the nonflammable, virtually odorless propellants of this type. These propellants have a low order of toxicity. Some fluorinated hydrocarbon propellants are flammable and will form explosive mixtures with air. These propellants are designated by the terms *Fluorocarbon, Propellant,* or the letters *FC* followed by the number. Thus, Propellant 152a and Propellant 142b are both fluorinated hydrocarbon propellants but both compounds are flammable. Unfortunately, the distinction becomes a little confusing in the scientific literature because the terms Propellant 12, Propellant 11, etc., are often used in a general sense in order to avoid the use of trade names, such as "Freon" 12 propellant, "Genetron" 12, "Ucon" 12, and "Isotron" 12, etc.

The "Freon" propellants and other fluorinated propellants are listed in Table 4–1 along with some of their physical properties. The compounds are listed in increasing order of their boiling points and, therefore, decreasing order of their vapor pressures. Pressures in the aerosol industry are commonly expressed as pounds per square inch gage (psig). Although the air pressure at sea level is 14.7 psi, most gages are set to read zero under these conditions. Therefore, these gages contain air at a pressure of about 14.7 psi although the pointer reads zero. In order to convert gage pressure into absolute pressure, it is necessary to add 14.7 to the gage reading.

$$\text{psig} + 14.7 = \text{psia, i.e., } 15.3 \text{ psig} + 14.7 = 30.0 \text{ psia.}$$

The data in Table 4–1 show that propellants are available with vapor pressures ranging from subatmospheric pressure up to 121.4 psig at 70°F. Propellants having any desired pressure within the available range may be obtained by using mixtures of propellants. Such mixtures are commonly used, not only because they are more economical in many cases, but also because there are certain pressure regulations governing aerosol products that must be considered. Some of the propellants have too high a vapor pressure to be used by themselves in certain aerosol products.

Although there is a considerable variety of propellants listed in Table 4-1, only Propellant 12, Propellant 11, and Propellant 114 are widely used in the aerosol industry. The other propellants, however, have special properties that may result in an increasing use of these propellants in the future.

The characteristics of the various propellants are summarized below:

"Freon" 22 Propellant. "Freon" 22 is used extensively as a refrigerant but very little as an aerosol propellant. As a result of its high vapor pressure, "Freon" 22 must be used in combination with a vapor pressure depressant, such as an organic solvent, or another propellant.

"Freon" 22 is stable in acidic or neutral medium but decomposes under alkaline conditions. This limits its usefulness in aerosol products. The instability under alkaline conditions, coupled with the fact that "Freon" 22 is not a particularly good solvent[3] and has a high vapor pressure, accounts for the minor use of this compound in the aerosol industry. There are some instances, such as in formulations with various glycols, where "Freon" 22–"Freon" 11 propellant mixtures are useful.[25] The high pressure and low-molecular weight of "Freon" 22 permits a high proportion of "Freon" 11 to be used in the propellant mixture. "Freon" 11 is a good solvent and propellants containing high proportions of "Freon" 11 have high compatibility with many materials, such as glycols.

"Freon" Food Propellant 115. "Freon" 115 is an extremely stable propellant with a low order of toxicity and is approved for use as a food propellant by the FDA. It is considerably more expensive than the common propellants, such as "Freon" 12 and "Freon" 11, and therefore, has found little use in nonfood aerosols. It undoubtedly has very poor solvent properties, judging by the number of fluorine atoms in the molecule.

"Freon" 12 Propellant. "Freon" 12 was the first, and is still the most widely used fluorinated hydrocarbon propellant. Next to "Freon" 11, it is the least expensive of the fluorinated propellants. "Freon" 12 is stable to hydrolysis under most acidic and alkaline conditions and thus has been used to a considerable extent in aqueous-based products. It is essentially odorless and has a low order of toxicity. "Freon" 12 has too high a vapor pressure to be used alone, and, therefore, is used in combination with other propellants, such as "Freon" 114 or "Freon" 11, or with organic solvents.

Propellant 152a. Propellant 152a has found little acceptance thus far by

TABLE 4-1 PROPERTIES OF THE "FREON" PROPELLANTS AND OTHER FLUORINATED PROPELLANTS

Propellant	Formula	Molecular Weight	Boiling Point (°F)	Vapor Pressure (psig at 70°F)	Density (g/cc) at 70°F	References
"Freon" 22	$CHClF_2$	86.5	−41.4	121.4	1.21	(3, 5, 8, 29, 30)
"Freon" 115	$CClF_2CF_3$	154.5	−37.7	103.0	1.31	(5, 11, 28)
"Freon" 12	CCl_2F_2	120.9	−21.6	70.2	1.33	(5, 8, 29, 30)
Propellant 152a	CH_3CHF_2	66.1	−11.0	63.0	0.91	5, 10, 29, 30
Propellant 142b	CH_3CClF_2	100.5	14.4	29.1	1.12	(9, 29, 30)
"Freon" C-318	C_4F_8(Cyclic)	200.0	21.1	25.4	1.51	(5, 6, 7, 29, 30)
"Freon" 114	$CClF_2CClF_2$	170.9	38.4	12.9	1.47	(5, 8, 29, 30)
"Freon" 21	$CHCl_2F$	102.9	48.1	8.4	1.32	(4, 5, 29, 30)
"Freon" 11	CCl_3F	137.4	74.8	(13.4 psia)	1.49	(5, 8, 29, 30)
"Freon" 113	CCl_2FCClF_2	187.4	117.6	(5.5 psia)	1.57	(5, 29, 30)

the aerosol industry. The main disadvantages of Propellant 152a are that it is flammable and is somewhat more expensive than the standard fluorinated propellants. However, Propellant 152a has a low-molecular weight, and therefore, less Propellant 152a is required to produce the same volume of gas during vaporization than the common propellants. This should be taken into account when considering cost comparisons with higher molecular weight propellants.

Propellant 152a has a density less than that of water, and it is possible to prepare propellant mixtures in combination with the "Freon" propellants which have a density approximating that of water. Such propellant mixtures would be particularly useful in the formulation of aqueous emulsion systems. When the density of the aqueous phase and the propellants are similar, the rate of creaming is reduced.

Propellant 142b.   Propellant 142b is presently used to a limited extent in some foam formulations, since it imparts quick breaking characteristics to some otherwise stable foams. Propellant 142b has a relatively low density which makes it effective in formulating emulsion systems. Propellant 142b has been reported to enhance the fragrance of certain perfumes.[13] It is slightly flammable, and somewhat more expensive than "Freon" 12 or "Freon" 11.

"Freon" Food Propellant C-318.   "Freon" C-318 propellant is the most chemically inert of the "Freon" propellants and has been approved by the FDA for use as a food propellant. It is generally used in blends with "Freon" 115 for foods. "Freon" C-318 is considerably more expensive than the common "Freon" propellants, and is, therefore, limited in its use in nonfood aerosols. It is also an extremely poor solvent.

"Freon" 114 Propellant.   "Freon" 114 is stable to hydrolysis and is commonly used with "Freon" 12 for the formulation of aqueous based aerosols. "Freon" 114 has little odor and is used extensively in the formulation of aerosol colognes and perfumes. It is somewhat more expensive than either "Freon" 12 or "Freon" 11 and, as a result, is not used as widely as the latter two propellants. "Freon" 114 is also used for the formulation of aerosol cosmetics and pharmaceuticals because of its high stability and low order of toxicity.

"Freon" 21 Propellant.   "Freon" 21 has aroused considerable interest in the aerosol industry because of its excellent solvent characteristics. "Freon" 21 has a low order of toxicity and should have application in the formulation of personal products and pharmaceuticals. It is stable under

acid or neutral conditions but will hydrolyze rapidly in alkaline systems. The use of "Freon" 21 is, therefore, limited to the former two systems.

"Freon" 11 Propellant. "Freon" 11 is the least expensive of the fluorinated hydrocarbon propellants and next to "Freon" 12, is the most widely used. "Freon" 11 has too low a vapor pressure to be used alone as a propellant and is generally used in combination with "Freon" 12. "Freon" 11 has excellent solvent properties and is useful in the formulation of many products where propellant compatibility might be a problem. "Freon" 11 propellant decomposes in the presence of water and certain metals and, for this reason, the use of "Freon" 11 in aqueous-based systems is limited. This is why "Freon" 114 propellant is used with "Freon" 12 in such systems in place of "Freon" 11.

"Freon" 113 Fluorocarbon. "Freon" 113 is not a propellant because it has too high a boiling point. However, "Freon" 113 is a good solvent and should be considered for aerosols where this property is important. It is used in combination with propellants, such as "Freon" 22 or "Freon" 12.

Other Liquefied Gas Propellants

The great popularity of the "Freon" fluorinated hydrocarbon propellants for aerosol products is due to the fact that the propellants are nonflammable, have a low order of toxicity, are essentially odorless, and have a range of boiling points which make them effective aerosol propellants. In recent years, other propellants, such as propane, isobutane, and butane, have also been used in some products. There was a combination of factors that led to the use of these liquefied petroleum gases, but one of the major reasons was the low cost of the hydrocarbons. Other propellants, which are used to a much lesser extent, are dimethyl ether and vinyl chloride.

The properties of the liquefied gas propellants are listed in Table 4–2. All of the propellants listed in Table 4–2 are very flammable. This is the major disadvantage of these products. Flammability is discussed in detail in Chapter 14. The propellants have vapor pressures that range from 16.4–110.3 psig at 70°F. They have the advantage of having much lower molecular weights than the standard fluorinated hydrocarbon propellants. In comparison with the fluorinated hydrocarbon propellants, lower weights of the hydrocarbons will provide the same volume of gas when the propellant is vaporized. This factor also decreases the cost of aerosols formulated with these compounds.

The liquefied petroleum gases are available in two grades, a pure grade and a special aerosol grade. The pure grade is designed for cosmetic and

TABLE 4-2 PROPERTIES OF THE HYDROCARBONS AND MISCELLANEOUS PROPELLANTS

Propellant	Formula	Molecular Weight	Boiling Point (°F)	Vapor Pressure (psig at 70°F)	Density (g/cc) at 70°F	References
Propane*	C_3H_8	44.1	−43.7	110.3	0.50	(12, 15)
Dimethyl Ether	$(CH_3)_2O$	46.1	−12.7	63.0	0.66	(14)
Vinyl Chloride	$CH_2=CHCl$	62.5	7.9	34.0	0.91	(14, 15, 16)
Isobutane*	$C_4H_{10}(iso)$	58.1	10.9	31.0	0.56	(12, 15)
n-Butane*	$C_4H_{10}(n)$	58.1	31.1	16.4	0.58	(12, 15)

* Pure Grade.

pharmaceutical uses. The special aerosol grade is less pure and is considered suitable for general aerosol applications. The properties of the hydrocarbons listed in Table 4–2 are those of the pure grade. Specific comments on the various propellants are given below.

Propane. Propane is used to a limited extent as the propellant in some paint products. It has a very high vapor pressure and must be used either in combination with another propellant, which will serve as a vapor pressure depressant, or with organic solvents, which also reduce the vapor pressure. Propane is a relatively poor solvent. As a general rule, the hydrocarbon propellants are quite stable and have a low order of toxicity.

Dimethyl Ether. To date, dimethyl ether has not been used alone as a propellant, but only as a component of propellant blends. These blends are discussed in the following section.

Vinyl Chloride. Vinyl chloride, a comparatively good solvent, has not been used by itself as a propellant in the United States but is a component of many propellant blends. For nonaerosol uses, it is normally shipped with an inhibitor, such as phenol, to prevent polymerization. However, most of the vinyl chloride used in aerosol products is uninhibited. Considering the known toxicity of phenol, usage of a phenol-inhibited product could, perhaps, pose problems for many aerosol products. Such usage should be checked to be certain that polymerization of the vinyl chloride did not occur during storage in the aerosol. Vinyl chloride has approximately the same vapor pressure as "Freon" 12–"Freon" 11 (50/50).

Isobutane. The major use of isobutane has been as a propellant for aqueous-based aerosol products and as a component of nonflammable propellant blends. Except for the flammability of isobutane, it has most of the properties necessary for use as a propellant. Isobutane has a low density and can be used in combination with the "Freon" propellants at such a ratio that the resulting propellant blend has a density close to one. These propellant blends are useful for aqueous emulsion systems.

n-Butane. n-Butane is normally present to a minor extent as an impurity in isobutane. n-Butane has not been used alone as a propellant since it has a fairly low vapor pressure.

Common Blends of Nonflammable and Flammable Propellants

A number of blends of nonflammable and flammable propellants are used at the present time. The purpose of blending the propellants is to reduce the cost of the nonflammable fluorinated propellants by the addition of the low-cost flammable propellants and still obtain a propellant blend that is relatively low in flammability characteristics. Some of the common propellant blends are listed in Table 4–3. Comments about the blends are as follows (flammability properties are discussed in Chapter 14):

Propellant 12–Propane (91/9). This blend has been used to a limited extent in paint products. The blend has a high vapor pressure but the solvents present in the paints serve as pressure depressants.

Propellant 12–Vinyl Chloride Blends. A number of these blends are used in coating compositions. Vinyl chloride has good solvent properties, and the blends are more compatible with many resins than Propellant 12 alone.

Propellant A. Propellant A (Propellant 12–Propellant 11–isobutane— 45/45/10) is used in a number of products as a substitute for Propellant 12–Propellant 11 (50/50). The addition of the isobutane reduces the cost of the blend slightly.

Propellant P. Propellant P (Propellant 12–Propellant 11–dimethyl ether —75/10/15) was developed primarily as a nonflammable paint propellant less expensive than Propellant 12 alone. Dimethyl ether is a good solvent and the blend is a better solvent than "Freon" 12 alone.

TABLE 4-3 PROPERTIES OF VARIOUS PROPELLANT BLENDS

Propellant Blend	Vapor Pressure (psig at 70°F)	Density (g/cc)	References
Propellant 12–Propane (91/9)	82	1.15	14
Propellant 12–Vinyl Chloride (80/20)	66	1.21	14
Propellant 12–Vinyl Chloride (65/35)	61	1.13	14, 19
Propellant P (Propellant 12–Propellant 11–Dimethyl ether—75/10/15)	62	1.21	18
Propellant A (Propellant 12–Propellant 11–Isobutane—45/45/10)	38	1.22	14, 17

Compressed Gas Propellants

Compressed gases, such as nitrogen, carbon dioxide, and nitrous oxide also are used as propellants for aerosol products. These gases have a number of limitations as propellants, and the number of products pressurized with the compressed gases is relatively small. In the nonfood field, the use of compressed gases has been limited to products with coarse sprays, such as windshield de-icers, furniture polishes, etc. The compressed gases have a low order of toxicity and are inexpensive.

Nitrogen was one of the first gases used as a propellant in nonfood aerosols. Nitrogen was essentially insoluble in the aerosol concentrates and, therefore, was confined to the vapor phase. Since very little nitrogen was dissolved in the liquid phase, it did not change the characteristics of the concentrate during discharge. The function of the nitrogen was to supply pressure to force the product out of the aerosol container. The concentrate had to be broken up into droplets by the use of mechanical breakup actuators since there was little propellant dissolved in the liquid phase. In a sense, nitrogen acted like a piston. Only relatively coarse sprays could be obtained with products pressurized with nitrogen.

The use of nitrogen as a propellant was greeted at first with considerable enthusiasm. Nitrogen was inert, low in toxicity, and inexpensive.[20,21,22] However, customer complaints soon forced a revision of the attitude regarding nitrogen. In the first place, the spray characteristics changed noticeably as the products were discharged. Since practically all of the nitrogen was concentrated in the vapor phase, the increase in volume of the vapor phase as the product was discharged caused a drop in the nitrogen pressure and this in turn caused the spray to become coarser.

There was another disadvantage. Since the nitrogen was concentrated in the vapor phase (head space), the volume of the vapor phase had to be increased considerably in order to provide enough propellant in the container to discharge all the product. The product fill, therefore, was appreciably less than that in liquefied gas aerosols.

Other objections arose as a result of misuse of the aerosol products. If the consumer accidentally inverted the container and discharged the vapor phase instead of the liquid phase, practically all of the nitrogen was lost immediately. This left a product without any propellant and the product could not be discharged from the container.

Nitrous oxide and carbon dioxide have been accepted to a greater extent than nitrogen. They have considerably more solubility in the liquid phase of aerosols than nitrogen and this is particularly true with chlorinated solvents. Propellant systems based upon combinations of methylene chloride, 1,1,1-trichloroethane and nitrous oxide (the "Aerothene" system) have been suggested by Anthony[23,24] for use with such products as hair sprays. The higher solubility of nitrous oxide in the chlorinated solvents provides more latitude in the event the container is discharged in the wrong position and also results in a smaller pressure drop as the product is used. Some of the potential problems to be considered with this type of system include the coarse spray inherent in any compressed gas system, the difficulties in perfuming products containing substantial concentrations of 1,1,1,-trichloroethane, the effect of high concentrations of chlorinated solvents upon the skin, valve components and nearby plastic articles, and the change in spray characteristics that occurs during use.

In addition, although the increased solubility of nitrous oxide in the liquid phase lessens the effect of misuse, the problem still exists and if the product is misused very often, most of the compressed gas will be lost and the product will not operate satisfactorily. The effect of misuse upon product performance has been investigated by Hinn and Webster.[20,25]

Nitrogen, since it was insoluble in the liquid phase, gave products that had the same characteristics as the original concentrates. Nitrous oxide and carbon dioxide, with their higher solubility in the liquid phase, tended to aerate the concentrates. With products such as tooth paste, the aerated product was not acceptable. According to Hinn,[20] the use of compressed gases as propellants is limited to products that discharge as foams, aerated streams, or residual sprays. Mechanical breakup devices are necessary to achieve a coarse spray, even though some of the gas is dissolved in the liquid phase. The pressure drops noticeably during use, partially because the gas dissolved in the liquid phase is released at a relatively slow rate. Equilibrium between the gas in the vapor phase and the liquid phase usually is not established while the product is being used.[20] Therefore, the pressure decreases more rapidly than would be expected because the drop in pressure, which results from an increase in the volume of the vapor phase, is not compensated for by the release of gas from the liquid phase.

At the present time, a number of products, such as window cleaners, furniture polishes, starch sprays, windshield de-icers, and engine starters are pressurized with compressed gases. The windshield de-icer is an interesting example of a product in which a compressed gas is a far better propellant than a liquefied gas. The windshield de-icers were originally formulated with Propellant 12. This product performed satisfactorily at room temperature but when the product was left in a car at subfreezing temperatures, as most of the windshield de-icers were, the pressure inside the container bcame too low to discharge the product. The compressed gases were ideal for the windshield de-icers because large variations in temperature caused relatively minor changes in pressure. The pressure–temperature curves for gases are comparatively flat, compared to those for the liquefied gas propellants. This is illustrated in Table 4–4 in which the pressures of a compressed gas are shown in comparison with the pressures of Propellant 12 at temperatures of 0°F, 32°F, and 70°F. It was assumed that the gas has a pressure of 70.2 psig initially at 70°F.

TABLE 4-4   COMPARATIVE PRESSURES OF LIQUEFIED AND COMPRESSED
GAS PROPELLANTS AT DIFFERENT TEMPERATURES

	Pressure (psig)	
Temperature (°F)	Compressed Gas	Propellant 12
70	70.2	70.2
32	64.1	30.1
0	59.1	9.2

The properties of nitrogen, carbon dioxide, and nitrous oxide and the solubilities of the gases in various solvents are given in References 20, 23, and 25. The solubilities of the compressed gases in the fluorinated hydrocarbon propellants are given in Reference 21.

Liquefied Gas–Compressed Gas Combinations

Several years ago there was considerable interest in combinations of liquefied gases and compressed gases as propellants for aerosols. This interest developed as a result of speculation that a compressed gas could be substituted for a portion of the liquefied gas propellant in an aerosol product without affecting the spray characteristics. If this were true, then propellant costs could be reduced because the compressed gases were less expensive than the fluorinated hydrocarbon propellants.

It was observed, for example, that when a residual type product formulated with a liquefied gas propellant was pressurized with carbon dioxide or nitrous oxide to about 90 psig, there was an initial increase in the fineness of the spray. Therefore, it was assumed that the concentration of liquefied gas propellant normally used in a given product could be decreased and that the change in spray characteristics resulting from the decreased percentage of liquefied gas propellant could be compensated for by pressurizing with a compressed gas.

While the idea had merit, there were several problems connected with the use of the liquefied gas–compressed gas combinations. Although the addition of a compressed gas to a liquefied gas results in a finer spray initially, the spray properties change during use of the product until they are about the same as that before the compressed gas was added. One of the major reasons the spray characteristics change during discharge is that the volume of the vapor phase increases and consequently the pressure exerted by the compressed gas decreases. This change is even more noticeable with a product pressurized only with a compressed gas.

For example, a residual product formulated with 20% "Freon" 12 propellant will have certain spray properties. If this product is pressurized to 90 psig with a compressed gas, such as nitrous oxide, the initial spray will be somewhat finer than that from the product formulated with "Freon" 12 alone as the propellant. However, as the product is used, the spray becomes increasingly coarser and ultimately has about the same characteristics as those from the product before it was pressurized with the compressed gas.

An aerosol formulation with a composition of 20% insecticide concentrate, 55% methyl chloroform, and 25% Propellant 12, and pressurized to 80 psig with nitrous oxide was investigated by Webster, Hinn, and Lychalk.[26] The product had to be classified as a pressurized spray rather than an aerosol insecticide because the particle size was not sufficiently small to meet the requirements for a space insecticide. A hair spray formulated with a combination of methylene chloride, 1,1,1-trichloroethane, hair spray concentrate, Propellant 12, and nitrous oxide has been suggested by Anthony.[24]

In the field of food aerosols, combinations of "Freon" Food Propellant C-318 with either nitrous oxide or "Freon" Food Propellant 115 have been found to possess advantages over any of the propellants by themselves. Many food aerosols need to be refrigerated and "Freon" C-318 by itself has too low a vapor pressure at refrigeration temperatures to be an effective propellant. The addition of "Freon" 115 or nitrous oxide provides enough pressure, however, so that a satisfactory discharge is obtained at the lower temperatures.[27,28]

An extensive discussion of the properties and functions of the various propellants used in the aerosol industry has been written by Reed.[29,30] These articles should be consulted for detailed information concerning aerosol propellants.

REFERENCES

1. "American Standard Number Designation of Refrigerants," ASHRAE Standard 34–57 and ASA Standard B 79.1 (1960).
2. "Freon" Aerosol Report FA-3, "Kauri-Butanol Numbers of 'Freon' Propellants and Other Solvents."
3. "Freon" Aerosol Report FA-7, "Propellant Solutions Containing "Freon" 22 Chlorodifluoromethane."
4. "Freon" Aerosol Report FA-28, "Freon" 21 Aerosol Solvent and Propellant."
5. "Freon" Technical Bulletin B-2, "Properties and Applications of the 'Freon' Fluorinated Hydrocarbons."
6. "Freon" Technical Bulletin B-18, " 'Freon' C-318."
7. "Freon" Technical Bulletin B-18B, "Physical Properties of 'Freon' C-318 Perfluorocyclobutane."
8. "Freon" Technical Bulletin D-6, "Comparative Stability of 'Freon' Compounds."

9. "Freon" Technical Bulletin D-60, "Properties of FC-142b, 1-Chloro-1, 1-Difluoroethane-CH$_3$CClF$_2$."
10. "Freon" Technical Bulletin D-62, "Properties of 1,1-Difluoroethane (DFE) CH$_3$CHF$_2$."
11. "Freon" Technical Bulletin T-115, "Thermodynamic Properties of 'Freon' 115 Monochloropentafluoroethane."
12. Phillips Hydrocarbon Aerosol Propellants, Technical Bulletin 519 (1961) Phillips Petroleum Company.
13. "Genetron" Aerosol Propellants, General Chemical Division, Allied Chemical and Dye Corporation (1955).
14. "Isotron" Technical Information File, "The Isotron' Blends," Vol. 2, No. 1. Pennsalt Chemical Company.
15. Matheson Gas Data Book, The Matheson Company, Inc., April, 1965.
16. The Dow Chemical Company Technical Bulletin, "Vinyl Chloride Monomer." 1954.
17. Union Carbide Chemicals Company Technical Bulletin, " 'Ucon' Propellant A."
18. Union Carbide Chemicals Company Technical Bulletin, "New 'Ucon' Paint Propellant."
19. "Genetron" Product Information Bulletin, " 'Genetron'–Vinyl Chloride Aerosol Propellants," General Chemical Division, Allied Chemical Company.
20. J. S. Hinn, Proc. 47th Mid-Year Meeting, CSMA (May 1961).
21. E. M. Datzler, F. C. Haller, and P. I. Smith, Aerosol Age 10, 70 (May 1965).
22. M. J. Root, Proc. Sci. Sect. Toilet Goods Assoc. 29 (June 1958).
23. T. Anthony, Soap Chem. Specialties 43, 185 (December 1967).
24. T. Anthony, Aerosol Age 13, 31 (September 1967).
25. R. C. Webster, Aerosol Age 6, 20 (June 1961).
26. R. C. Webster, J. S. Hinn, and P. A. Lychalk, Proc. 49th Mid-Year Meeting, CSMA (May 1963).
27. "Freon" Aerosol Report A-61, "New Food-Aerosol Propellant."
28. "Freon" Aerosol Report A-63. "Food Propellant 'Freon' 115."
29. F. T. Reed, "Propellants," in H. R. Shepherd, "Aerosols, Science and Technology," Interscience Publishers, Inc., New York, N.Y., 1961.
30. F. T. Reed, "Fluorocarbon Propellants," in A. Herzka, "International Encyclopaedia of Pressurized Packaging" (Aerosols), Pergamon Press, Inc. New York, N.Y., 1966.

5

CONTAINERS

TINPLATE CONTAINERS

The first aerosol containers produced in quantity were the high-pressure, cumbersome cylinders that were developed during World War II for packaging insecticides. These were expensive and had little sales appeal. In 1943, however, the Continental Can Company initiated a research program to develop a low-cost unit suitable for aerosol insecticides. After four years of research, an acceptable container became available to the industry. This was a modified 12-oz round can with a concave top and bottom. The valve, with a metal dip tube attached, was soldered to a hole in the center of the top. The cans were shipped with the top and valve attached and the customers were supplied with machines to double seam the bottom.

The container was known as a 211 × 413 concave style aerosol can. In the container industry, can sizes are always stated so that the last two numbers are sixteenths of an inch. Any preceding numbers are inches. The diameter is given first followed by height. A 211 × 413 container is $2^{11}\!/_{16}$ in. in diameter and $4^{13}\!/_{16}$ in. in height, measured outside and over the seam.

The Crown Can Division of the Crown Cork and Seal Company started development work on aerosol containers in 1945,[1] and the containers resulting from this work were made available in 1947. These were drawn, two-piece containers manufactured without a side seam. They were originally prepared for capping with the valve staked into the top, but in 1948 the Crown containers were manufactured with the now familiar 1-in. curl opening.

The 12-oz concave type was discarded by the Continental Can Company in 1951, and a can with a special dome top was marketed. The 1-in.

curled opening, which had been adopted earlier by the Crown Can Company, had become popular and the Continental cans were then supplied with this opening.

The American Can Company manufactured 12-oz aerosol containers during 1948 but did not enter the field actively until about 1950. The 1-in. curl opening was adopted by the American Can Company in 1955 and this opening has since, with a few exceptions, become standard for the industry. Other companies that now manufacture aerosol containers include the Heekin Can Company, the National Can Corporation, and the Sherwin-Williams Co.

Tinplate is one of the most popular metals for aerosol containers. It is a carbon steel alloy and is available in various gages with different quantities of tin on the surface. The tin is first deposited on the surface by electrolytic methods and then melted so that it forms a continuous film bonded to the surface of the steel. The alloy is available with different thickness of tin on the surface and the differences are indicated by the terms ½-lb, 1-lb, and 1½-pound tinplate. The weight designation of tinplate is used in the tinplate industry with the following meaning: there is a standard area over which the varying weights of tin are spread. This standard area is known as a base box and is the area of both sides of 112 sheets of steel with dimensions of 14 × 20 in. The thickness of the steel is indicated by the weight per base box which runs from about 85–112 lb or about 0.010 to a little over 0.012-in. thick. The thickness of the tin on the surface varies and is not uniform. For example, the thickness of tin on samples of 1-lb tinplate varied from 0.00003–0.00010 in. with averages of about 0.00005–0.00008 in. The theoretical thickness, assuming 1 lb of tin distributed evenly over the surface would be 0.000065 in.[2]

The increase in the number of aerosol products in the early years was almost matched by the increase in the variety of aerosol containers. Containers became available with tinplate, lacquer-lined black iron, and lacquer-lined tinplate construction. In 1962 there were available three different style bottoms—concave, concave with stacking feature, and flat; ten plate thicknesses; nine tinplate coatings; various interior enamels—epons, epoxys, phenolics, vinyls, etc. (these were roller coated or spray applied, and single or double coatings were available); two different side-seam structures—inside tabs where maximum pressure resistance was dictated, and outside tabs where enamel coverage was the major concern; four side-seam solders; three side-stripe enamels to protect the side-seam; and two end-seam compounds.

Aerosol side-seam tinplate containers are available in a variety of sizes —from 3–24 oz in capacity. Two-piece drawn cans are available in both 6- and 12-oz sizes. Also, a variety of aluminum containers are available in sizes ranging from ½ oz on up. These may be obtained both coated and uncoated.

Some common aerosol containers and their approximate capacity 100% full are given in Tables 5–1, 5–2, and 5–3.[14,15]

TABLE 5-1 CAPACITIES AND SIZES OF SIDE-SEAM AEROSOL CONTAINERS

American Can Company, Continental Can Company, Crown Cork and Seal Company

		Capacity	
Can Size	*Common Industry Designation*	*Fluid Ounces*	*cc (approximate)*
202 × 214	3 oz	5.2	154
202 × 314	4 oz	7.0	207
202 × 406	6 oz	7.9	236
202 × 509	8 oz	10.0	296
202 × 708	10 oz	13.6	403
211 × 413	12 oz	14.1	418
211 × 510	14 oz	16.5	488
211 × 604	16 oz	18.3	542
211 × 713	20 oz	22.8	592

TABLE 5-2 CAPACITIES AND SIZES OF "REGENCY LINE" OF AEROSOL CONTAINERS

American Can Company

		Capacity	
Can Size	*Common Industry Designation*	*Fluid Ounces*	*cc (approximate)*
207.5 × 413	10 oz	11.4	337
207.5 × 509	12 oz	13.4	393
207.5 × 605	14 oz	15.2	450
207.5 × 701	16 oz	16.9	500

The choice of a particular metal container for any given aerosol product depends upon many factors. The two-piece, lacquered-lined tinplated containers are slightly more expensive that the side-seam containers but are generally used for aqueous-based aerosol foam proucts, where corrosion

TABLE 5-3   CAPACITIES AND SIZES OF "SPRATAINER" DRAWN CONTAINERS

Crown Cork and Seal Company

| Can Size | Capacity | |
	Common Industry Designation	cc (approximate)
202 × 411	6 oz	225
214 × 411	12 oz	375

resistance to aqueous products is desired. These containers also are used where the pressures of the aerosol products are higher than are permitted in the standard side-seam containers.

The side-seam containers are used with aerosol products where corrosion resistance is not too great a problem and where the pressures of the formulations permit the use of these containers. Actually, the tinplate on the side-seam containers has been reported to be more uniform than that on the drawn tinplate containers because the drawing operation may increase the uneven distribution of the tinplate on the steel. However, the lacquer lining on the side-seam containers is not as effective in preventing corrosion as in the drawn cans because of the difficulty of obtaining a uniform lacquer coating over the soldered side seam.

The technical aspects of metal containers have been covered in considerable detail by Johnsen[1] in a series of four articles in *Aerosol Age.*

ALUMINUM CONTAINERS

Aluminum containers have not, as of yet, been used extensively in the United States in spite of the fact that containers with high strength and excellent appearance are available. This is partially due to the cost of the aluminum containers which is still higher than that of the tinplate containers. In countries where the price differential between tinplate and aluminum containers is not significant, aluminum containers are used to a much greater extent. Practically all of the aerosols in Italy and most of those produced in Switzerland are packaged in aluminum.[3]

In the United States, aluminum containers have been used mostly for specialty and drug products where container costs are not as important as for some of the more common aerosol items. The estimated output of aluminum aerosol containers in the United States was 25 million in 1961 and 45 million in 1963.[4]

The two aluminum cans commercially available in the United States are the single-piece, seamless container made by extrusion, and the two-piece unit manufactured by drawing or extruding the shell and double-seaming the shell to the bottom. The single-piece container with the drawn-in neck has the advantage that it allows considerable latitude in basic shape and permits continuous internal coatings. These containers may also be embossed with decorative and functional designs during extrusion. A further advantage of aluminum containers is that they are lighter than tinplate containers with the same capacity. An aerosol container of 6-oz capacity requires 1 oz of aluminum as against 2 oz for the tinplate containers.

One of the disadvantages of aluminum is that the strength of aluminum is less than that of steel of the same gage. Therefore, in order to achieve containers of equal strength, it is necessary to fabricate aluminum containers with thicker walls. Since aluminum is more expensive than tinplate at the same metal thickness, this difference in strength also increases the cost of the aluminum containers compared to the tinplate containers. However, aluminum containers with considerable strength are available. A two-piece Swiss container is guaranteed for pressures up to 170 psi and an aluminum can with a bursting strength in excess of 250 psi is available in the United States.[5]

GLASS CONTAINERS

The introduction of glass aerosol bottles in the early 1950's had a considerable impact upon the aerosol market. Glass bottles could be supplied in a variety of colors, sizes and shapes. The combination of the low-cost metal containers and the low-pressure propellants was responsible for the rapid expansion of the aerosol industry, but the initial metal containers were not particularly attractive for packaging products such as colognes and perfumes. The aesthetic appeal of the glass bottles stimulated the development of a whole series of personal products.

Aerosol laboratories now had low-cost test tubes available. Prior to this time, if an aerosol chemist wanted to observe an aerosol formulation and determine its compatibility, it was necessary to prepare the formulation in a glass compatibility tube.[6] In view of the cost of the compatibility tubes, most aerosol laboratories had relatively few in stock. Formulation work was limited since each tube had to be cleaned and reassembled after use. With the advent of the aerosol glass bottles, innumerable formulations could be prepared inexpensively, observed over a period of time, and discarded after use. The availability of glass bottles for formulation work in the laboratory undoubtedly hastened the introduction of many new aerosol products.

One of the first glass bottle products to appear on the market was "Larvex," an aqueous-based mothproofer formulated with "Freon" 114 propellant. This product appeared in 1952 in a 16-oz uncoated glass bottle. Because of the possibility of breakage and the potential hazard involved with products packaged under pressure in the glass bottles, a considerable amount of research was carried out in an attempt to find a suitable means for protecting glass bottles. In 1952, the Bristol-Myers Company announced the details of a successful method of coating aerosol bottles with plastic.[7] In the same year, the Wheaton Glass Company acquired the rights to this process and made plastic-coated bottles available to the industry in 1953. The first product in plastic-coated bottles, a cologne, appeared on the market in 1954.

Shortly after the appearance of the first coated glass bottles, a great controversy arose concerning the safety of the glass bottle products packaged under pressure. Innumerable drop tests were carried out to determine what hazards might result if a consumer happened to drop a bottle in a bathroom, for example. Many years have passed since these tests were first carried out and the excellent safety record that has been established for glass bottle products has demonstrated the relatively low order of the hazards that exist with the use of these products.

The plastic coating used for the bottles is a thermoplastic polyvinyl chloride. The coating functions in three ways:

1. It protects the bottle from scratches that would weaken the bottle.
2. The plastic coating protects the bottle when the bottle is dropped and thus reduces the chance of the bottle breaking.
3. If the bottle does break, the plastic coating is designed to retain the glass fragments so that they do not fly through the air and provide a hazard to the consumer.

Today, glass bottles can be obtained in a wide variety of shapes, colors and capacities, varying from 10–355 cc or ⅓–12 fl oz.[8,9] Bottles are available with or without the plastic coatings. Bottles with coatings adhered to the glass or bottles with unbonded coatings may be obtained. Coatings with vent holes may also be obtained. These vent holes allow for the release of pressure in the event the bottles do break. In testing bottles, the water bath temperatures should not be above 100°F. Otherwise, absorption of water by the coating will cause the coating to become opaque. Glass bottles are more expensive than metal containers but they have a sales appeal that metal containers do not have. In addition, products that are corrosive may be packaged in the corrosion-resistant glass. As a result of these properties, glass bottles have become firmly established in the aerosol industry.

Detailed information concerning the properties and testing of glass bottles is given in Reference 10.

PLASTIC CONTAINERS

Plastic containers have been under consideration for packaging aerosols for many years.[10-12] However, there are very few products marketed in plastic containers as yet. Although plastic containers have many advantages, they also have certain disadvantages and each type of plastic container has to be thoroughly tested before a product can be marketed. Some of the materials of construction that have been considered for plastic containers are melamine resins, phenolic resins, nylon, acetal copolymers, polypropylene, and linear polyethylene.[10]

The strength of the plastic containers provides a freedom in design and style that is not possible in glass bottles. They are light in weight and are extremely resistant to corrosion and breakage. These properties have stimulated continued interest in more effective development of plastic containers.

One of the major disadvantages of the plastic containers is their permeability to gases. Permeation may move in either direction. Moisture may escape from an aqueous solution or an emulsion, propellant vapors may escape, or various solvents may diffuse through the plastic. On the other hand, oxygen may permeate from the outside and attack sensitive components. In products formulated with ethyl alcohol, water may be absorbed from the outside. The rate of diffusion of acetone or methylene chloride through polyacetal containers is too large for these compounds to be used in the containers.[13]

Another problem with plastic containers is migration of the plasticizer from the plastic. This may cause unacceptable changes in the product itself or discoloration on the outside of the container.

These disadvantages, in conjunction with the high cost of plastic containers, have been sufficient to prevent any large-scale use to date.

PRESSURE AND SIZE LIMITATIONS

For many years the Federal Government has been concerned with the safe transportation of explosives and other dangerous materials by common carrier within the United States. Regulations are in effect which cover the packing, marking, loading, and handling of these materials and these have been published in the Code of Federal Regulations. The regulations are also available in Agent T. C. George's Tariff No. 19, issued August 5, 1966, effective September 5, 1966. This tariff is published by the Bureau of Explosives which is connected with the Association of American Railroads. A copy of the tariff may be obtained from the Bureau of Explosives, 63 Vesey Street, New York City.

Until the last two years, the regulations were administered by the Interstate Commerce Commission, but on October 15, 1966, a law passed by Congress established a new Department of Transportation (DOT) which then assumed the responsibility for the safety laws relating to railroads, motor carriers, etc. that previously had been the function of the ICC. This law became effective April 1, 1967.

Among the many substances that are classified dangerous materials in the Code of Federal Regulations are compressed gases. Many aerosol products are defined as compressed gases and are subject to the regulations. The regulations cover the types of containers, and conditions used for such products. Violation of the regulations can result in a fine or imprisonment or both.

Some of the DOT regulations concerned with pressure and size limitations in aerosol products are discussed in the following sections. It should be emphasized that these excerpts of the regulations are presented merely to illustrate the type of limitations that exist. The DOT regulations are much more extensive than the excerpts indicate and the regulations themselves must be consulted when there is any question of legality of the product. Other DOT regulations pertaining to aerosols are concerned with the flammability and testing of loaded products. These are discussed in the appropriate sections.

According to Paragraph 73.300 of the regulations, a compressed gas is defined as follows:

"The term *compressed gas* shall designate any material or mixture having in the container an absolute pressure exceeding 40 psi at 70°F or, regardless of the pressure at 70°F, having an absolute pressure exceeding 104 psi at 130°F, or any liquid flammable material having a vapor pressure exceeding 40 psi absolute at 100°F, as determined by the Reid method covered by the American Society for Testing Materials Method of Test for Vapor Pressure of Petroleum Products (D-323)."

In Paragraph 73.300 (d), a liquefied compressed gas is defined essentially as follows:

"A liquefied compressed gas is a gas which, under the charged pressure, is partially liquid at a temperature of 70°F."

Compressed gases must be shipped in cylinders, tank cars, cargo tanks, or portable tank containers (see Paragraphs 73.308, 73. 314, and 73.315). Many aerosol products are classified as compressed gases and, if there were no exemptions to the regulations, these aerosol products would have to be packaged in cylinders, etc. However, exemptions to the regulations for compressed gases (Agent T. C. George's Tariff No. 19) have been made so that aerosol products can be packaged and shipped in low-pressure containers. Those concerned with pressure and size limitations are as follows:

Paragraph 73.306

Exemptions from compliance with regulations for shipping compressed gas.

"(a) General exemptions. Compressed gases, except poisonous gases as defined by Paragraph 73.326 (a) and except those for which no exemptions are provided as indicated by the "No exemption" statement in Paragraph 72.5, when, in accordance with one of the following subparagraphs are, unless otherwise provided, exempt from specification packaging, marking, and labeling requirements, except that marking the name of contents on outside container is required for shipments via carrier by water. . . .

(1) When in containers of not more than 4 fluid oz water capacity (7.22 cu in. or less).

(3) When inside nonrefillable metal containers charged with a solution of materials and compressed gas or gases which is nonpoisonous, provided all of the following conditions are met:

(i) Capacity must not exceed 50 cu in. (27.7 fl oz).

(ii) Pressure in the container must not exceed 180 psig at 130°F. If the pressure exceeds 140 psig at 130°F but does not exceed 160 psig at 130°F, a specification DOT 2P inside metal container must be used; if the pressure exceeds 160 psig at 130°F, a specification DOT 2Q inside metal container must be used. In any event, the metal container must be capable of withstanding without bursting a pressure of one and one-half times the equilibrium pressure of the content at 130°F."

Exemption No. (1) includes glass bottles, plastic containers, etc. If the capacity of the containers does not exceed 4 oz, then there are no pressure restrictions. However, even if the capacity does exceed 4 oz, the product still will not be regulated if the pressure is sufficiently low so that the product is not classified as a compressed gas according to the definition in Paragraph 73.300. If the product is packaged in a container with a capacity larger than 4 oz and the pressure results in the product being classified as a compressed gas, then the product must meet the other conditions required for exemption before it can be shipped in aerosol containers.

Exemption (3) (ii) defines the types of metal containers that may, depending upon the pressure of the products, be used. For products in which the pressure does not exceed 140 psig at 130°F, all of the usual metal aerosol containers, either sideseam or drawn, can be used. For products in which the pressure exceeds 140 psig at 130°F but does not exceed 160°F, specification DOT 2P containers must be used. The requirements for a container to be classified as a 2P container are listed in Paragraph 78.33, Specification 2P. These specifications include the type and size of the container, the material of construction, the method of manufacture, wall thickness, and the following tests:

Paragraph 78.33–8—Tests

(a) One out of each lot of 25,000 containers or less, successively produced per day shall be pressure tested to destruction and must not burst below 240-psi gauge pressure. The container tested shall be complete with end assembled.

(b) Each such 25,000 containers or less, successively produced per day, shall constitute a lot and if the test container shall fail, the lot shall be rejected or ten additional containers may be selected at random and subjected to the test under which failure occurred. These containers shall be complete with ends assembled. Should any of the ten containers thus tested fail, the entire lot must be rejected. All containers constituting a lot shall be of like material, size, design, construction, finish and quality.

The containers must also be identified, as indicated in the following paragraph:

Paragraph 78.33–9—Marking

(a) On each container by printing, lithographing, embossing, or stamping "DOT 2P" and manufacturer's name or symbol. If the symbol is used, it must be registered with the Bureau of Explosives.

Until recently, the only containers that met 2P specifications were the 6-oz drawn containers and the 12-oz drawn containers that had been manufactured with a sufficiently heavy bottom. However, some side-seam containers that meet the 2P specifications are now available.

If the pressure exceeds 160 psig at 130°F but does not exceed 180 psig, then specification DOT 2Q metal containers must be used. The requirements for a container to be classified as a 2Q container are listed in Paragraph 78.33a Specification 2Q. Besides the specifications concerning the type, size of container, etc., the container must pass the following tests:

Paragraph 78.33a–8—Tests

(a) One out of each lot of 25,000 containers or less, successively produced per day, shall be pressure tested to destruction and must not burst below 270-psi gauge pressure. The container tested shall be complete with end assembled.

(b) Each such 25,000 containers or less, successively produced per day, shall constitute a lot and if the test container shall fail, the lot shall be rejected or ten additional containers may be selected at random and subjected to the test under which failure occurred. These containers shall be complete with ends assembled. Should any of the ten containers thus tested fail, the entire lot must be rejected. All containers constituting a lot shall be of like material, size, design, construction, finish and quality.

The containers must also be identified, as shown in the following paragraph:

Paragraph 78.33a–9—Marking

(a) On each container by printing, lithographing, embossing, or stamping "DOT 2Q" and manufacturer's name or symbol. If the symbol is used, it must be registered with the Bureau of Explosives.

The present regulations may be summarized as follows:

Glass Bottles.   If the capacity is 4 oz or less, there are no pressure regulations. Most loaders limit the pressures to about 25 psig at 70°F for coated bottles and about 15 psig for uncoated bottles. If the capacity is over 4 oz, and the product is classified as a compressed gas, i.e., a pressure greater than 40 psia at 70°F, the product cannot be shipped in interstate commerce.

Metal Containers.   For containers with capacities not over 4 oz, the above pressure regulations apply. For those over 4 oz,

(a) pressures may not exceed 140 psig at 130°F for the low-pressure containers.

(b) for pressures between 140 psig and 160 psig at 130°F, a specification 2P container must be used.

(c) for pressures between 160 psig and 180 psig at 130°F, a specification 2Q container must be used.

(d) products with pressures over 180 psig at 130°F are classified as compressed gases and must be considered nonexempt under the standard regulations.

The exemptions for foodstuffs, soap, cosmetics, beverages, biologicals, etc., are covered separately in Paragraph 73.306 (b).

REFERENCES

1. M. Johnsen, *Aerosol Age* **7**, 20, 29, 39 (June–September 1962).
2. W. C. Beard, "Valves," in H. R. Shepherd, "Aerosols: Science and Technology," Interscience Publishers, Inc., New York, N.Y., 1961.
3. P. W. Sherwood, *Chem. Age India* **11**, 90 (1961).
4. "Aluminum Cans Setting Records," *Light Metal Age* **11** (February 1963).
5. A. Taranger, *Soap Chem. Specialties* **33**, 109 (July–August 1957).
6. "Freon" Aerosol Report FA-14, "Handling 'Freon' Fluorinated Hydrocarbon Compounds in the Laboratory."
7. R. H. Thomas, Proc. 38th Mid-Year Meeting, CSMA (June 1952).
8. "Plastic-Coated Glass Pressure Containers," Brochure by Owens-Illinois.
9. Brochure on "Aerosol Glass Bottles," Wheaton Plastic-Cote Corporation.
10. R. H. Thomas, "Glass and Plastic Containers," in H. R. Shepherd, "Aerosols: Science and Technology, Interscience Publishers, Inc., New York, N.Y., 1961.
11. J. N. Owens, *Aerosol Age* **4**, 20, 53 (April–May 1959).
12. J. C. Pizzurro and R. Abplanalp, *Aerosol Age* **2**, 37 (June 1957).
13. J. J. May, *Soap Chem. Specialties* **40**, 135 (August 1964).
14. "Aerosol Handbook," Metal Division, Continental Can Company (April 1964).
15. Technical Bulletins, American Can Company, Crown Cork and Seal Company.

6

VALVES
AND ACTUATORS

Aerosol valves have several important functions—they keep the product in the container when the aerosol is not being used, and they allow the product to be discharged when desired. In addition, valves and actuators can have a considerable influence upon the spray or foam properties of the aerosol.

AEROSOL VALVES

There is a wide variety of valves available, and it is usually possible to obtain a valve which will be satisfactory for most aerosol products, regardless of the application. Besides the so-called standard valves used for products such as hair sprays, deodorants, insecticides, etc., metering valves, which deliver specific quantities of product at each discharge, are available. Other valves allow the product to be used in both the upright and inverted positions. Some valves are designed only for foam products while others can be used for either spray or foam products, depending upon the actuator. Valves are available for metal containers, glass bottles, or plastic containers.

Details on all the variations available in valves and actuators may be obtained from the valve companies. Beard[1,2] and Harris and Platt[3] have written excellent review articles on valves.

Standard Spray Valves

Although the various common spray valves may differ somewhat in appearance and construction, basically they operate in about the same way. The stem of the valve, which is open to the atmosphere, contains a small orifice. In the closed position, this orifice is sealed by an elastomeric gasket. When the valve stem is pushed down or tilted to the side, the orifice is moved away from the gasket and down into the aerosol container. The aerosol product is then free to flow through the orifice and out into the atmosphere.

A typical spray valve designed for use with metal containers is shown in both the open and closed positions in Figure 6–1. The major components of the valve are the valve stem or core, valve seat gasket, spring, and valve housing or valve body which holds the spring. These components are assembled into a mounting cup which is crimped into the standard 1-in. opening in a metal container. The mounting cup contains a flowed-in gasket which prevents leakage between the cup and the container. The constriction at the bottom of the housing is called the tail piece. The dip tube is connected to the tail piece.

The stem is hollow from the top down to a point slightly below the inlet orifice and is sealed at the bottom. The valve housing is constructed so that a shoulder on the housing fits tightly against the valve seat gasket. The stem extends down into the spring as shown. When the valve is not actuated and is in the off or closed position, the spring inside the housing forces a shoulder on the stem up against the valve seat gasket. There is a groove in the stem which contains the inlet orifice. When the valve is in the off position, the inlet orifice is closed by the valve seat gasket which fits around and into the groove. When the stem is depressed by pushing down on the actuator, the inlet orifice is moved down and away from the gasket so that the inlet orifice is open to the interior of the aerosol. This is shown in Figure 6–1. The aerosol product is then free to leave the container. The pressure in the aerosol forces the product up the dip tube

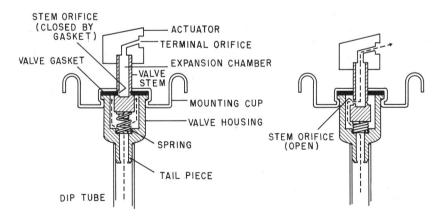

Figure 6–1   Spray valve.

and into the valve housing. From here it enters the exposed inlet orifice in the stem, travels up the stem and out the orifice in the actuator. Many valves are constructed so that the product is discharged when the valve is tilted to one side. Tilting of the valve uncovers the inlet orifice and allows the product to be discharged.

The inlet orifice in the stem is also called the metering orifice since its size determines the rate of discharge of the aerosol product. Valves are available with inlet orifices varying from about 0.013–0.030 in.

Valve stems while they are usually constructed from nylon may also be made from acetal resin or metal. Brass stems are used for formulations that might soften plastic.

The expansion chamber is formed by the hollow portion of the stem. Although the value of an expansion chamber in promoting small particle size in aerosol sprays was recognized as early as 1901, the first high-pressure insecticides developed during World War II were equipped with an oil burner type nozzle without an expansion chamber. Beard[1] has suggested that the reason for this was that the cooling produced by the vaporization of the high pressure propellant precipitated the insecticide when the propellant was allowed to vaporize before leaving the nozzle. Also, the nozzles on the insecticides were constructed so that they imparted a swirling motion to the product and some breakup of the spray was obtained by mechanical action.

Subsequently, exemptions to the ICC regulations allowed the packaging of aerosol insecticides in the low pressure beer cans if the pressure did not exceed 40 psig at 70°F. It was found that valves without expansion chambers did not provide as fine a spray as desired. In order to obtain a valve which would produce a spray with the most uniform particle size distribution, Fulton[4] and his coworkers investigated the effect of variations in orifice and expansion chamber dimensions upon particle size. They found that the most effective valve had an expansion chamber with a capacity of 0.008 cu in. and inlet and outlet orifices with diameters of 0.015 and 0.021 in., respectively. The most uniform sprays were obtained when the ratio of the areas of the inlet and outlet orifices was about 2:3. The difference in pressure in the expansion chamber and in the container was about 5 psi.

Valve seat gaskets are generally made of neoprene or Buna N. Specially compounded gaskets are available for formulations that contain an appreciable concentration of liquids with high solvent power, such as methylene chloride.

The housing or valve body serves to hold the spring and to provide a connection for the dip tube. The valve body may be made from metal or from plastics such as "Zytel" Type 101 nylon or "Delrin" acetal resin. The lower end of the valve body or *tail piece* usually has an i.d. of about

0.060–0.080 in. Valves with additional orifices added to the tail piece are available. The purpose of this is to obtain a further reduction in spray rate or to add another expansion chamber. Clogging may be a problem with the orifice in the tail piece because there is no wiping action of the valve components to clean the orifice.

The dip tubes are normally constructed of polyethylene. Considerable attention must be directed to the behavior of dip tubes in aerosol formulations. If the product causes stress cracking or the dip tube does not fit the tail piece properly, the dip tube may fall off during aging or during pressure filling. Some formulations cause excessive swelling and growth of dip tubes and this should be checked before a formulation is marketed. Dip tubes are usually attached to the outside of the tail piece. However, capillary dip tubes are sometimes used, particularly for water based products, and these may be inserted inside the tail piece. A detailed discussion of dip tubing, which covers the historical development, manufacture, materials of construction, and problems associated with dip tubes, has been published by Seckel.[5]

The mounting cup is usually made from one pound tinplate and may be obtained either unlined or coated with epoxy or vinyl resins. The sealing or flowed-in gasket in the mounting cup is added by applying liquid neoprene compound to the side and in the lip of the rotating cup. The liquid neoprene is dried and vulcanized in place.

Metering Valves

Metering valves are designed to deliver specific quantities of a product each time the valve is actuated. Metered quantities are necessary for many pharmaceutical products where the dosage of the drug must be controlled. Metering valves are also useful for products such as the more expensive perfumes where it is desirable to limit the quantity of material sprayed in order to avoid waste. Metering valves that deliver quantities ranging from a fraction of 1 cc up to 15 cc with each depression of the valve are available.

Metering valves are generally used with glass bottles or small metal containers. The principles involved in the construction of glass bottle valves are essentially the same as those for valves for metal containers. However, glass bottle valves are attached to the bottles by crimping a metal ferrule around the outside of the lip of the bottle while valves for metal containers are attached by expanding a cup inside the 1-in. opening in the container. The valve seat gasket in the glass bottle valve rests on the top of the bottle; therefore, the only metal that comes in contact with the aerosol formulation is the spring in the valve cup. In valves for metal containers, the metal of the mounting cup is exposed to the aerosol formulation.

The construction of one type of metering valve for glass bottles is illustrated with a simplified diagram in Figure 6–2. In this valve, the valve body or housing is the metering chamber, and its size determines the quantity of product that is discharged. While the aerosol is being discharged, the metering chamber is sealed off from the rest of the product in the container so that only the material in the metering chamber is released. When the product is not being used, the metering chamber is closed to the atmosphere and open to the interior of the container.

As shown on Figure 6–2, the valve stem is constructed with a shoulder, below which is a spring. An elastomeric gasket is located at the bottom of the spring and rests on the opening from the metering chamber to the tail piece. The inlet orifice in the valve stem is located in the upper portion of the stem and when the valve is in the off position, the orifice is above the valve seat gasket and outside the valve cup. The lower end of the valve stem is flattened and sealed at the bottom. In the off position, the flattened portion of the valve stem extends into the elastomeric gasket at the bottom of the metering chamber. Because the stem is flattened at this point, the gasket does not make a complete seal around the valve stem. The aerosol product fills the metering chamber through the opening between the gasket and the flattened portion of the valve stem.

When the valve is depressed, the rounded portion of the valve stem, which is located above the flattened section, ultimately passes through the gasket at the bottom of the metering chamber. When this occurs, the

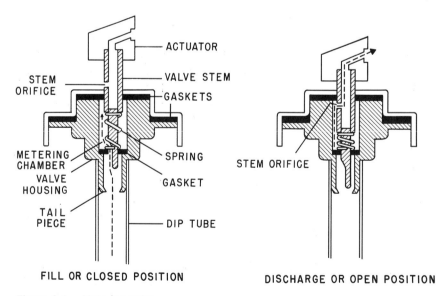

FILL OR CLOSED POSITION   DISCHARGE OR OPEN POSITION

Figure 6–2   Metering valve.

metering chamber is sealed off from the rest of the product in the aerosol container because the bottom gasket makes a complete seal around the rounded part of the valve stem. This takes place while the inner orifice is still located above the valve seat gasket in the upper part of the valve. Upon further depression of the valve stem, the part of the stem containing the inlet orifice is pushed down past the valve seat gasket and into the metering chamber. The product in the metering chamber is then discharged by passage through the inlet orifice, into the hollow portion of the valve stem, and out through the outer orifice in the actuator. When the valve is released, the portion of the stem containing the inlet orifice rises through the valve seat gasket and closes off the metering chamber to the outside. This occurs before the lower flattened portion of the valve stem reaches the lower gasket at the bottom of the metering chamber. After the flattened stem section reaches the lower gasket, the product in the container fills the empty metering chamber through the opening between the gasket and the valve stem in preparation for the next discharge.

Vapor Tap Valves

Vapor tap valves (valves with holes of varying sizes in the lower portion of the spring cup) are available from the valve companies. When these valves are actuated, propellant vapor passes through the hole in the spring cup and mixes with the product that is passing up through the dip tube. This feature produces a softer spray. It has also been reported that the sprays are less chilling than those from valves without vapor taps.

The diameter of the holes in the vapor tap and in the tail piece can be adjusted so that aerosol products can be sprayed either upright or inverted. When the container is inverted, the product passes through the hole in the spring cup and the propellant vapor passes through the end of the dip tube.

A disadvantage of the vapor tap valves is that the loss of propellant through the vapor tap reduces the concentration of propellant in the aerosol. In products formulated with low concentrations of propellants, this may cause a noticeable change in spray characteristics during use of the product. In products that contain combinations of propellants, the composition of the propellants in the vapor phase is usually different from that in the liquid phase. Loss of the propellant from the vapor phase may change the composition of the remaining propellant sufficiently so that a change in spray characteristics during use occurs.

The Gulf SA Valve

The Gulf SA (Spray Anyway) valve is an unusual valve designed to spray in either the upright or inverted position.[6] Essentially, the valve body is equipped with a shoulder, above which is a vapor tap. A sliding washer of metal or plastic is also attached to the valve body. When the container is in the upright position, the washer rests on the shoulder of the valve body and closes off the vapor tap. The valve then functions as a normal valve. In the inverted position, the washer slides away from the shoulder and exposes the vapor tap. The formulation enters the vapor tap and is discharged through the valve. There has been no widespread use of this valve as yet.

Foam Valves

Foam valves differ principally from spray valves in that the foam valves have a large diameter delivery spout. This delivery spout serves as the expansion chamber, and the end of the delivery spout is the outer orifice. The inner orifice is usually large. Some valves are manufactured so that either a spray actuator or a foam actuator can be used. Other valves may be used only for foam products. When a spray actuator is used for a foam product, the product usually discharges as a foamy stream. With a foam actuator, expansion takes place in the delivery spout and the product discharges as a foam.

Actuators or Buttons

Actuators are available in a wide variety of sizes and shapes. Actuators, which spray in either a horizontal or vertical position or at a 45° angle, may be obtained. The usual material of construction is polyethylene. Actuators normally contain the outlet or terminal orifice and are manufactured with orifices varying in diameters from 0.016–0.60 in. A diameter of about 0.016–0.018 in. appears to be standard. Actuators with the orifices having a somewhat larger diameter than that of the metering orifice in the valve stem are usually chosen in order to obtain a more uniform particle size in the spray. Two typical actuators are illustrated in Figure 6–3. One has a straight or standard taper while the other has a reverse taper. The reverse taper produces a somewhat finer spray with a wider cone.

Many aerosol products in which coarse sprays are desired, such as window cleaners or furniture polishes, are formulated either with low concentrations of liquefied gas propellants or with compressed gases. In these products, the concentration of propellant usually is not sufficient to break up the concentrate into a spray when a standard actuator is used and the product is discharged as a stream. In order to obtain a spray from these products, it is necessary to break up the stream by mechanical action, just as a stream of water from a garden hose is broken up by the nozzle. Most of the valve companies manufacture actuators that function by breaking up the stream mechanically. Although these actuators usually have specific

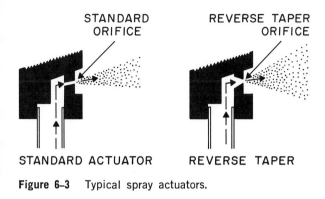

Figure 6–3   Typical spray actuators.

tradenames, they are generally referred to as mechanical breakup actuators. They function by imparting a swirling motion to the product so that it leaves the orifice in the form of a conical sheet. The film breaks up into droplets a short distance from the actuator. The swirling motion of the liquid stream is obtained by a tangential arrangement of the feed channels with respect to the terminal or outlet orifice.

In products that contain sufficient liquefied gas propellant giving a fairly good spray with a standard actuator, a mechanical breakup actuator will generally produce a somewhat finer spray with a wider angle.

Some actuators contain both inlet and outlet orifices. These actuators are manufactured with a stem like tube. The tube fits into the valve and has a slit at the bottom which serves as the internal orifice. These valves are particularly useful for products such as aerosol paints because both the inlet and outlet orifices can be cleaned by removing the actuator. The orifices in the stem are usually round, although they may be rectangular. They vary in diameter from 0.0135–0.055 in.

REFERENCES

1. W. C. Beard, "Valves," in H. R. Shepherd, "Aerosols: Science and Technology," Interscience Publishers, Inc., New York N.Y., 1961.
2. W. C. Beard, *Aerosol Age* 11 (April, May 1966).
3. R. C. Harris, and N. E. Platt, "Valves," in A. Herska, "International Encyclopaedia of Pressurized Packaging (Aerosols)," Pergamon Press, Inc., New York, N.Y., 1966.
4. R. A. Fulton, A. Yeomans, and E. Rogers, Proc. 36th Mid-Year Meeting, CSMA, 51 (June 1950).
5. P. H. Seckel, Proc. 49th Ann. Meeting, CSMA, 34 (December 1962).
6. A. H. Samuel, R. V. Sharpless and A. C. Miller, Proc. 43rd Ann. Meeting, CSMA, 42 (December 1957).

7

LOADING
METHODS

All aerosol propellants are gases at room temperature and atmospheric pressure. However, the only propellants loaded into aerosol containers as gases are the compressed gases, nitrogen, carbon dioxide, and nitrous oxide. All other propellants are liquefied first and then loaded into the containers. The propellants can be liquefied either by cooling them below their boiling point or by applying sufficient pressure at room temperature. In the former case the cold, liquefied propellants are metered directly into open aerosol containers. This method of loading is referred to as cold filling or refrigeration filling. Propellants that are liquefied by pressure at room temperature are forced into the container through the valve. This method of loading is referred to as pressure filling.

The equipment and machinery designed for loading aerosol products varies all the way from simple laboratory apparatus to very elaborate and sophisticated production lines. There are a number of suppliers of both laboratory and production equipment and their names and addresses may be obtained by consulting the Aerosol Buyers Guide, *Aerosol Age* **13** (October, 1968).

LIQUEFIED GAS PROPELLANTS

Cold Filling

The arrangement of a more or less typical commercial cold filling line is illustrated in Figure 7–1. The following steps are involved in the filling process:

STEP 1. The empty aerosol containers are removed from their shipping cartons and placed in the unscrambler. The unscrambler sepa-

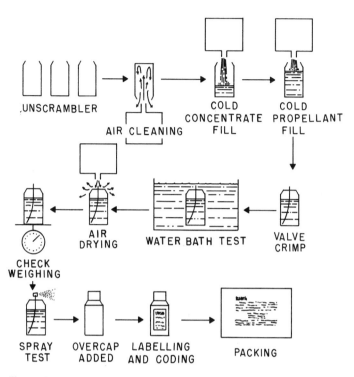

Figure 7-1   Arrangement of a typical cold filling line.

rates the containers and sends them into the line at predetermined intervals.

STEP   2. After the cans leave the unscrambler, they are inverted and cleaned by an air blast which removes dirt and other foreign matter. In some cases, the air cleaning is carried out in the presence of a continuous vacuum.

STEP   3. The concentrate is cooled to a temperature of about $0°- -10°F$, depending upon the product and propellant, and added to the open container. If the concentrate were not cooled, propellant losses would be excessive.

STEP   4. The propellant is cooled to a temperature somewhat below its boiling point, usually between $-20°- -40°F$, and metered into the open container.

STEP   5. The valve is placed on the container and crimped. The can may be coded at this point for identification purposes.

STEP   6. The filled containers are heated to a temperature of $130°F$ in a water bath as required by the DOT regulations. If cans leak or burst in the water bath, this is an indication of a faulty loading operation that must be corrected immediately. Two of the possible causes are improperly crimped valves or overfilled containers that become liquid full at 130°F.

STEP   7. After passing through the water bath, the containers are air dried. This prepares the containers for labelling and coding and also removes water from areas where corrosion could occur, such as the valve cup.

STEP   8. The containers may be weighed at this point to be certain that the product fill is correct. In many plants the check weighing operation is automatic

STEP   9. Actuators are placed on valves if necessary and the product may be spray tested. Other tests, such as those for foam products, may be carried out at this point.

STEP 10. Protective caps or overcaps are placed on the containers.

STEP 11. For the purpose of identification the cans are labelled and coded if this has not been done previously.

STEP 12. The complete aerosol products are packed in shipping cartons.

The cold-fill procedure has the following advantages and disadvantages:

Advantages

(a) Fairly high production rate, compared to the conventional pressure filling method. Large quantities of the cold liquefied propellant can be added rapidly through the 1-in. opening in the metal container. Therefore, the method is advantageous for products formulated with high concentrations of propellant. A typical rate of fill is sixty cans per minute.

(b) Most of the air is removed from the container during the loading. Usually enough of the liquefied propellant vaporizes during the loading to displace the air remaining in the container after addition of the concentrate.

Disadvantages

(a) Expensive refrigeration equipment is required for cooling the concentrate and propellant.

(b) The method is suitable only for products whose concentrates can be cooled without freezing, without precipitation of ingredients or do not become excessively viscous. These points can be checked first by loading the concentrate into glass bottles and observing the concentrate after it has been cooled. In some cases where the problems of precipitation or viscosity of the cold concentrate occur, and where "Freon" 12–"Freon" 11 blends are used for the propellant, it is possible to eliminate these problems by premixing some of the "Freon" 11 propellant with the concentrate. "Freon" 11 is a good solvent, and its presence in the concentrate may prevent precipitation of the active ingredients or prevent excessive increases in viscosity during cooling. The ratio of "Freon" 12 to "Freon" 11 in the propellant blend is adjusted to compensate for the "Freon" 11 that is added to the concentrate.

(c) Ice may form on the cold filling nozzles and subsequently contaminate the product. Excessive moisture in aerosols can contribute to corrosion.

(d) The temperatures of the concentrate and propellant must be adjusted with care in order to obtain adequate removal of air from the container without incurring excessive propellant losses.

In the laboratory, cold filling is carried out by leading the propellant through a set of copper coils cooled below the boiling point of the propellant. The coils can be cooled by enclosing them in a commercial deep freeze unit or in an insulated box. In any case, the coils are equipped with suitable valves and connections for attaching the unit to a propellant cylinder and for allowing the liquefied propellant to flow into an aerosol container. If an insulated box is used, the coils may be cooled with a mixture of dry ice and a solvent. It is advantageous to use a nonflammable solvent, such as "Freon" 11, for safety reasons.

Conventional Pressure Filling

Pressure filling was developed initially for products that could not be cold filled. This included aqueous based products that froze, and products in which the active ingredients were not soluble at the cold-fill temperatures.

Pressure filling differs from cold filling mainly in the way the propellant is added. In the cold fill process, the cold concentrate and cold propellant are both loaded into the container before the valve is added and crimped. In the pressure fill method, the concentrate is added to the container at room temperature, and this is followed by the crimped valve. The propellant is then loaded through the valve under pressure.

A typical conventional pressure loading line is illustrated in Figure 7–2 and involves the following steps:

STEP 1. Unscrambling—Same as for cold filling.
STEP 2. Air cleaning—Same as for cold filling.
STEP 3. The concentrate, at room temperature, is added to the aerosol container.
STEP 4. The air is removed from the vapor phase in the container either

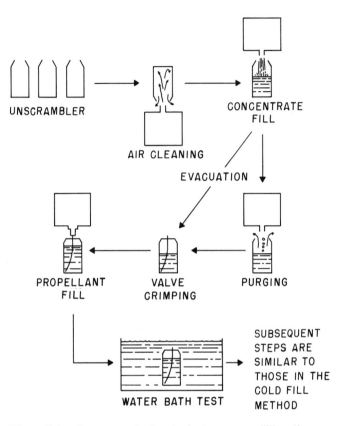

UNSCRAMBLER

AIR CLEANING

CONCENTRATE
FILL

EVACUATION

PROPELLANT
FILL

VALVE
CRIMPING

PURGING

WATER BATH TEST

SUBSEQUENT
STEPS ARE
SIMILAR TO
THOSE IN THE
COLD FILL
METHOD

Figure 7-2   Arrangement of a typical pressure filling line.

by purging with propellant or by evacuation. If the air is removed
by evacuation, the valve is placed on the container and crimped
in the same operation.

STEP 5. If the air is removed by purging, the valve is placed on the container and crimped as soon as the purging operation is completed.

STEP 6. The propellant is pressure loaded through the valve. The total
pressure in the system for loading the propellant is generally about
50 psi higher than the vapor pressure of the propellant.

STEP 7. The filled containers are then heated to a temperature of 130°F
in the water bath.
From here on, the steps are the same as those described for the
cold loading process.

The advantages and disadvantages of the conventional pressure loading
method are as follows:

Advantages

(a) No refrigeration equipment is required.
(b) The concentrate does not have to be cooled; therefore, the problems of insolubility of the active ingredients or high viscosity at low cold-fill temperatures are minimized.
(c) The method is suitable for loading hydrocarbons, or other flammable propellants.

Disadvantages

(a) The method is comparatively slow because the propellant has to be forced through the small valve orifices. This is compensated to some extent by the fact that most of the pressure-loaded products have relatively low concentrations of propellant. A typical rate of loading is about fifteen cans per minute.
(b) Air must be removed either by purging or evacuation.

Rotary Pressure Filling

In recent years, rotary pressure filling units with a multiple number of filling heads have been developed. These have high production rates and this eliminates one of the main disadvantages associated with the conventional pressure filling procedure. For example, one rotary pressure filling unit with fifteen filling heads is reported to have a production rate of 270 cans per minute.[1] Units with a smaller number of heads are available.

Undercap Filling

One of the latest techniques is referred to as undercap filling. The propellant filling head performs three functions. It evacuates the container to remove air, loads the propellant, and crimps the valve.[2] This is illustrated in Figure 7–3. After the concentrate has been added at room temperature,

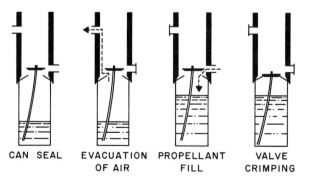

CAN SEAL   EVACUATION   PROPELLANT   VALVE
             OF AIR   FILL   CRIMPING

Figure 7–3   Undercap filling. (*Courtesy of the Kartridg Pak Co.*)

the valve is placed on the container as in the conventional pressure fill procedure. The filling head then contacts the top of the container, forming an airtight seal. The air is removed from the container by evacuation and this holds the valve slightly above the 1-in. opening in the container. The propellant flows in through another port and passes underneath the valve cup and into the container through the 1-in. opening. After the propellant has been loaded, the valve is crimped immediately.

Rotary machines with a multiple number of undercap filling heads are available. The unit with eighteen filling heads is reported to have a production rate of 300 cans per minute.

Other Loading Procedures

Certain perfume products contain impurities that are insoluble in the final aerosol formulation. In this case, loaders have found it advantageous to premix the concentrate and propellant, cool the mixture to a predetermined temperature, and allow the mixture to stand several days. The mixture of propellant and concentrate is then filtered and cold filled. This procedure gives products that are sparkling clear.

Bower and Appenzeller[3] have suggested a pressure loading procedure which involves premixing a portion of the propellant with the concentrate. This provides a concentrate with a slight positive pressure that is self-purging. The concentrate is loaded into the containers at room temperature and the remainder of the propellant is pressure loaded. The method eliminates the need for purging or evacuation steps during pressure loading.

COMPRESSED GASES

Compressed gas aerosols are produced either by loading the propellant through the valve to a predetermined pressure, usually about 80–100 psi, or by saturating the aerosol solvent system with the compressed gas prior to addition to the container. The saturated solvent is loaded into the container by the undercap filling technique. The arrangement of the subsequent equipment is similar to that in the production lines for products with liquefied gas propellants.

When the compressed gas is added through the valve, the concentrates are loaded in the container first, air is removed either by evacuation or by purging with nitrogen, and the valve is added and crimped. If the aerosol is formulated with nitrogen, which is almost insoluble in a concentrate, then the gas is forced through the valve using a piece of equipment called a gasser. The quantity of compressed gas added usually is small. Weight checks are not satisfactory as a means of determining if the correct quantity of propellant has been added. For this reason, pressure checks are made after the gassing operation.

If the partially soluble compressed gases, carbon dioxide and nitrous oxide, are employed as propellants, it is necessary to add a shaking step during the loading of the gas in order to establish equilibrium as rapidly as possible between the gas dissolved in the liquid phase and that in the vapor phase. If equilibrium is not achieved, the amount of gas added will be insufficient and the ultimate pressure will be lower than desired. The filling head in this case is called a gasser-shaker. Rotary undercap fillers for the partially soluble compressed gases are also available.

The latest technique for filling compressed gas aerosols formulated with organic solvents involves impregnating the solvent system with the gas prior to filling. This is accomplished with a piece of equipment called the Saturator. The saturated solvent is then pumped to an undercap filler.

A typical saturator combined with an undercap filler is illustrated in Figure 7–4. The aerosol concentrate is added to the container which is led to the undercap filler. The container is evacuated under the filler head and the desired volume of impregnated solvent is pumped into the container from the saturator. The valve is lowered and crimped. The same system may be used for products formulated with combinations of liquefied and compressed gas propellants. In some instances the entire product can be

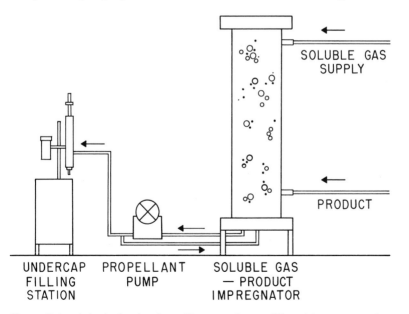

Figure 7–4   A typical saturator with an undercap filler. (*Courtesy of the Katridg Pak Co.*)

passed through the saturator and this eliminates the necessity for adding concentrate to the container prior to undercap filling.

The solvent saturator is discussed in detail by Anthony.[4]

REFERENCES

1. Technical Brochure, "Aerosol Pressure Filler,' The Kartridg Pak Company.
2. Technical Brochure, "Undercap Filling," The Kartridg Pak Company.
3. F. A. Bower and R. G. Appenzeller, Proc. 44th Ann. Meeting, CSMA, December, 1957 (also available as "Freon" Aerosol Report A-47).
4. T. Anthony, *Soap Chem. Specialties* **43,** 185 (December, 1967).

8

SPRAY CHARACTERISTICS

One of the most important objectives in developing an aerosol formulation is to obtain a product with a satisfactory spray. If the product is for space applications, such as a room deodorant or space insecticide, then a fine spray with a small particle size (droplet size) is necessary so that the particles will remain suspended in the air for as long as possible.

At the other extreme are products for residual applications, such as insecticides for crawling insects, where a coarse, wet spray is desirable so that as much of the product remains on the surface as possible. This is desirable, not only for reasons of efficiency, but also to avoid inhalation of the toxic ingredients of the insecticide. In between these two classes of products with very wet or very fine sprays, there is a whole series of aerosols in which sprays with medium particle size are the most suitable, such as personal deodorants, hair sprays, coating compositions, etc. A knowledge of the factors that affect spray characteristics is essential for any chemist involved in formulation studies.

MECHANISM OF SPRAY FORMATION

Formation of a liquid-concentrate spray by liquefied gas propellants disrupting the cohesive molecular state of the concentrate has been discussed to some extent in Chapter 3. However, at this point, a more detailed treatment of some theoretical aspects necessary for comprehending certain variances in spray characteristics seems judicious.

A common misconception about an aerosol product is that all of the propellant vaporizes as soon as the product leaves the outer orifice in the actuator. An investigation by York[1] and Wiener[2] indicated that only a minor proportion of the propellant flashes immediately into vapor. York calculated that if "Freon" 12–"Freon" 11 (50/50) propellant were discharged from an aerosol container, approximately 14% of the propellant would flash into vapor. Wiener carried out a similar series of calculations for mixtures of "Freon" 12 with "Freon" 11 and "Freon" 114. The results are presented in a series of graphs showing the percent of flashing as a function of temperature for each propellant or propellant blend. According to the calculations of Wiener, the percentage of "Freon" 12–"Freon" 11 (50/50) propellant that flashed when sprayed into a room at 70°F would be about 19%.

The process of flashing when the propellant is discharged is so rapid that the energy (heat) required to actuate a liquid–vapor conversion has to be drawn from the propellant itself and not from the surrounding air. The withdrawal of energy cools the remaining liquid-propellant droplets. Flash vaporization and resultant cooling continue until the spray droplets reach their boiling point at atmospheric pressure. This happens almost immediately after the propellant leaves the outer orifice.

The percentage of propellant that flashes during the cooling stage is comparatively low, i.e., about 14%–19% for "Freon" 12–"Freon" 11 (50/50). Further evaporation of the liquid droplets occurs at atmospheric pressure during passage through the air. The energy for this evaporation is acquired from the surrounding atmosphere, and the rate of evaporation is low when compared to that during the initial flash vaporization.

This general picture can be extended to an aerosol product in which flash vaporization in the expansion chamber of the valve, as well as at the outer orifice, takes place. As the liquid-aerosol stream enters the expansion chamber, a portion of the propellant vaporizes. The aerosol spray is then cooled to an equilibrium temperature corresponding to the expansion-chamber pressure. When the resulting vapor–liquid mixture reaches the outer orifice, additional flashing of propellant occurs until the droplets are cooled to a temperature corresponding to the boiling point of the mixture at atmospheric pressure. After these two flash vaporizations, a slower evaporation of droplet liquid transpires with the vapor pressure essentially equal to atmospheric pressure. Once again, the energy necessary for this evaporation is supplied by the surrounding atmosphere. The rate of evaporation depends upon the rate of heat transfer from the atmosphere to the droplets.

If the flashing of the propellant is primarily responsible for the initial breakup of the liquid-aerosol stream into droplets, then the higher the proportion of propellant that flashes, the smaller will be the particle size of the droplets in the spray. According to Wiener, some of the factors that contribute to a high degree of flashing of the propellants are a high specific heat, a low heat of vaporization, and a low boiling point. Propellants with a high specific heat have a comparatively high reservoir of energy readily available for the vaporization; those with a low heat of vaporization require less energy for vaporization; and propellants with a low boiling point (a high vapor pressure) have to be cooled to a greater extent to reach the temperature at which the vapor pressure of the spray equals atmospheric pressure.

Many other factors (the effect of surface tension, effects of drag forces on the droplets, coalescence of the droplets, etc.) are involved in the formation of a spray. However, relatively little is known about the influence of these factors.[1]

From the preceding discussion, it can be determined that the particle size depends upon the distance the particles are from the valve. After the flash vaporizations in the expansion chamber and the outer orifice, the droplets continue to decrease in size as they travel through the air owing to secondary evaporation at atmospheric pressure. The secondary evaporation can be very rapid. Thus, when a solution of "Freon" 12–"Freon" 11 (50/50) propellant is sprayed against the hand at a distance of 1 in. from the valve, a considerable amount of the cold liquid propellant is deposited on the hand. At a distance of 2 ft, all of the propellant has evaporated, and there is no sensation of cold.

When high boiling solvents with a low rate of evaporation are present, the change in droplet size after evaporation of the propellant will be less than that in aerosols formulated with lower boiling solvents, such as ethyl alcohol. Rapid evaporation of the alcohol occurs so that the particles may completely evaporate when only a few feet from the valve.

MEASUREMENT OF PARTICLE SIZE (DROPLET SIZE)

The actual measurement of the droplet sizes from an aerosol spray is a difficult and sometimes almost impossible task. A tentative method for measuring particle sizes of products (insecticides) formulated with high boiling solvents has been developed by the U. S. Department of Agriculture.[3] This method involves spraying the aerosol into a wind tunnel. The droplets in the spray are collected on a silicone-coated microscope slide and the diameters of 200 droplets are determined with a calibrated microscope. The particle-size distribution is indicated by a cumulative weight-percent method that shows the weight percent of the spray with a particle size smaller than or equal to a given diameter in microns. One of the terms used is the *mass median diameter*, which is the droplet size that divides the spray in half so that 50% of the particles (by weight) have a diameter less than the mass median diameter while the remainder consists of particles with equal or larger diameters. The results may also be plotted as a typical frequency distribution curve.

The wind tunnel has sufficient length so that practically all of the propellant flashes off and evaporates from the droplets before they reach the microscope slide. The droplets that reach the slide, therefore, are assumed to have decreased to their ultimate size. This method is satisfactory only for aerosols formulated with high boiling solvents that produce a spray with a relatively fine particle size. The method is not satisfactory for aerosols that have been formulated with lower boiling solvents such as water or ethyl alcohol because these solvents evaporate completely in the wind tunnel, and there is no deposition of droplets on the slide. Thus far, no simple method has been developed for measuring the particle size distribution of sprays of this type. Also, sprays that have very coarse particles cannot be measured satisfactorily in the wind tunnel because many of the droplets fall to the bottom of the tunnel before reaching the slide. Only the finer particles are deposited on the slide, and this does not provide a correct picture of the particle size distribution.

A recent article by Lefebvre and Tregan[4] describes a method for determining particle size distribution in which the aerosol droplets are caught on a thin film of magnesium oxide which has been deposited on a glass slide. According to the authors, a perfectly circular hole is obtained where the droplet impacts the slide, and the diameter of the hole is essentially the same as that of the droplet. Examination of the various holes on the slide indicates the particle size distribution of the spray. Lefebvre and Tregan also discuss the use and limitations of high speed microphotography in investigations of particle sizes in aerosol sprays.

For residual products with very coarse sprays, measurement of the actual particle size is considered to be impractical and difficult.[5] An indication of the particle size of residual products may be obtained, however, by determining the pickup efficiency of the product. Pickup efficiency is defined as the percentage of the low volatile components deposited on a surface and is considered to be a function of the particle size.[6] Pickup efficiency is determined by spraying an aerosol product against a blotter target, clamped to an aluminum sheet, which is suspended from a balance. The amount of low volatile material sprayed from the container can be calculated from the weight of aerosol sprayed, assuming that the composition of the aerosol is known. The weight of low volatile material that reaches the blotter can be determined by the increase in weight of the blotter.

Another procedure for evaluating spray patterns has been described by Root.[7] This method involves spraying the aerosol against a piece of paper that has been treated with a dye-talc mixture. An advantage of this method is that the paper can be photographed after it has been sprayed, thus providing a permanent record of the spray.

Other methods of measuring particle size of aerosols are discussed in References 8–14. These methods include collecting and weighing particles on filters of known pore size and also precipitating the particles with an electrical precipitator in a tube of known weight. These methods are not generally used in the aerosol industry.

Most aerosol laboratories are not equipped with specific apparatus for determining spray characteristics. By standing behind the aerosol as it is sprayed, visual observations of the spray pattern can be made. If the spray appears to rise or disappears shortly after leaving the outer orifice, the spray is considered to be fine or very fine. If the spray travels out in a horizontal path, the spray is classified as medium, and if the droplets fall to the ground, it is a coarse spray. Another method for discerning the type of spray would be to spray the product against a sheet of paper or the hand and note how wet or dry the spray is.

For consistency in visual classification of sprays it is helpful to prepare a series of standard aerosols. For example, a series of "Freon" 12–ethyl alcohol combinations can be prepared in which the "Freon" 12–ethyl alcohol ratio varies from 90/10–10/90 in increments of 10% of the components. The sprays from the control samples vary from very fine to a stream and can be classified by visual observation as very fine, fine, medium-fine, etc. Whenever it is necessary to determine the type of spray of an actual aerosol product, the product can be compared with samples in the control series. The aerosol product is then given the same classification as the control sample it most closely resembles.

FACTORS AFFECTING SPRAY CHARACTERISTICS

The spray characteristics of aerosols may be varied by changing the concentration or type of propellant, valve and actuator, or solvent. Major changes, however, are usually obtained by varying the concentration or type of propellant.

Variation in Propellant Concentration

An increase in the concentration of propellant decreases the particle size. There are several ways of explaining this effect. It requires energy to break up the liquid aerosol stream into particles or droplets, and since the propellant is the source of the energy, it would be expected that as the concentration of propellant was increased, the particle size would decrease.

At high concentrations of propellant, there is a comparatively large quantity of energy available to break up a relatively low quantity of concentrate and this gives small particles. At lower concentrations of propellant, however, there is proportionately less energy available and this has to be distributed throughout a much larger quantity of concentrate.

Another factor is the dilution effect of increasing quantities of propellant. As the proportion of propellant is increased, the molecules of the concentrate become more and more separated from each other. This weakens the forces of attraction between the molecules of the concentrate. Consequently, when the aerosol is sprayed, the propellant evaporates and the concentrate molecules are too far apart to form large droplets.

The concept of propellant flashing can also be helpful in understanding why increasing the propellant concentration decreases particle size. Wiener[2] showed that as the percentage of propellant flashing into vapor increased, the particle size decreased. Any factor, therefore, that influences the extent to which the propellant flashes will also affect the particle size. One of the factors that determined the amount of propellant that flashed was the vapor pressure (or boiling point) of the propellant. Propellants with high vapor pressure (low boiling points) flashed to a greater extent than propellants with low vapor pressures (high boiling points). The same principle applies when propellants are mixed with solvents of the concentrate such as ethyl alcohol. For example, assume that an aerosol is formulated with mixtures of "Freon" 12–"Freon" 11 (50/50) propellant and ethyl alcohol. At low concentrations of propellant the solution will have a low vapor pressure because ethyl alcohol functions as a vapor pressure depressant for the propellant. The extent to which the propellant flashes during discharge is low, and consequently, the particle size is large. As the concentration of propellant is increased and that of the alcohol is decreased, the vapor pressure increases and the extent to which the propellant flashes also increases. Therefore, the particle size decreases.

TABLE 8-1   EFFECT OF VARIATION IN PROPELLANT/ETHYL ALCOHOL RATIO UPON SPRAY CHARACTERISTICS

| Composition of Mixture (wt %) | | | |
"Freon" 12–"Freon" 11 (50/50) Propellant	Ethyl Alcohol	Vapor Pressure (psig at 70°F)	Spray Characteristics
25	75	10.0	Coarse
50	50	25.5	Medium
75	25	33.5	Medium to Fine
90	10	36.5	Very Fine

Th effect of varying the concentration of propellant upon spray characteristics is illustrated by the data in Table 8–1. The pressures of the mixtures listed in Table 8–1 were obtained on samples prepared by cold filling.

Variation in Propellant Type

Another method for changing the spray characteristics of aerosols is to vary the type of propellant. Low boiling propellants with high vapor pressures give finer sprays than higher boiling propellants with lower vapor pressures. The principle involved is the same as that discussed in the previous section with respect to propellant flashing. Formulating a product with a propellant that has a high vapor pressure will result in a greater degree of flashing than when a propellant with a lower vapor pressure is used.

TABLE 8-2   EFFECT OF VARIATION IN TYPE OF PROPELLANT UPON THE SPRAY CHARACTERISTICS OF PROPELLANT/ETHYL ALCOHOL MIXTURES

Composition of mixtures—75% Propellant–25% Ethyl Alcohol

Propellant	Spray Characteristics	Pressure of Mixtures (psig at 70°F)
"Freon" 12	Very Fine	65.5
"Freon" 12–"Freon" 11 (50/50)	Medium to Fine	33.5
"Freon 12–"Freon" 11 (30/70)	Medium	21.5
"Freon" 11	No Spray	—

The data in Table 8–2 show the effect upon the spray characteristics of mixtures formulated with 75% propellant and 25% ethyl alcohol as the ratio of "Freon" 12 to "Freon" 11 varies. "Freon" 12 propellant has a high vapor pressure and a low boiling point, and gives a fine spray. "Freon" 11 boils at about room temperature and has a low vapor pressure. The product formulated with "Freon" 11 gives no spray. The particle size of the spray becomes finer as the proportion of "Freon" 12 in the propellant increases.

Effect of Valves and Actuators

The effect of valves and actuators upon spray characteristics was discussed briefly in Chapter 6. Because of the wide variety of valves available, no attempt is made to correlate valve design and particle size of the sprays.

Variations in valves have a pronounced effect upon spray rate as shown in Table 8–3. The data were obtained by packaging a 50% "Freon" 12–50% ethyl alcohol mixture with four different valves and determining the spray rate at 70°F.

TABLE 8-3   EFFECT OF VARIATION IN VALVE AND ACTUATOR UPON SPRAY
RATE-50% "FREON" 12 PROPELLANT—50% ETHYL ALCOHOL

Pressure = 54–55 psig *at 70°F*

Type of Valve			Type of Actuator		
Body Orifice	Stem Orifice	Comments	Comments	Orifice	Spray Rate (g/sec)
0.013	0.018	Low spray rate	Standard	0.018	0.67
0.080	0.018	Standard	Standard	0.018	1.25
0.080	0.030	Vapor tap (0.020)	Standard	0.018	1.39
0.080	0.030	Standard	Standard	0.018	1.94

The following conclusions can be drawn:
1. Decreasing the body orifice from 0.080 to 0.013 decreased the spray rate.
2. Increasing the stem orifice from 0.018 to 0.030 increased the spray rate.
3. The addition of the vapor tap decreased the spray rate.

As previously mentioned, mechanical breakup actuators generally have a very marked effect upon spray properties in comparison with the so-called standard actuators. Mechanical breakup actuators generally give a finer spray with a wider cone than the standard actuators.

Solvents

The type of solvent or solvents used in an aerosol product will also influence the particle size of the spray. When aerosols are formulated with a relatively volatile solvent such as ethyl alcohol, the alcohol may evaporate completely after the spray has traveled only a short distance from the valve. The particle size will then be very fine and the spray will feel dry to the hand. On the other hand, high boiling solvents, such as kerosene, evaporate much more slowly in the air and the spray from an aerosol with kerosene as the solvent will have a larger particle size and the spray will feel wetter to the hand.

Temperature

The temperature at which an aerosol product is used must also be considered in formulation studies. As the temperature increases, the spray becomes finer. The extent to which an aerosol propellant flashes is a function of the difference between the temperature of the propellant in the aerosol container and the boiling point of the propellant. At higher temperatures, more of the propellant will have to flash before the droplets are cooled to their equilibrium temperature.

Most aerosols are formulated and used at room temperature so that temperature is not a very important factor. However, there are exceptions, such as suntan sprays that will be used on hot beaches, and windshield de-icers, which are used in cold weather.

COMPARING THE SPRAY PROPERTIES OF AEROSOL PRODUCTS

Quite often it is desirable to replace the original propellant in an aerosol product with a different propellant while maintaining the original spray properties of the aerosol. An empirical method has been developed[12] for determining what concentrations of different propellants will produce about the same spray properties with a given concentrate. The method is based upon a modification of the method for determining pickup efficiencies of residual products; it may be used for comparing the spray properties of aerosols with fine or coarse sprays. It assumes that within certain limits different propellants appear to give approximately the same spray characteristics with the same solvent (or concentrate) when the propellant–solvent ratios are adjusted to give the same pickup efficiency.

As an example, it was determined that mixtures of various propellants with odorless mineral spirits that gave a pickup efficiency of 70% had the following compositions as shown in Table 8–4.

TABLE 8-4 PROPELLANT–ODORLESS MINERAL SPIRITS MIXTURES WITH PICKUP EFFICIENCIES OF 70%

Propellant	Propellant–Odorless Mineral Spirits Ratio (wt %)
"Freon" 12	63/37
"Freon" 12–"Freon" 11 (70/30)	71/29
"Freon" 12–"Freon" 11 (50/50)	81/19

As an illustration of the use of the data, assume that an aerosol product has been formulated with odorless mineral spirits as the solvent and

"Freon" 12 propellant and that it is desirable to use "Freon" 12–"Freon" 11 (50/50) as the propellant in place of "Freon" 12. The data in Table 8–4 indicate that the ratio of propellant to concentrate would have to be increased from 63/37 to about 81/19 in order to maintain the same spray properties.

Data that illustrate the effect of variation in pickup efficiency of combinations of "Freon" 12–"Freon" 11 and "Freon" 12–"Freon" 114 propellants with odorless mineral spirits and ethyl alcohol are available in Reference 12.

REFERENCES

1. J. L. York, *J. Soc. Cosmetic Chemists* **7**, 204 (1956).
2. M. V. Wiener, *J. Soc. Cosmetic Chemists* **9**, 289 (1958)
3. "A Tentative Method for Determination of the Particle Size Distribution of Space Insecticide Aerosols," Proc. 43rd Ann. Meeting, CSMA (December 1956) ("Freon" Technical Bulletin A-45).
4. M. Lefebvre and R. Tregan, *Aerosol Age* **10**, 31, 32 (July, August 1965).
5. A. H. Yeomans, Proc. 40th Ann. Meeting, CSMA (December 1953).
6. "A Tentative Method for Determining Pickup Efficiency of Residual Aerosol Insecticides," Proc. 43rd Mid-Year Meeting, CSMA (May 1957) ("Freon" Technical Bulletin A-46).
7. M. J. Root, *Aerosol Age* **2**, 21 (August, 1957).
8. R. A. Fulton, *Aerosol Age* **4**, 22 (July 1959).
9. K. R. May, *J. Sci. Instr.* **22**, 187 (1945).
10. J. M. Pilcher, R. I. Mitchell, and R. E. Thomas, Proc. 42nd Ann. Meeting, CSMA, 74 (December 1957).
11. W. B. Tarpley, *Aerosol Age* **2**, 38 (December 1957).
12. P. A. Sanders, *Aerosol Age* **11** (January, February 1966) ("Freon" Aerosol Report A-65).
13. R. R. Irani and C. F. Callis, "Particle Size, Measurement, Interpretation and Application," John Wiley and Sons, Inc., New York, N.Y., (1963); see also, *Drug Cosmetic Ind.*, **93**, 567 (October 1963).
14. J. H. Burson et al. *Rev. Sci. Instr.* **34**, 1023 (1963).

9

VAPOR
PRESSURE

The vapor pressure of a liquid is defined as the pressure of the vapor that is in equilibrium with the liquid. The vapor pressure of any single liquid is constant at a given temperature regardless of the quantity of liquid that is present. This means that in most aerosol products formulated with liquefied gases, the pressure remains essentially constant during the life of the package. This is one of the advantages of a liquefied gas as a propellant compared to a compressed gas.

The vapor pressure of an aerosol product is an important property. It must be determined because of the restrictions that the DOT regulations place upon the pressure of aerosol products. It is also useful for quality control. Pressure measurements on a product during loading serve as a check on such factors as the quantities of materials added to the containers, effectiveness of purging to remove air, etc.

RAOULT'S LAW

Practically all aerosol products consist of a mixture of propellant with other components, such as solvents and active ingredients. If the components are soluble in the propellant, then these materials will lower the vapor pressure of the propellant. The reason for this is that the vapor pressure of a propellant is determined by the frequency with which the molecules escape from the surface of the propellant and enter the vapor phase. When some other material is dissolved in the propellant, the concentration of molecules of propellant in the surface is decreased and therefore the rate of escape of the molecules of the propellant is decreased. The material dissolved in the propellant could be solid, liquid, or gas.

The lowering of the vapor pressure of a liquid by the addition of another substance is known as Raoult's Law. Raoult's Law can be stated as follows, where the term *solvent* refers to the original liquid and the term *solute* refers to the material that is added to the solvent:

The depression of the vapor pressure of a solvent upon the addition of a solute is proportional to the mole fraction of solute molecules in the solution.

It is also proportional to the mole fraction of the solvent. This can be expressed by the following equation, where P_s is the vapor pressure of the solution, P is the vapor pressure of the pure solvent, and x is the mole fraction of the solvent.

$$P_s = xP \qquad (9\text{-}1)$$

If the material that is added to the liquid has an appreciable vapor pressure itself, then its vapor pressure will also be lowered as a result of dilution by the initial solvent. Thus, the addition of "Freon" 11 propellant to "Freon" 12 lowers "Freon" 12's vapor pressure; and conversely, the addition of the "Freon" 12 will decrease "Freon" 11's vapor pressure. The lowering of the vapor pressure of the solute is known as Henry's Law. Raoult's Law applies to the solvent and Henry's Law to the solute.[1]

DALTON'S LAW OF PARTIAL PRESSURES

The total pressure in any system is equal to the sum of the individual or partial pressures of the various components. The basis for this is that when two or more gases occupy the same space, neither of the gases interferes with the pressure of the other. In 1802 John Dalton observed that in any mixture of gases, each gas exerted the same pressure as if it were alone in the volume occupied by the gases. Therefore, the total pressure of the mixture is equal to the sum of the partial pressures of the gases in the mixture. This is known as Dalton's Law.

CALCULATION OF VAPOR PRESSURE

When combined, Raoult's and Dalton's laws, provide a means of calculating the vapor pressure of any solution if the composition of the solution and the molecular weights and vapor pressures of the pure components are known. In carrying out the calculation, the partial pressures are obtained first by multiplying the mole fraction of each component by its vapor pressure. The total pressure of the solution is obtained by adding together the partial pressures of the components.

For example, the vapor pressure at 70°F of a mixture of 30% "Freon"

12 and 70% "Freon" 11 can be calculated as shown below. The vapor pressures and molecular weights of "Freon" 12 and "Freon" 11 are as follows:[4]

"Freon" 12

$$\text{Vapor pressure at } 70°F = 84.9 \text{ psia}$$
$$\text{Molecular weight} = 120.9$$

"Freon" 11

$$\text{Vapor pressure at } 70°F = 13.4 \text{ psia}$$
$$\text{Molecular weight} = 137.4$$

1. Assume there are 100 g of solution. The moles of the two components in 100 g of the propellant blend are obtained by dividing the weights of the components by their molecular weights.

$$\text{Moles "Freon" 12} = \frac{30}{120.9} = 0.248$$

$$\text{Moles "Freon" 11} = \frac{70}{137.4} = 0.509$$

$$\text{Total moles in mixture} = 0.757$$

2. The mole fraction of each component is obtained by dividing the moles of each component by the total moles in the mixture.

$$\text{Mole fraction "Freon" 12} = \frac{0.248}{0.757} = 0.33$$

$$\text{Mole fraction "Freon" 11} = \frac{0.509}{0.757} = 0.67$$

3. The partial pressure of each component at a specific temperature is obtained by multiplying the absolute vapor pressure of the pure component by its mole fraction in the liquid phase.

Partial pressure of "Freon" 12 = 0.33 × 84.9 psia = 28.0 psia
Partial pressure of "Freon" 11 = 0.67 × 13.4 psia = 9.0 psia

4. The total pressure is obtained by adding the partial pressure of the components.

$$\text{Total pressure of "Freon" 12/"Freon" 11 (30/70)}$$
$$= 28.0 + 9.0 = 37.0 \text{ psia at } 70°F \text{ or } 22.3 \text{ psig}$$

The same method of calculating vapor pressures is used regardless of whether the added component is a solid, liquid or gas. If a mixture were prepared consisting of 30% "Freon" 12 and 70% of a soluble solid with a molecular weight of 137.4, the vapor pressure of the solution would be calculated as follows, assuming the vapor pressure of the solid was essentially zero:

Moles "Freon" 12 $= \dfrac{30}{120.9} = 0.248$

Moles solid $\quad = \dfrac{70}{137.4} = 0.509$

Total moles $\quad\quad\quad = 0.757$

Mole fraction of "Freon" 12 $= 0.33$
Mole fraction of solid $\quad = 0.67$
Partial pressure of "Freon" 12 $= 0.33 \times 84.9 = 28.0$ psia
Partial pressure of solid $\quad = 0.67 \times 0 = 0$ psia
Total pressure of solution $= 28.0 + 0 = 28.0$ psia or 13.3 psig at 70°F

CALCULATION OF THE COMPOSITIONS OF BLENDS WITH SPECIFIC VAPOR PRESSURES

Sometimes a propellant blend with a specific vapor pressure is desired. If the composition of the blend is not known, it can be calculated if the vapor pressures and the molecular weights of the individual components are known.

Two Component Blends

The composition of a two-component blend with a specified vapor pressure can be calculated using the following equation:

$$P_1 (1 - X_2) + P_2 X_2 = P \qquad (9\text{–}2)$$

where
    $P_1 =$ Vapor pressure (psia) of component 1
    $P_2 =$ Vapor pressure (psia) of component 2
    $P \;=$ The specified vapor pressure of the blend that is desired.
    $X_1 =$ Mole fraction of component 1 in the liquid phase of the blend.
    $X_2 =$ Mole fraction of component 2 in the liquid phase of the blend.

Since P_1, P_2 and P are known, X_2 can be calculated. X_1 is obtained by subtracting X_2 from 1 since the sum of X_1 and X_2 must equal 1.

The preceding equation is derived as follows:
    Let $Pa_1 =$ partial pressure of component 1 in the blend
           $Pa_2 =$ partial pressure of component 2 in the blend

1. $Pa_1 + Pa_2 = P$
    This follows from Dalton's Law of partial pressures.

2. $P_1X_1 = Pa_1$
    $P_2X_2 = Pa_2$
    This follows from Raoult's Law.

3. Therefore,
    $$P_1X_1 + P_2X_2 = P \tag{9-3}$$

4. $$X_1 + X_2 = 1, \text{ or } X_1 = 1 - X_2 \tag{9-4}$$

5. Substituting the value of X_1 from Equation 9–4 into Equation 9–3 gives
    $$P_1 (1 - X_2) + P_2X_2 = P \tag{9-2}$$

The composition of the blend in weight percent can be obtained by multiplying the mole fractions by the molecular weights. This gives the weight in grams of each component in one mole of the blend. Thus, if Mwt_1 and Mwt_2 are the molecular weights respectively of components 1 and 2,

X_1Mwt_1 = grams of component 1 in one mole of blend
X_2Mwt_2 = grams of component 2 in one mole of blend

$$\text{Wt \% of component 1 in blend} = \frac{\text{grams component 1}}{\text{total grams}} \times 100$$

$$\text{Wt \% of component 2 in blend} = \frac{\text{grams component 2}}{\text{total grams}} \times 100$$

As a specific example, assume that the composition of a blend of "Freon" 12 and isobutane with a vapor pressure of 40 psig (54.7 psia) at 70°F is desired.

1. Let X_1 = Mole fraction of "Freon" 12 in the liquid phase
        X_2 = Mole fraction of isobutane in the liquid phase
        P_1 = Vapor pressure of "Freon" 12 at 70°F = 84.9 psia
        P_2 = Vapor pressure of isobutane at 70°F = 45.7 psia
        P = Desired vapor pressure of blend = 54.7 psia

2. Substituting in Equation 9–2 gives
    $$84.9 (1 - X_2) + 45.7X_2 = 54.7$$

3. Solving for X_2 gives a value of 0.77 for the mole fraction of isobutane. The mole fraction of "Freon" 12 is, therefore, 0.23.

4. The composition of the blend in weight percent is determined as follows:

Wt in grams of "Freon" 12 in one mole = 0.23 × 120.9 = 27.8 g
Wt in grams of isobutane in one mole = 0.77 × 58.0 = 44.6 g
Total weight of one mole in grams   = 72.4 g

$$\text{Wt \% "Freon" 12 in blend} = \frac{27.8}{72.4} \times 100 = 38.4\%$$

$$\text{Wt \% isobutane in blend} = \frac{44.6}{72.4} \times 100 = 61.6\%$$

Therefore, the composition of the "Freon" 12–isobutane blend that has a calculated vapor pressure of 54.7 psia at 70°F is 38.4% "Freon" 12 and 61.6% isobutane.

Three Component Blends

When a three-component blend with a specific vapor pressure is desired, the problem becomes a little more complicated because there are many different combinations of the three components that have the same vapor pressure. One approach to the problem of three-component mixtures is to calculate the compositions of several different blends with the specified vapor pressure and plot the data on a triangular coordinate chart. The resulting plot will show all possible combinations of the three components that have the desired vapor pressure. Since there are a number of different three-component blends with the same vapor pressure, the selection of a final single blend will of necessity depend upon some additional factor such as cost, flammability, density, etc.

In order to construct the vapor pressure plot on the triangular coordinate chart, it is not necessary to calculate the compositions of three-component blends. The two points at each end of the vapor pressure plot fall on the base lines for two-component blends. Therefore, all that is required are the compositions of the two different two component blends that have the same vapor pressure as that specified for the three-component blend. These can be calculated by using Equation 9–2. These compositions can be plotted on the triangular coordinate chart, and a straight line can be drawn between the points. All possible combinations of the three components with the same vapor pressure will fall on this line. The points at the end of the line represent the compositions of the two-components blends with the specified vapor pressure; all other points represent the combinations of three components with the same vapor pressure.

As an example, assume that the various compositions of "Freon" 12, "Freon" 11, and isobutane that have a vapor pressure of 54.7 psia (40 psig) at 70°F are desired. The vapor pressures of the three components are as follows:

Vapor pressure of "Freon" 12 at 70°F = 84.9 psia
Vapor pressure of "Freon" 11 at 70°F = 13.4 psia
Vapor pressure of isobutane   at 70°F = 45.7 psia

Using these values, the compositions of "Freon" 12–"Freon" 11 and "Freon" 12–isobutane blends having a vapor pressure of 54.7 psia can be calculated using Equation 9–2. The "Freon" 12/"Freon" 11 blend with the specified vapor pressure of 54.7 psia at 70°F has a composition of 54.8% "Freon" 12 and 45.2% "Freon" 11. The "Freon" 12/isobutane blend with a vapor pressure of 54.7 psia has a composition of 38.4% "Freon" 12 and 61.6% isobutane. These two compositions can be plotted on a triangular coordinate chart and a line drawn between them as shown in Figure 9–1. The compositions of all three-component blends with the vapor pressure of 54.7 psia will fall on this line.

Equation 9–5 can be used to calculate the compositions of three-components blends with a given vapor pressure, assuming that the vapor pressures and molecular weights of the individual components are known:

$$P_1 + X_2(P_2 - P_1) + X_3(P_3 - P_1) = P \qquad (9\text{–}5)$$

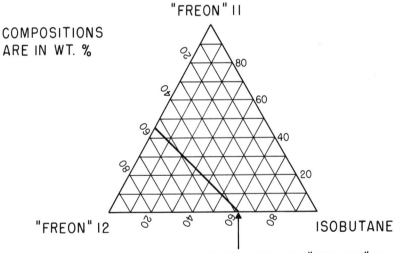

THE COMPOSITIONS OF ALL MIXTURES OF "FREON" 12, "FREON" 11, AND ISOBUTANE WITH A VAPOR PRESSURE OF 54.7 psia AT 70°F FALL ON THIS LINE.

Figure 9–1   Mixtures of "Freon" 12, "Freon" 11, and isobutane with a vapor pressure of 54.7 psia at 70°F.

P_1, P_2, and P_3 are the vapor pressures of the individual components and P is the specified or desired vapor pressure of the three-component blend. X_1, X_2, and X_3 are the mole fractions of the components 1, 2, and 3 respectively. Since there are two unknowns in Equation 9–5, it is necessary to assign a value to one of the mole fractions, i.e., X_2 and then solve the equation for the other mole fraction. The reason that X_1 does not appear in Equation 9–5 will become clear from the derivation of Equation 9–5 which is given below:

Equation 9–5 can be derived as follows, where Pa_1, Pa_2, and Pa_3 represent the partial pressures of components 1, 2, and 3 respectively:

1. $Pa_1 + Pa_2 + Pa_3 = P$ (Dalton's Law)

2. $P_1X_1 = Pa_1$, $P_2X_2 = Pa_2$, and $P_3X_3 = Pa_3$ (Raoult's Law)

3. Therefore,
$$P_1X_1 + P_2X_2 + P_3X_3 = P \qquad (9–6)$$

4. $X_1 + X_2 + X_3 = 1$, and therefore,
$$X_1 = 1 - X_2 - X_3 \qquad (9–7)$$

5. Substituting the value of X_1 from Equation 9–7 into Equation 9–6 gives,
$$P_1(1 - X_2 - X_3) + P_2X_2 + P_3X_3 = P$$

Simplifying the above equation gives Equation 9–5.

As an example, assume again that the various compositions of "Freon" 12, "Freon" 11, and isobutane that have a vapor pressure of 54.7 psia (40 psig) at 70°F are desired. Let X_1, X_2, and X_3 represent the mole fractions of "Freon" 12, "Freon" 11, and isobutane, respectively. Also, let P_1, P_2, and P_3 represent the vapor pressures of "Freon" 12, "Freon" 11 and isobutane. The vapor pressures at 70°F of the three components are as follows:

"Freon" 12—Vapor pressure (P_1) at 70°F = 84.9 psia
"Freon" 11—Vapor pressure (P_2) at 70°F = 13.4 psia
Isobutane —Vapor pressure (P_3) at 70°F = 45.7 psia

Substituting the values for the various vapor pressures in Equation 9–5 gives

$$84.7 + X_2(13.4 - 84.9) + X_3(45.7 - 84.9) = 54.7$$

Now assume that X_2 has a value of 0.3. By substituting this value for X_2 and solving for X_3, it is found that X_3 is 0.22. Therefore, X_1 has a value of 0.48. Again, assume that X_2 has a value of 0.1. Under these conditions, X_3 is found to have a value of 0.59 and X_1 a value of 0.31. The mole fractions, X_1, X_2, and X_3, in each of the two blends, can be converted into weight percent as usual. The results are shown in Table 9–1.

TABLE 9-1 BLENDS WITH A VAPOR PRESSURE OF 54.7 PSIA

	Composition of Blend (wt %)	
Component	$X_2 = 0.3$	$X_2 = 0.1$
"Freon" 12	51.5	44.3
"Freon" 11	36.9	16.0
Isobutane	11.6	39.7

By referring to Figure 9–1, it can be seen that these compositions fall on the plot for compositions with a calculated vapor pressure of 54.7 psia.

DEVIATION FROM RAOULT'S LAW

In many cases the vapor pressure of a solution calculated by Raoult's Law differs from the actual measured pressure. There are several reasons for this. If the solute associates with the solvent so that molecular complexes are formed, the calculation of the mole fraction of the components, if it does not take into account the molecular weight and concentration of the various complexes, will be incorrect. Generally, it is difficult to determine the extent of complex formation and often the presence of complexes is known only because of the deviation from Raoult's Law. Equation 9–7, which describes the compositions of three-component mixtures with the same vapor pressure, gives a straight line plot. However, the actual measured pressures for three-component blends of "Freon" 12, "Freon" 11, and isobutane fall on a curve when plotted on a triangular coordinate chart because of deviations from Raoult's Law.

Another reason for a deviation from Raoult's Law is that either the solvent or solute may, itself, be associated in the liquid state and thus is considered to be polymeric. Ethyl alcohol, for example, is known to be highly associated. This is the explanation for the high boiling point of ethyl alcohol compared to isomeric dimethyl ether, which is not associated. The experimentally determined vapor pressures of "Freon" propellant–ethyl alcohol mixtures are generally found to be higher than vapor pressures calculated from Raoult's Law because the calculation assumes that ethyl alcohol is monomeric in the liquid state.

In other instances, disassociation may occur or chemical reactions may take place. Again, calculations of the mole fractions of the components will give incorrect results if they do not take into account the disassociation or chemical reactions. In cases where a deviation from Raoults' Law occurs, it is usually not because Raoult's Law has failed but only because the mole fractions of the components in the solution were incorrectly calculated. Therefore, this is only an apparent deviation. There are some cases involving ionization where the deviation is considered to be real.[2]

PRESSURES OF AEROSOL FORMULATIONS

The vapor pressure of any aerosol formulation (excluding the effect of air) depends upon the pressure of the propellant and the effect that the other components in the formulation have upon the pressure of the propellant. If the concentrations of the solvents and active ingredients are sufficiently high so that the mole fraction of the propellant in the aerosol is appreciably reduced, then the vapor pressure will be noticeably reduced. In many products, the concentrations of the active ingredients are so low that the effect upon the vapor pressure is negligible. Usually, it is the solvents in the concentrate that have the most noticeable effect in depressing the vapor pressure of the propellant. In some cases, the components themselves may have appreciable vapor pressures, and this will contribute to the total pressure.

Effect of Air

Up to the present, we have been concerned only with the vapor pressure of the aerosol formulation itself, i.e., the vapor pressure of the solution that consists of the liquefied propellant and concentrate. This vapor pressure will be referred to in subsequent discussions as the *vapor pressure of the aerosol formulation* (P_f). Another factor that may have a major effect upon the pressure in the container is the amount of air that is present.

When air is present in an aerosol, it increases the pressure in the container. This follows from Dalton's Law of partial pressures that when two or more gases are in the same space, each gas exerts the same pressure as if it were alone. If the pressure of air in the vapor phase of an aerosol container is 10 psi, then the total pressure in the aerosol container will be equal to the vapor pressure of the aerosol formulation, P_f, plus the pressure of the air, 10 psi. This can be expressed as follows:

$$P_t = P_f + P_a, \text{ or } P_t = P_f + 10. \tag{9-8}$$

P_t = total pressure in the aerosol container
P_f = pressure of the aerosol formulation (propellant and concentrate)
P_a = pressure due to air

The total pressure in the aerosol container, P_t, will be referred to as the *pressure in the container*, the *pressure of the aerosol product*, or the *total pressure* in order to differentiate it from the vapor pressure of the aerosol formulation P_f, alone.

All aerosol products contain air because all empty aerosol containers initially are full of air at atmospheric pressure, i.e., about 14.7 psi. What happens to this air during the loading of the containers depends upon the method of filling, type of purging, etc. Air may also be present in the solvents, etc., that are used in the preparation of the aerosols.

Many times it is desirable to know how much pressure the air is contributing to the total pressure in the container. This is usually difficult to determine from a pressure measurement alone since the pressure of the combination of propellant and concentrate (the aerosol formulation) may not be known because of the complexity of the mixture. In these cases, a determination of the air concentration in the vapor phase of the product by gas chromatography coupled with a measurement of the pressure of the product will give both the pressure due to air alone and the vapor pressure of the aerosol formulation.

For example, assume that the pressure of an aerosol product at 70°F was found to be 50.3 psig (65 psia) and that the concentration of air in the vapor phase had been determined by gas chromatography to be 20 vol % (this is the same as mole %). Therefore, the pressure due to air is 0.20×65.0 psia = 13 psi and the vapor pressure of the aerosol formulation is $65 - 13 = 52$ psia or 37.3 psig at 70°F.

In order for the calculations to be accurate, the concentration of air in the vapor phase by gas chromatography should be determined at the same temperature at which the pressure was measured. This is because the solubility of air in the liquid phase of the aerosol is a function of the temperature, as is the pressure of the formulation. The solubility of air in the liquid phase of the aerosol determines the concentration of air in the vapor phase.

If the vapor pressure of the aerosol formulation, P_f, is known, then the measurement of the pressure in the product after loading will indicate how much air is present. The actual pressure in the container minus the pressure of the aerosol formulation without air will give the pressure due to air.

The effect of air upon the pressure of aerosol products is discussed in detail in Reference 3. The vapor pressures of combinations of propellants is given in Reference 4. The vapor pressures of mixtures of propellants and various solvents are given in References 5 and 11.

Effect of Method of Filling

The method of filling aerosols has a considerable influence on the amount of air entrapped in the container. Products that are cold filled normally contain relatively little entrapped air. In these products, the pressure due to air usually does not exceed 3–4 psi. The reasons for the low amount of entrapped air are:

The concentrate and the propellant physically displace most of the air in the container when they are added. The combined liquid volumes of the concentrate and propellant can amount to as high as 85% of the volume in the container. Also, the propellant, which is added to the cold concentrate, generally vaporizes sufficiently so that most of the air remaining after addition of the concentrate and the propellant is displaced by the vaporizing propellant.

In the pressure fill procedure, the concentrate added to the container displaces the amount of air corresponding to the volume of the concentrate. The container is then capped, thus trapping the air remaining in the container. When the propellant is pressure filled through the valve, the propellant will compress the air already in the container and increase the pressure even further.

PRESSURE MEASUREMENT

If an aerosol product did not contain air, a pressure measurement would be comparatively simple. A pressure gage would be connected to the aerosol (vapor phase) with a shutoff valve between the gage and the aerosol product. After the aerosol had been brought to the desired temperature, the valve between the gage and the aerosol would be opened and the pressure read on the gage. When the valve between the aerosol and the gage was opened, the vapor phase of the aerosol would expand into the gage because the vapor pressure of an aerosol is higher than atmospheric pressure while a gage is usually at atmospheric pressure. The loss of propellant vapor into the gage might be expected to decrease the pressure in the vapor phase of the aerosol. However, sufficient liquefied gas propellant would vaporize to replace that lost to the gage and this would maintain a constant pressure in the aerosol. Therefore, the correct pressure could be read on the gage almost immediately after the valve to the gage had been opened.

However, if air is present in the aerosol, as it is in practically all aerosols, then a different situation exists that increases the complexity of the pressure measurement. This may be seen more easily by first considering a closed aerosol container filled with *nothing except air* at a pressure of 50 psi. If the valve of this container is opened so that some of the air escapes, the pressure in the container will decrease just as the pressure in a tire decreases as a result of loss of air from a puncture.

If an attempt is made to obtain the pressure of the container with 50 psi of air, using a gage at atmospheric pressure, some of the air in the container will enter the gage and this will continue until the pressures in the container and gage are equal. However, since some of the air left the container, the pressure in the container will decrease because now there is less air present than there was originally. The pressure in the container will decrease whether the air in the container escapes into a room or into a gage. The pressure that is recorded on the gage will be less than that which was present in the aerosol container initially and is not an accurate value for the original pressure in the container. The extent to which the pressure in the container decreases depends upon the initial pressures in the container and the gage and upon the volumes of the container and gage.[7]

A reverse situation would occur if the pressure in the gage were higher than that in the container. If the air pressure in the gage were 100 psi and that in the container were 50 psi, then when the valve between the gage and the container was opened, some of the air in the gage would enter the container and increase the pressure in the container. The final pressure recorded on the gage, would be higher than the initial pressure in the container. Again, the gage reading would not give an accurate value for the original pressure in the container.

The most accurate measurement would be obtained when, before the valve between the aerosol and gage was opened, the pressure in the gage was adjusted so that it was the same as that in the aerosol container, i.e., 50 psi. When the gage containing 50 psi of air was attached to the aerosol container and the valve between the two was opened, there would be no loss of air from the aerosol container into the gage nor any transfer of air from the gage into the aerosol container. The pressure recorded on the gage would be a true indication of the pressure in the container.

The previous considerations can be applied to the pressure measurement of an *aerosol product containing air*. The total pressure of an aerosol product is equal to the sum of the vapor pressure of the aerosol formulation and the pressure of the air in the vapor phase (Dalton's Law).

$$P_t = P_f + P_a$$

If the pressure of the aerosol product is measured with a gage at atmospheric pressure, the vapor phase of the aerosol product will expand into the gage until the final pressure in the gage and the aerosol are equal. During the expansion of the vapor phase into the gage, the pressure of the formulation, P_f, remains relatively constant because additional liquefied propellant vaporizes to replace that lost to the gage. The final pressure recorded on the gage is lower than the initial pressure in the aerosol product because P_a decreases during the measurement.

The fact that taking the pressure of an aerosol product with a gage at atmospheric pressure gives a lower reading than the original pressure was recognized early in the aerosol industry. A committee was established in the Chemical Specialties Manufacturers Association to develop a method for measuring the pressure of an aerosol which would take into account the air in the product. A procedure, which achieves essentially the ideal situation where the adjusted gage pressure is about equal to that in the aerosol,[6] was adopted. This method involves prepressurizing the gage before the pressure is measured and is carried out with a modified gage which has additional valves for adding air to the gage or for releasing pressure from the gage while it is attached to the aerosol container.

The pressure measurement is carried out as follows: assume that the pressure in the container is 50 psi and that the pressure in the gage is 14.7 psi, or atmospheric pressure. The gage is attached to the container but the valve between the gage and the container is closed. When the valve between the gage and the container is cracked, or barely opened, the pointer on the gage will start to move upwards because the pressure in the container is higher than that in the gage. The valve between the gage and the container is now closed and the pressure in the gage is increased with air to a higher value, for example, 30 psi, through the additional air valve on the gage. After closing the air pressure valve, the valve between the gage and the container is barely opened to determine if the pressure of 30 psi in the gage is lower or higher than that in the container. Again the pointer on the gage will start to move upwards because the pressure in the gage is still lower than that in the container.

The process is repeated. The valve between the gage and container is closed and the pressure in the gage again increased to 60 psi. When tested, the pointer on the gage will now start to move downwards, because the gage pressure is higher than the container pressure. After closing the valve between the gage and container, some of the gage pressure is released through the valve to the outside and the pressure in the container rechecked. Ultimately, a point will be reached where the pressure in the gage is the same as that in the container and when the valve between the gage and the container is opened, the pointer on the gage will not move because the pressures in the container and gage are the same.

This procedure sounds somewhat complicated and time consuming but actually a pressure measurement of this type can be carried out by a trained technician in a matter of a few minutes after the aerosol has reached the specified temperature. During the measurement, there is very little loss of air from the container to the gage or little transfer of air from the gage to the container. Since the pressures of most aerosol products range from about 30–60 psig at 70°F, the gage is usually prepressured initially to about 40 psig.

When pressure measurements are used as a quality control check for products being loaded commercially, it is not necessary to obtain the exact pressure. It is assumed that the correct pressure of the product has already been determined in the laboratory using a prepressurized gage. As far as the products on the loading line are concerned, the important factor to check is whether or not the pressure of the product changes at some time during the loading. This would indicate a change in such factors as the quantity of propellant or concentrate that was being added.

The presence of air in an aerosol causes more problems in the measurement of pressure than are realized initially. In an aerosol product, part of the air is present in the vapor phase and part of the air is dissolved in the liquid phase. The distribution ratio of the air between the liquid and vapor phase is constant when the system is at equilibrium and depends upon such factors as the volume fill in the container, the solubility of air in the concentrate, and the temperature of the aerosol. For example, in a container filled 70% volume full with "Freon" 12/"Freon" 11 (50/50) propellant alone, at 70°F, 56% of the air is in the liquid phase and 44% in the vapor phase. In a container filled only 50% volume full with "Freon" 12/"Freon" 11 (50/50), 35% of the air is in the liquid phase and 65% is in the vapor phase.[3]

The fact that air distributes itself between the liquid and vapor phases shows why it is necessary to shake an aerosol container at intervals before a pressure measurement. The agitation is necessary in order to establish equilibrium between the air in vapor phase and that in the liquid phase at the temperature at which the pressure measurement is being carried out.

There are a number of processes that theoretically could take place during a pressure measurement with a nonprepressurized gage. These are discussed in detail in Reference 7. For example, one question that arises is whether the expanding vapor phase of the aerosol mixes with the air in the gage or merely compresses it. The two different processes would give different pressures. The data indicate that the expanding vapor phase of the aerosol merely compresses the air in the gage. The air in the gage, therefore, acts as a medium for transmittal of pressure.

The use of a prepressurized gage is necessary in order to obtain an accurate pressure. There is also another advantage to prepressurizing the gage. If vapor pressures are taken through the dip tube with a nonprepressurized gage, the aerosol product could enter the gage. Many aerosol products, such as aerosol paints, could contaminate the gage and cause it to become inoperative. Prepressurizing the gage will avoid this.

In the laboratory, the simplest method for taking pressures vapor phase is to prepare the samples without dip tubes. Pressures can then be taken directly through the valve. Another procedure is to use a can puncturing device that will puncture either the bottom or the side of the container. Can puncturing devices allow pressures to be taken vapor phase on cans equipped with dip tube valves.

REMOVAL OF AIR FROM AEROSOL CONTAINERS

The presence of air in an aerosol container can be undesirable for several reasons. Air may increase the pressure sufficiently so that the DOT limits are exceeded. The increased pressure will also increase the spray rate. Air can cause deterioration of aerosol products by reacting with oxygen sensitive components, such as perfumes, or by catalyzing free radical reactions. In addition, oxygen is often considered to be a factor in many cases where corrosion occurs.[12,13] For these reasons, attempts are usually made to reduce the air content to a minimum during the loading of aerosol products. The amount of air remaining in aerosols prepared by cold filling usually is relatively low and in most cases there is no need to take additional steps to remove air from these products. The problems that result from the presence of air occur with pressure-filled aerosols.

One of the first methods considered for removing air from pressure-filled products was to invert the container and spray enough material so that the vapor phase was removed. This procedure removes the air initially present in the vapor phase and thus reduces the air concentration. However, it will not remove all of the air in the aerosol because a considerable proportion of the air is dissolved in the liquid phase. If the vapor phase is removed by venting and the container is shaken, air will leave the liquid phase and enter the vapor phase in order to establish equilibrium between the liquid phase and the vapor phase. This procedure can be continued and while the air is constantly reduced it will never be eliminated entirely because of the air in the liquid phase. This procedure is impractical because of the loss of propellant that occurs during the venting.

One effective air removal method is to purge the aerosol vapor phase with propellant vapor or liquid propellant before the valve is attached and crimped. In order to obtain efficient purging, certain precautions must be observed. For example, if liquid propellant is dropped into the container, the temperature of the concentrate must be high enough so that the propellant will vaporize after it contacts the concentrate surface.

The technique of purging is important. In one successful procedure, a slow stream of propellant vapor is started through the purging tube and the tube is lowered until it almost touches the surface of the concentrate. The purging tube is then raised slowly out of the container, allowing the propellant vapor to displace the air in the vapor phase. Purging with too vigorous a flow of vapor may create turbulence and result in ineffective purging.

The container should be capped with the valve as soon as possible after purging has been completed. If the interval between purging and capping is too long, the propellant in the vapor phase may dissolve in the concentrate, and air will reenter the container.

Attempts have been made to eliminate air by first purging the empty container with propellant and then adding the concentrate. Usually this is not a very effective procedure because the turbulence caused by the addition of the concentrate results in some air reentering the aerosol container. In other cases, propellant vapor purging with the purging tube outside and above the container has been attempted but the effectiveness is minimal.

Another procedure is to premix some of the propellant with the concentrate before the concentrate is added to the container.[9] The concentration of propellant in the concentrate should be such that the mixture has a slight positive pressure. Under these conditions, the concentrate will be self purging after it has been added to the container.

Most of the recent commercial pressure-filling equipment is capable of removing air from the containers by evacuation before crimping the valve. This probably is the simplest method for eliminating air from pressure-filled products.[10]

In laboratory work it is sometimes desirable to reduce the air to a concentration below that obtained with either cold filling or purging. This may be accomplished by adding an excess of the propellant to the concentrate and boiling off the excess.[8] For example, if "Freon" 12/"Freon" 11 propellant ethyl alcohol mixtures with very low air content are desired, they may be obtained by loading the ethyl alcohol into the container, adding an excess of the "Freon" 11, boiling off the excess, capping, and pressure loading the "Freon" 12. The container should be capped while vapors of the "Freon" 11 are still overflowing the container.

REFERENCES

1. F. H. Getman and F. Daniels, "Outlines of Theoretical Chemistry," p. 163, John Wiley & Sons, Inc., New York, N.Y., 1931.

2. MacDougall, F. H., "Physical Chemistry," p. 244, The Macmillan Company, New York, N.Y., 1936.
3. "Freon" Aerosol Report FA-15, "The Effect of Air on Pressure in Aerosol Containers."
4. "Freon" Aerosol Report FA-22, "Vapor Pressure and Liquid Density of 'Freon' Propelants."
5. L. Flanner, *Aerosol Age* **9,** 27 (March 1964).
6. "Freon" Aerosol Report FA-11, "A Method for Measuring the Vapor Pressure of Aerosol Products."
7. P. A. Sanders, *Aerosol Age* **11** (August, September 1966) ("Freon" Aerosol Report A-68).
8. "Freon" Aerosol Report A-51, " 'Freon' 11 S Propellant."
9. F. A. Bower and R. G. Appenzeller, Proc. 45th Meeting, CSMA (December 1957) ("Freon" Aerosol Report A-457).
10. Technical Bulletin, "Undercap Filling," The Kartridge Pak Company.
11. P. A. Sanders, *Aerosol Age* **11** (January, February 1966).
12. F. L. LaQue and R. R. Copson, "Corrosion Resistance of Metals and Alloys," 2nd ed., Reinhold Publishing Corporation, New York, N.Y., 1963.
13. P. A. Sanders, *Soap Chem. Specialties* **42,** 74, 135 (July, August 1966) ("Freon" Aerosol Report A-66).

10

SOLUBILITY

In formulation studies where a homogeneous system is desired, i.e., one in which all of the components are mutually soluble, a number of unnecessary experiments can often be avoided if the solubility properties of the aerosol propellants and common aerosol solvents are known. When an aerosol propellant is added to a concentrate (which usually consists of active ingredients dissolved in a solvent or mixture of solvents) the addition of the propellant results in the formation of a new solvent system. The new solvent system consists of the propellant plus the solvent or solvents in the concentrate. The new solvent system may be a better or a poorer solvent for the active ingredients than that in the original concentrate. It is possible for an active ingredient to be quite soluble in a solvent, such as ethyl alcohol, but relatively insoluble in a mixture of propellants and ethyl alcohol. In some cases, it is necessary to change the propellant in order to obtain a homogeneous system. "Freon" 11 propellant is a fairly good solvent, for example, but "Freon" 12 and "Freon" 114 propellants are relatively poor solvents, as are the hydrocarbons. A propellant containing a high proportion of "Freon" 11 may be required in order to obtain sufficient solubility for the active ingredients.

The "Freon" propellants are soluble in a wide variety of organic solvents and they are also solvents for a wide variety of compounds. A considerable amount of miscellaneous data has been accumulated on the solubility of the propellants in a number of compounds and is available in Reference 1. The "Freon" propellants are not miscible with water nor with many glycols.[2] The immiscibility of the propellants with water has resulted in many intensive investigations of emulsion systems prepared with the "Freon" propellants and water.[3-6]

There are a number of different physical properties of propellants and solvents which are useful for comparing the solubility properties of a compound with other compounds or in predicting the miscibility or solubility properties of the compound itself. Two of these, the Kauri–Butanol values and the solubility parameters, will be discussed in the following sections. Kauri–Butanol values provide an indication of the relative solubility properties of a liquid in comparison with other liquids. Solubility parameters are useful for predicting whether two liquids will be miscible or whether an amorphous solid or polymer will dissolve in a solvent.

KAURI–BUTANOL (K–B) VALUES

Kauri–Butanol values are used to indicate the relative solvent powers of liquids. The test initially was designed for evaluating the hydrocarbon solvents used in paint and lacquer formulations and was restricted to solvents with a boiling point over 104°F (40°C).[7] However, its use has been extended to include the lower boiling hydrocarbon propellants as well as the fluorinated propellants and it appears to be applicable to these compounds as well. Although the test has been criticized because of its artificial nature,[8] there appears to be a correlation between Kauri–Butanol values and the more fundamental solubility parameters indicating that K–B values may be more significant than has been assumed.

The Kauri–Butanol value of a solvent is the number of milliliters required to produce a certain degree of turbidity when added to 20 g of a standard solution of Kauri resin in n-butyl alcohol at 25°C. The concentration of Kauri resin in butyl alcohol has been standardized so that 20 g of the Kauri–Butanol solution require the addition of 105 cc of toluene to reach the end point. Toluene, with its K–B value of 105, is the standard for solvents with Kauri–Butanol values over 60 while a n-heptane–toluene (75/25) blend with a K–B value of 40 is the standard for solvents with K–B values under 60.

Because of the higher vapor pressure of the liquefied gas propellants, the test procedure has to be modified to include the use of pressure equipment. In carrying out the test, the 20 g of standard Kauri–Butanol solution is placed in a suitable pressure vessel, such as a 4-oz aerosol bottle. The bottle is capped with a standard valve without dip tube and placed over a sheet of No. 10 print (normal newspaper print is satisfactory). The solution is then titrated with the propellant using a pressure burette. The end point is reached when the sharp outlines of the 10-point print, as viewed through the liquid in the bottle, become obscured or blurred due to precipitation of the Kauri resin from solution. Details of the Kauri–Butanol test are given in Reference 7.

The K–B values for the fluorinated propellants are listed in Table 10–1

TABLE 10-1   SOLUBILITY PROPERTIES OF THE FLUORINATED
HYDROCARBON PROPELLANTS

Propellant	Formula	Kauri–Butanol Value	Solubility Parameter
"Freon" 21	$CHCl_2F$	102	8.0
"Freon" 11	CCl_3F	60	7.5
"Freon" 113	CCl_2FCClF_2	31	7.2
"Freon" 22	$CHClF_2$	25	6.5
Propellant 142b	CH_3CClF_2	20	6.8
"Freon" 12	CCl_2F_2	18	6.1
"Freon" 114	$CClF_2CClF_2$	12	6.2
Propellant 152a	CH_3CHF_2	11	7.0(?)
"Freon" C-318	C_4F_8	10	5.6

and those for various other propellants are listed in Table 10–2. The data
in Table 10–1 indicate that "Freon" 21 fluorocarbon, with a K–B value of
102, has the best solvent properties of the "Freon" compounds. This is
one reason why this particular compound has been of interest for the
formulation of pharmaceutical products. "Freon" 11, with a K–B value
of 60, is next in solvent power and is followed by "Freon" 113 with a
K–B value of 31. The rest of "Freon" compounds have lower K–B values
and are relatively poor solvents.

TABLE 10-2   SOLUBILITY PROPERTIES OF THE HYDROCARBONS,
MISCELLANEOUS PROPELLANTS, AND PROPELLANT BLENDS

Propellant	Formula	Kauri–Butanol Value	Reference
Propane	C_3H_8	ca 25	15
Vinyl Chloride	$CH_2 = CHCl$	58	16
Isobutane	C_4H_{10}	ca 25	15
n-Butane	C_4H_{10}	ca 25	15
Propellant 12– Propane (91/9)		18–25	16
Propellant 12–Vinyl Chloride (80/20)		25	16
Propellant 12–Vinyl Chloride (65/35)		33	16
Propellant P [Propellant 12– Propellant 11– Dimethyl Ether (75/10/15)]		31	17

Vinyl chloride (Table 10–2) has a K–B value of 58 and has fairly good solvent properties. It is comparable to "Freon" 11. The hydrocarbon propellants have K–B values of about 25 and, therefore, have relatively poor solvent properties.

The K–B values of a wide variety of organic solvents are available.[8,9]

SOLUBILITY PARAMETERS

The use of solubility parameters provides a method for predicting whether mixtures of liquids will be miscible or whether a given amorphous solid will dissolve in a solvent or mixtures of solvents. Solubility parameters are physical constants for compounds, just as boiling points and vapor pressures. Solubility parameters have been determined for a large number of organic solvents, propellants, and resins and polymers and are available in tables. If two solvents have approximately the same solubility parameters, they will be miscible. Therefore, by finding the solubility parameters for any liquid in the tables and comparing it with the solubility parameters of other liquids, it is possible to predict the miscibility of the various solvents. If a resin has about the same solubility parameter as that of a given solvent, the resin usually will dissolve in the solvent. The method is not exact and it fails in certain cases, but a comprehension of solubility parameter theory makes possible a better understanding of much of the solubility phenomena that are encountered in the aerosol field.[10]

The concept of solubility parameters was originated by Hildebrand.[11] Burrell,[12,13] in an excellent series of articles, demonstrated how solubility parameters can be used in applications involving coating compositions. His articles should be consulted because they provide a comprehensive and yet very readable discussion of solubility parameters.

The derivation of solubility parameters starts with the free-energy equation from thermodynamics which is written as follows:

$$\Delta F = \Delta H - T\Delta S \qquad (10\text{–}1)$$

This equation states that the change in free energy, (ΔF), in any process is equal to the heat of mixing, (ΔH), minus the product of the absolute temperature, (T), and the change in entropy (ΔS). The process involved could be a chemical reaction, the mixing of two liquids, the dissolving of a solid in a liquid, etc. The idea of entropy may be a little difficult to visualize. Burrell explains entropy as a measure of disorder or randomness or it may be considered as a measure of the freedom of movement. As an example, a molecule tightly bound in a crystal lattice has low entropy while a gas molecule, which is essentially unrestricted in its movement, has a high entropy. Processe tend to proceed so that the system will increase in entropy.

One of the most important applications of the equation is that it can be used to determine whether any process will take place. If the change in the free energy, ΔF, is negative, then the process will occur. For example, assume that it is desirable to know whether or not a given material is soluble in one of the "Freon" propellants and that it is not possible to determine this experimentally. If the heat of mixing, ΔH, the temperature, T, and the change in entropy, ΔS, of the system were known, then ΔF, the free energy, could be calculated. If the heat of mixing, ΔH, were less than $T\Delta S$, then the change in free energy, ΔF, would be negative. Therefore, the process would take place, i.e., the material would dissolve in the "Freon" propellant.

In order to arrive at the point where solubility parameters come into the picture, it is necessary to rearrange Equation 10–1 and substitute a term containing solubility parameters for the heat of mixing. Hildebrand has shown, that when two components are involved in a process, the heat of mixing, ΔH, is equal to the following term:

$$\Delta H_m = V_m \left[\left(\frac{\Delta E_1}{V_1} \right)^{1/2} - \left(\frac{\Delta E_2}{V_2} \right)^{1/2} \right]^2 \phi_1 \phi_2 \qquad (10\text{--}2)$$

where ΔH_m = over-all heat of mixing, calories
V_m = total volume in cc of the mixture
ΔE = energy of vaporization of component 1 or 2, calories
V = molar volume of component 1 or 2, cc
ϕ = volume fraction of component 1 or 2

The term $\Delta E/V$ is the energy of vaporization per cc and has been described as the *cohesive energy density*. It indicates the amount of energy that has to be absorbed by one cc of a liquid to overcome the intermolecular forces which hold the molecules together.

The equation is then rearranged as follows:

$$\frac{\Delta H_m}{V_m \phi_1 \phi_2} = \left[\left(\frac{\Delta E_1}{V_1} \right)^{1/2} - \left(\frac{\Delta E_2}{V_2} \right)^{1/2} \right]^2 \qquad (10\text{--}3)$$

This shows that at a given concentration, the heat of mixing per cc is equal to the square of the difference between the square roots of the cohesive energy densities of the two components. The symbol δ, which Hildebrand termed the solubility parameter, was assigned to this quantity, so that the following equation results:

$$\delta = \left(\frac{\Delta E}{V} \right)^{1/2} \qquad (10\text{--}4)$$

Therefore, if the solubility parameters are substituted for the cohesive energy densities, Equation 10–3, it then becomes:

$$\frac{\Delta H_m}{V_m \phi_1 \phi_2} = (\delta_1 - \delta_2)^2 \qquad (10\text{--}5)$$

The term $(\delta_1 - \delta_2)^2$ is, therefore, proportional to ΔH. When ΔH, in the original free energy equation, is replaced with the term $(\delta_1 - \delta_2)^2$, the following is obtained:

$$\Delta F \propto (\delta_1 - \delta_2)^2 - T\Delta S \qquad (10\text{--}6)$$

Now, it can be seen that as the solubility parameters, δ_1 and δ_2, approach each other, the term $(\delta_1 - \delta_2)^2$ approaches zero.

$$(\delta_1 - \delta_2)^2 \longrightarrow 0$$

Therefore, this term drops out of Equation 10–6 leaving it essentially in the form as follows:

$$\Delta F = 0 - T\Delta S \qquad (10\text{--}7)$$

If the solubility parameters approach each other, ΔF is usually negative and the process under consideration will take place. This is why two solvents with about the same solubility parameters will usually be miscible, or why an amorphous solid will dissolve in a liquid having about the same solubility parameter as that of the solid. To repeat, when the solubility parameters are about the same, the term $(\delta_1 - \delta_2)^2$ becomes essentially zero and ΔF is usually negative.

The solubility parameter concept explains many of the puzzling features of solvency. For example, there are known instances where a mixture of two solvents will dissolve a given solid, although neither solvent by itself will dissolve the solid. The explanation for this is that one of the liquids has too high a solubility parameter to dissolve the solid while the other liquid has too low a solubility parameter. The average solubility parameter of the mixture of liquids, however, is close enough to that of the solid so that the mixture is a solvent for the solid. The solubility parameter concept was invaluable in explaining some of the phenomena observed with the formation of foams from aqueous ethyl alcohol aerosol systems.[10]

Burrell lists a number of methods used to calculate solubility parameters and also provides an extensive list of the solubility parameters of various solvents and polymers. The solubility parameters of the propellants are given in Table 10–1.[14] It is interesting that with hydrocarbon solvents there is a relationship between the K–B values of the liquids and the solubility parameters.[12] There also appears to be a rough correlation between the K–B values and solubility parameters for the fluorinated propellants, as shown in Table 10–1.

SWELLING EFFECT OF PROPELLANTS UPON ELASTOMERS

Gaskets manufactured from elastomers are used in practically all aerosol valves. If an aerosol formulation causes excessive swelling of the gaskets, the aerosol valve may fail to operate properly. Therefore, the effect of propellants and various solvents upon the elastomers has been a subject of considerable interest to the aerosol industry.[18]

The elastomers most widely used in aerosol valves are based on various types of neoprene and buna. In studying the effect of the propellants upon the elastomers, the extent of swelling is usually determined by measuring the increase in length that occurs when the elastomers are tested in a particular liquid.[18] The extent of swelling of Neoprene GN, Buna N, and natural rubber by a variety of propellants is shown in Table 10–3. Carbon tetrachloride and chloroform, although not used in aerosol formulations as a result of their toxicity, have been included in the list for comparative purposes. They are excellent solvents and are known to have a pronounced swelling effect upon elastomers.

The compounds in Table 10–3 have been listed in decreasing order of their solubility parameters in order to show the relationship between the solubility parameters and extent of swelling. Compounds with solubility parameters closest to those of the elastomers have the most effect upon the elastomers. The solubility parameters of the three elastomers range from 8.3–9.4.[12] Compounds with solubility parameters below 7.2 have very little effect upon the elastomers. Compounds with solubility parameters much higher than those of the elastomers, such as ethyl alcohol (12.7) and water (23.4), would also have little effect.

Kauri–Butanol values have been included in Table 10–3. There is a fairly consistent relationship between the extent of swelling and the K–B values of the propellants. Compounds with the highest K–B values have the most pronounced effect upon the elastomers; as a general rule, propellants with the highest solubility parameters also have the highest K–B values.

EFFECT OF "FREON" PROPELLANTS UPON PLASTICS

Since some of the components of aerosol valves are constructed from plastic materials, the effect of the propellants upon plastics has been of much interest. The potential use of plastic aerosol bottles has also stimulated interest in the suitability of these materials for use with the propellants.

A considerable amount of information on the effect of the propellants on plastics is available.[19,20] A brief summary on a limited number of plastics is given in Reference 21 and is abstracted below.

TABLE 10-3 SWELLING OF ELASTOMERS BY VARIOUS COMPOUNDS

Compound	Solubility Parameter	Kauri–Butanol Value	Percent Increase in Length at Room Temperature		
			Neoprene GN	Buna N	Natural Rubber
Methylene chloride	9.5	136	37	52	34
Chloroform	9.1	208	43	54	45
Carbon tetrachloride	8.6	113	36	11	44
"Freon" 21	8.0	102	28	48	34
"Freon" 11	7.5	60	17	6	23
"Freon" 113	7.2	31	3	0.5	17
Propellant 152a	7.0(?)	11	2	1	2
Propellant 142b	6.8	20	3	3	5
"Freon" 22	6.5	25	2	26	6
"Freon" 114	6.2	12	0	0	2
"Freon" 12	6.1	18	0	2	6
"Freon" C-318	5.6	10	0	0	0

Solubility Parameter of Neoprene GN = 9.2.
Solubility Parameter of Buna N = 9.4.
Solubility Parameter of Natural Rubber = 8.3.

"Teflon" Tetrafluoroethylene Resin

No swelling was observed when submerged in "Freon" liquids but some diffusion with "Freon" 12 and "Freon" 22 was found.

Polyvinyl Alcohol

This material is not affected by the "Freon" propellants and therefore is very suitable for use with the propellants.

Vinyl Polymers

There is so much variation in the types of polymers and plasticizers that generalizations are difficult. Samples should be tested before use.

Nylon

Nylon is generally suitable for use with the "Freon" compounds. Many valve components are constructed from nylon.

Polyethylene

This is suitable for many applications, and dip tubes are usually constructed of polyethylene. However, it can be affected by strong solvents.

"Lucite" Acrylic Resins (Methacrylate polymers)

This type of polymer is affected by many of the "Freon" propellants and its use with them is doubtful. Aerosol coating compositions with these polymers have been formulated, which shows that the propellants have a solvent action.

Polystyrene

Polystyrene is not suitable for use with the Freon propellants since it is generally affected adversely by the propellants.

Phenolic Resins

As a general rule, phenolic resins are not affected by the "Freon" propellants. However, there is enough variety in the resins so that any special application should be tested. Some of the linings in aerosol containers are prepared from phenolic resins.

EFFECT OF STRUCTURE UPON SOLUBILITY PROPERTIES

The following two conclusions can be drawn from available swelling data:
1. Generally compounds containing fluorine cause less swelling than the corresponding compounds containing chlorine.
2. Halogenated compounds with one carbon atom and one hydrogen atom

cause more swelling than completely halogenated compounds or those containing two or more hydrogen atoms.

Using Kauri–Butanol values as a guide, the following conclusions can be drawn:

1. The increasing substitution of fluorine atoms for chlorine atoms decreases the Kauri–Butanol value.
2. Substitution of one hydrogen atom for a chlorine atom causes an increase in the Kauri–Butanol value. Further substitutions cause the K–B value to decrease as illustrated below. The K–B value for each compound is shown in parenthesis after the compound:

$$CCl_4 \ (113) \longrightarrow CCl_3F \ (60) \longrightarrow CCl_2F_2 \ (18)$$
$$CHCl_3 \ (208) \longrightarrow CHCl_2F \ (102) \longrightarrow CHClF_2 \ (25)$$
$$CH_2Cl_2 \ (136) \longrightarrow CH_2ClF \ (\text{not available}) \longrightarrow CH_2F_2 \ (\text{not available})$$

DETERMINATION OF MISCIBILITY OF THREE-COMPONENT SYSTEMS

There are a number of three-component solvent systems used in the aerosol industry in which only certain proportions of the three components are miscible and form clear solutions. Examples of systems of this type are "Freon" propellant–glycol–alcohol mixtures and "Freon" propellant–water–ethyl alcohol mixtures.

The most effective way of showing which particular compositions are miscible and form a single liquid phase and which compositions are not miscible and form two liquid layers is to determine the miscibility of various compositions in these mixtures and plot the data on a triangular coordinate chart. Extensive use of this method for illustrating the miscibility of "Freon" propellant/glycol/ethyl alcohol mixtures is shown in Reference 2 and for "Freon" propellant–water–ethyl alcohol mixtures in Reference 10.

A typical solubility chart for a "Freon" 12–propylene glycol–ethyl alcohol system is shown in Figure 10–1. Compositions to the left of the phase boundary are not miscible and form two layers. Compositions to the right of the boundary are completely miscible and form a single liquid phase.

There are several procedures which can be followed in order to obtain the data required for drawing a phase boundary, such as that illustrated in Figure 10–1. Two miscible components, such as ethyl alcohol and propylene glycol, can be weighed into glass bottles and "Freon" 12 propellant added until the solubility limit of "Freon" 12 in the mixture is

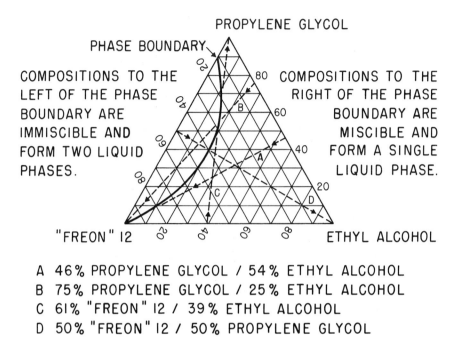

PROPYLENE GLYCOL

PHASE BOUNDARY

COMPOSITIONS TO THE LEFT OF THE PHASE BOUNDARY ARE IMMISCIBLE AND FORM TWO LIQUID PHASES.

COMPOSITIONS TO THE RIGHT OF THE PHASE BOUNDARY ARE MISCIBLE AND FORM A SINGLE LIQUID PHASE.

"FREON" 12   ETHYL ALCOHOL

A 46% PROPYLENE GLYCOL / 54% ETHYL ALCOHOL
B 75% PROPYLENE GLYCOL / 25% ETHYL ALCOHOL
C 61% "FREON" 12 / 39% ETHYL ALCOHOL
D 50% "FREON" 12 / 50% PROPYLENE GLYCOL

Figure 10–1 Determination of the solubility characteristics of "Freon" 12- propylene glycol-ethyl alcohol systems by titration.

exceeded. At this point the solution either becomes cloudy or immiscible propellant droplets are noticeable. Since the weights of the original mixture are known, as is the weight of the "Freon" 12 added, it is simple enough to calculate the composition at the point where immiscibility occurred. An example of this procedure is shown in Figure 10–1 using Composition B, consisting of 75% propylene glycol and 25% ethyl alcohol, as the original mixture. As "Freon" 12 is added, the composition of the three-component mixture will change continually along Line B as shown by the arrows. Immiscibility occurs where Line B intersects the phase boundary.

Another procedure is to start with an immiscible mixture, such as Composition D, consisting of 50% "Freon" 12 and 50% propylene glycol. Ethyl alcohol is added to the mixture by means of air pressure until the initially cloudy, immiscible mixture forms a clear solution. This end point is reached where Line D crosses the phase boundary. Since Line D crosses the phase boundary at about a right angle, the end point is quite sharp.

The problem that may arise in obtaining these data is shown on Figure 10–1 with a mixture, such as Composition A, consisting of 46% propylene glycol and 54% ethyl alcohol. When "Freon" 12 is added to this mixture, the solution remains clear until the point where Line A contacts the phase boundary. However, Line A is essentially tangential to the phase boundary at the point of contact and the end point is not clear cut. Actually it may be difficult to decide whether or not the mixture is single phase or whether two layers are present. A similar situation occurs if a mixture, such as Composition C, consisting of 61% "Freon" 12 and 39% ethyl alcohol, is titrated with propylene glycol. As the propylene glycol is added, the composition will change along Line C in the direction shown by the arrows. Ultimately, Line C contacts the phase boundary but again the end point will not be definite. If additional propylene glycol is added, a homogeneous system will result because the compositions will now fall to the right of the phase boundary.

The titrations may be carried out in typical aerosol glass bottles using a magnetic stirring device for agitation of the mixture during the addition of the third component. The third component is added from a pressure burette through the valve on the bottle. The valve should not be equipped with a dip tube. The proportions of the three components should be adjusted so that the end point is reached when the bottle is nearly full. Otherwise, the results may be in error because of the amount of propellant that is in the vapor phase of the bottle. This is particularly true when the solubility of the propellant in the mixture is very low.

REFERENCES

1. "Freon" Technical Bulletin B-7, "Solubility Relationship of the 'Freon' Fluorocarbon Compounds."
2. P. A. Sanders, *Aerosol Age* **5**, 26 (February 1960) ("Freon" Aerosol Report FA-25).
3. "Freon" Aerosol Report FA-21, "Aerosol Emulsions with "Freon" Propellants."
4. P. A. Sanders, *J. Soc. Cosmetic Chemists* **9**, No. 5, (September 1958) ("Freon" Aerosol Report A-49).
5. P. A. Sanders, *Aerosol Age* **8**, 33 (July 1963) ("Freon" Aerosol Report A-58).
6. P. A. Sanders, *J. Soc. Cosmetic Chemists* **17**, 801–830 (December 1966) ("Freon" Aerosol Report A-67).
7. 1966 Book of ASTM Standards, Part 20 (January).
8. "Freon" Aerosol Report FA-3, "Kauri-Butanol Numbers of 'Freon' Propellants and Other Solvents."
9. "Freon" Aerosol Report FA-26, "Solvent Properties Comparison Chart."
10. P. A. Sanders, *Drug Cosmetic Ind.* **99**, 56, 57 (August, September 1966) ("Freon" Aerosol Report A-59).
11. J. Hildebrand and R. Scott, "The Solubility of Nonelectrolytes," 3rd ed., Reinhold Publishing Corp., New York N.Y., (1949).
12. H. Burrell, *Offic. Dig. Federation Paint Varnish Prod. Clubs* **27**, 726 (1955).
13. H. Burrell, *Offic. Dig. Federation Paint Varnish Prod. Clubs* **29**, 1159 (1957).

14. "Freon" Technical Bulletin D-73, "Solubility Parameters."
15. Phillips Hydrocarbon Aerosol Propellants, Technical Bulletin 519 (1961). Phillips Petroleum Company.
16. "Genetron" Product Information Bulletin, " 'Genetron'–Vinyl Chloride Aerosol Propellants," General Chemical Division, Allied Chemical Company.
17. Union Carbide Chemical Company's Technical Bulletin, "New 'Ulcon' Paint Propellant."
18. "Freon" Technical Bulletin B-12A, "Effect of 'Freon' Fluorocarbons on Elastomers."
19. "Freon" Technical Bulletin B-41, "Effect of 'Freon' Compounds on Plastics."
20. J. A. Brown, Proc. 46th Ann. Meeting, CSMA (December 1959).
21. "Freon" Technical Bulletin B-2, "Properties and Applications of the 'Freon' Fluorinated Hydrocarbons."

11

VISCOSITY

Viscosity has been defined as the resistance of a liquid to flow or as the resistance experienced by one portion of a liquid when it moves over another portion.[1] Most people are familiar with viscous liquids, like molasses, and it is apparent that in these liquids there is a considerable resistance of the liquid to flow. In spite of the vast amount of experimental and theoretical work that has been carried out on viscosity, the theory of liquid viscosity is still incomplete.[2] Several theories on the structure of liquids[3] have been advanced to explain viscosity. These attribute a *pseudocrystalline* or a *microcrystalline* structure to liquids and the liquids thus possess some properties analogous to those of solids. These structures result from the strong intermolecular forces and close packing of the molecules in a liquid.

In aerosol formulations, viscosity problems occur mostly with products designed for surface coating applications, such as paints, lacquers, adhesives, etc. The essential ingredient in these products is a polymeric material, such as a resin or elastomer. Polymers consist of chains of molecules linked together and can have a variety of molecular weights, ranging from relatively low to very high molecular weights. For any homologous series of polymers with the same molecular structure, the viscosity of solutions of the polymer increases as the molecular weight of the polymer increases.

It is possible to visualize this effect by considering polymer chains in solution to exist as strands of barbed wire. The longer and larger the number of strands, the more entangled they will be in solution. The difficulty of moving one layer of the solution over another layer due to strand positions is also the result of entrapment of the solvent molecules by the coils and segments of the polymer chains. This may serve as a partial explanation for the abnormally high viscosity of polymer solutions.

Such solutions are difficult to formulate as a fine-spray aerosol. When a propellant is added to a polymer solution with high viscosity, the product discharges as a stream and not as a spray. If the solution of the polymer is diluted by the addition of more solvent, the polymer chains will start to separate since the chains are in constant motion in the solution. If the solution is dilute enough, the polymer chains will untangle sufficiently so that the polymer molecules may be separated. Under these circumstances, the product may be formulated as an aerosol.

The viscosity of any polymer solution is a function of the concentration of the polymer in the solution and increases as the concentration of polymer is increased. Therefore, dilution of the polymer solution by the addition of solvent will decrease the viscosity of the polymer solution because the concentration of the polymer in the solution will be decreased accordingly. The extent to which the viscosity is reduced by the addition of solvent depends not only upon the particular polymer that is involved, but also the type of solvent that is used for the dilution.[4-6] If the added liquid is a relatively poor solvent for the polymer, the polymer chains will tend to curl up and the viscosity will decrease rapidly.

VISCOSITY OF THE CONCENTRATE

It has been found that in many cases, the viscosity of the polymer solution, i.e., the concentrate, is a good indication of whether a given polymer solution can be formulated directly as an aerosol or whether the concentrate will have to be diluted by the addition of more solvent. An effective and simple method for determining the relative viscosity of a polymer solution is to measure the viscosity with a #4 Ford cup. The #4 Ford cup consists essentially of a stainless steel cup with a small orifice in the bottom of the cup. The time required for the polymer solution to flow through the orifice and drain out of the cup is measured with a stop watch. The viscosity is expressed as a number of seconds required to empty the cup. It has been observed experimentally with certain enamels,[7] that a concentrate with a #4 Ford Cup viscosity of about 20 sec may be formulated as an aerosol with a satisfactory spray. Most nonaerosol paints and lacquers sold for brush application have viscosities considerably higher than 20 sec and have to be diluted by the addition of solvents until the viscosity is low enough to produce an acceptable spray.

The experimental work involved in obtaining a concentrate with the correct viscosity for aerosol packaging can often be minimized by using the following procedure: first determine the #4 Ford cup viscosity of the solution of the polymer that is to be formulated. Then dilute the concentrated solution of the polymer with successive portions of a suitable solvent and determine the #4 Ford cup viscosity after each addition of solvent. The viscosity can then be plotted as a function of the concentration of the polymer in the solvent or as a function of the percent of solvent added to the concentrate. The results obtained with a hypothetical Polymer X might appear as shown in Table 11–1, assuming that the concentration of Polymer X in the initial concentrate was 40% by weight.

TABLE 11-1 EFFECT OF VARIATION IN CONCENTRATION UPON VISCOSITY

Concentration of Polymer X in Solvent (wt %)	#4 Ford Cup Viscosity (Sec)
40	120
30	82
20	48
10	20
5	10

These data can be illustrated graphically as shown in Figure 11–1. From the graph in Figure 11–1 it is possible to determine what concentra-

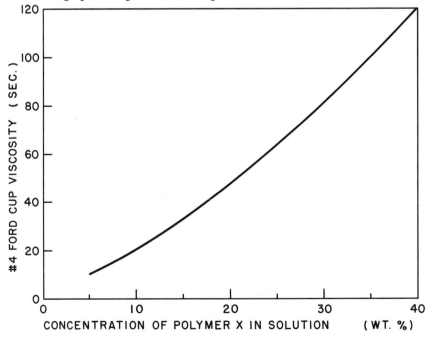

Figure 11-1   Relationship between solution viscosity and Polymer X concentration.

tion of polymer in the solvent will have any given viscosity. If the concentration of the polymer in the concentrate is not known, then the graph will show what concentration of added solvent is necessary to produce any given viscosity.

As far as the proportion of propellant required is concerned, it has been found experimentally that about 45–50% "Freon" 12 by weight in a coating composition will usually give a satisfactory spray if the viscosity of the concentrate is satisfactory. Another limiting factor in the formulation of an aerosol coating composition besides viscosity is the compatibility of the concentrate with the propellant. "Freon" 12 is a relatively poor solvent, and the addition of "Freon" 12 to the concentrate may create a solvent system in which the polymer is not soluble. In order to determine the concentrations of the polymer in the solvent that have sufficient compatibility with the propellant, the various solutions listed in Table 11-1 can be titrated with "Freon" 12 propellant until precipitation of the polymer occurs. The precipitation of the polymer can be considered the end point of the titration. The composition of the solution at the end point can be calculated if the composition of the initial concentrate is known and also the weight of "Freon" 12 that is added. The results that might be obtained with the hypothetical Polymer X are shown in Table 11–2.

TABLE 11-2   COMPATIBILITY OF "FREON" 12 PROPELLANT WITH
SOLUTIONS OF POLYMER X

% Polymer X In Concentrate	% "Freon" 12 In Mixture at End Point
40	2
30	11
20	26
10	50
5	63

The data in Table 11–1 and Table 11–2 can be combined on the same graph as illustrated in Figure 11–2. It is then possible to select reasonable compositions to evaluate. For example, a mixture containing 7.5% Polymer X in the concentrate has a #4 Ford cup viscosity of about 14 sec. This particular concentrate also may be formulated with up to 56% "Freon" 12 before precipitation of the polymer occurs. Therefore, a concentrate containing 7.5% Polymer X would be a logical product to test initially for spray properties.

Besides the viscosity of the concentrate, other factors are involved in

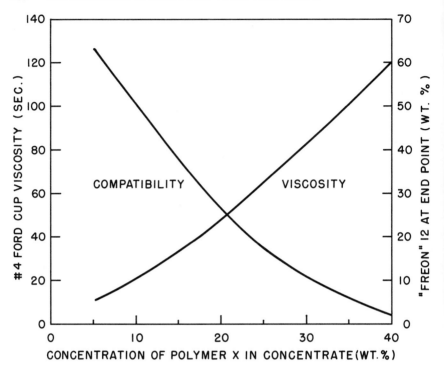

Figure 11–2   Relationship between solution viscosity, propellant compatibility and Polymer **X** concentration.

obtaining a satisfactory spray. Thus, with some high molecular weight polymers, a product with a satisfactory spray cannot be obtained even with #4 Ford cup viscosities below 20 sec. It is possible that viscosity changes in the valve during spraying and that these changes occur more readily with high molecular weight polymers than with polymers with lower molecular weight.

VISCOSITY OF THE AEROSOL PRODUCT

Up to now, we have discussed the viscosity of the polymer solution (the concentrate); however, one of the most important factors is the viscosity of the aerosol product itself. The measurement of aerosol formulation viscosity is somewhat more complicated than the measurement of concentrate viscosity because of the problems introduced by the pressure of the aerosol. An apparatus and procedure for measuring the viscosity of aerosol compositions has been described by Morrow and Palmer.[4] The viscometer (consistometer) developed by Morrow and Palmer used the falling ball principle to measure the viscosity of aerosol products under pressure. By using this apparatus, they were able to determine the relationship between the viscosity of an aerosol product and the pressure required to obtain a satisfactory spray. As would be expected, the higher the viscosity, the higher was the pressure required.

The effect upon viscosity of adding additional solvent to the polymer solution has been discussed. The same effect is obtained when the propellants are added to the concentrate. The addition of the propellant reduces the viscosity of the concentrate and in this respect the propellants act like standard thinners used in the paint industry within the limits of their compatibility.

The extent to which the viscosity is reduced by the addition of the propellants is a function of the particular propellant that is used.[5,6] For example, a 50/50 solution of "Freon" 22–"Freon" 12 is a more effective thinner than "Freon" 12 alone.

EFFECT OF PROPELLANT CONCENTRATION

For many years it has been known that there was a minimum concentration of propellant necessary to provide a satisfactory spray for any polymer solution. There also may be an upper limiting concentration of propellant at a given concentration of the polymer.[8] Thus, it was observed with an acrylic ester resin dissolved in toluene, that 25% "Freon" 12 was necessary in order to obtain a coarse, wet spray when the concentration of resin in the aerosol was 12%. When the concentration of resin was maintained at 12% and the concentration of "Freon" 12 increased above 25%, the spray became increasingly finer. Ultimately, the spray changed to a stream as the concentration of "Freon" 12 was increased above 50%. This effect occurs because as the propellant increases at a constant resin concentration, the solvent concentration decreases proportionately, and the viscosity of the concentrate increases. Also, as the proportion of propellant in the aerosol is increased, the extent to which the propellant flashes during discharge of the product increases. The cooling resulting from the increased flashing increases the viscosity of the product so that at higher propellant concentrations, the product is so viscous that a stream rather than a spray is obtained.

REFERENCES

1. F. H. Getman and F. Daniels, "Outlines of Theoretical Chemistry," 5th ed. 57, John Wiley & Sons, Inc., New York, N.Y., 1931.
2. H. Mark and A. V. Tobolsky, "Physical Chemistry of High Polymeric Systems," 2nd ed., Vol. 2, p. 282, Interscience Publishers, Inc., New York, N.Y., 1955.

3. S. Glasstone, "Textbook of Physical Chemistry," 2nd ed., p. 511, Macmillan Company, Ltd., London, England, 1956.
4. R. W. Morrow and F. S. Palmer, Proc. 39th Ann. Meeting, CSMA (December 1952) ("Freon" Aerosol Report FA-13).
5. F. S. Palmer and R. W. Morrow, Proc. 39th Mid-Year Meeting, CSMA (May 1953) ("Freon" Aerosol Report FA-13).
6. F. S. Palmer, "Freon" Aerosol Report FA-13, "The Formulation of Aerosol Coating Compositions with 'Freon' Fluorocarbon Propellants."
7. F. A. Bower, *Aerosol Age* 3 (December 1958) ("Freon" Aerosol Report FA-13).
8. P. A. Sanders, *Amer. Paint J.* (October, 1966) ("Freon" Aerosol Report FA-13A).

12

DENSITY

Aerosol propellant and formulation densities are significant because they determine the weight of propellants that can be shipped in cylinders or tank cars, or stored at loading plants, and the weight of aerosol products that can be packaged in an aerosol container. Quite often, there is some confusion concerning the terms *density* and *specific gravity*. The density of a substance is defined as the mass per unit volume at any given temperature. Thus, $d = m/v$, where d = density, m = mass, and v = volume. In the aerosol industry, density is usually expressed in g/cm^3 or in lb/ft^3.

Specific gravity is the ratio of the weight of a given volume of a substance at one temperature (t_2) to the weight of the same volume of a reference material, such as water. The temperature of the reference material (t_1) may be the same as the temperature of the substance whose specific gravity is under consideration, or it may be different. The specific gravity of any material is then reported using the symbol $d_{t_1}^{t_2}$, where as previously mentioned, t_1 is the temperature of the reference material and t_2 is the temperature of the substance. For example, iron has a specific gravity of 7.90 at 20°C compared to water at 4°C. Therefore, the specific of gravity of iron is described as $d_{4°}^{20°} = 7.90$. In the metric system, the density of water is taken as one. Under these conditions, the density and specific gravity of a substance have the same number. However, density has the dimensions of mass/volume, such as g/cc, while specific gravity is a pure number and has no dimensions.

The numerical values of the densities of the various propellants were given previously in Chapter 4 and need not be repeated here. The fluorocarbon propellants (with the exception of Propellant 152a) have densities greater than one. Therefore, in comparison with the hydrocarbon propellants which have densities less than one, a greater weight of products formulated with the fluorocarbon propellants can be packaged in a given volume than products formulated with the hydrocarbon propellants, assuming that the weight percentage of the propellants is the same in both cases.

CALCULATION OF DENSITIES OF MIXTURES

The densities of the common blends of various propellants are usually available from the literature. However, quite often densities of various mixtures, particularly of three-component blends, either have to be calculated or determined experimentally. The calculation of the density of a given mixture assumes that the volumes of the individual components are additive and that the total volume of the mixture is equal to the sum of the volumes of the individual components. The calculation of the density of a three-component mixture is illustrated by the following example.

Assume that the density of a "Freon" 12–"Freon" 114–"Freon" 11 (50/30/20) mixture at 70°F is desired. The density is calculated as follows:

1. First, the densities of the three propellants are obtained from a suitable source, such as Reference 1.

$$\text{"Freon" 12} \quad — \quad 1.32 \text{ g/cc}$$
$$\text{"Freon" 114} \quad — \quad 1.47 \text{ g/cc}$$
$$\text{"Freon" 11} \quad — \quad 1.48 \text{ g/cc}$$

2. Assume that the weight of the three-component blend is 100 g. The volume in cc contributed by each of the components is calculated by dividing the weight of the component by its density.

$$\text{Volume of 50.0 g of "Freon" 12} \ = \ \frac{50}{1.32} = 37.9 \text{ cc}$$

$$\text{Volume of 30.0 g of "Freon" 114} = \ \frac{30}{1.47} = 20.4 \text{ cc}$$

$$\text{Volume of 20.0 g of "Freon" 11} \ = \ \frac{20}{1.48} = 13.5 \text{ cc}$$

3. The total volume of 100 g of the blend is obtained by adding together the volumes of the individual components. The total volume, therefore, is:

$$37.9 \text{ cc} + 20.4 \text{ cc} + 13.5 \text{ cc} = 71.8 \text{ cc}$$

4. The density of the blend is equal to m/v and is obtained by dividing the weight of the blend, 100 g, by the volume, 71.8 cc.

$$\text{Density of the blend} = \frac{100.0 \text{ g}}{71.8 \text{ cc}} = 1.39 \text{ g/cc}$$

A rather common error in calculating densities is to assume that the density of a mixture of liquids can be obtained merely by multiplying the percentage of each component by its density and adding up the resulting values for the various components. In some cases, where the densities of the various components are fairly close to each other, this procedure will give approximately the same result as that obtained by the correct method. However, when the densities differ appreciably, a considerable error will result when the density of the mixture is calculated by multiplying the percentage of each component by its density. The difference that results with the two methods is illustrated by the following example. Assume that the density of a "Freon" 12–propane (50/50) mixture is desired. The density of this mixture is calculated correctly as follows:

1. The densities of the two components are:

$$\text{Density of "Freon" } 12 = 1.32 \text{ g/cc}$$
$$\text{Density of propane} \quad = 0.50 \text{ g/cc}$$

2. The volumes of the individual components in 100 g of the blend are:

$$\text{Volume of 50 g of "Freon" } 12 = \frac{50.0}{1.32} = 37.9 \text{ cc}$$

$$\text{Volume of 50 g of propane} \quad = \frac{50}{0.50} = 100.0 \text{ cc}$$

3. The total volume of 100 g of the "Freon" 12–propane (50/50) blend is:

$$37.9 \text{ cc} + 100.0 \text{ cc} = 137.9 \text{ cc}$$

4. The calculated density of the blend, therefore, is

$$\frac{100.0 \text{ g}}{137.9 \text{ cc}} = 0.72 \text{ g/cc}$$

Now, if the density of the blend is calculated erroneously by multiplying the percentage of each component by its density, the following result is obtained:

$$\text{Contribution of "Freon" } 12 = 0.50 \times 1.32 \text{ g/cc} = 0.66 \text{ g/cc}$$
$$\text{Contribution of propane} \quad = 0.50 \times 0.50 \text{ g/cc} = 0.25 \text{ g/cc}$$
$$\text{Total} = 0.91 \text{ g/cc}$$

The density of 0.72 g/cc obtained by the first procedure differs considerably from the value of 0.91 g/cc obtained by the second method.

CALCULATIONS OF COMPOSITIONS OF BLENDS WITH A SPECIFIC DENSITY

It is sometimes desirable to determine the composition of a propellant blend that has a specific density. For example, in the formulation of emulsion systems, it is advantageous to use a propellant blend having the same density as that of water, since this decreases the rate of creaming. The composition of a blend with a particular density can be calculated if the densities of the individual components are known.

Two-Component Blends

The composition of a two-component blend with a specific density can be calculated using the following equation:

$$X = \frac{100 \ (d_1 d_2 - d_1 D)}{D(d_2 - d_1)} \tag{12-1}$$

X = grams of Component 1 in 100 g of the propellant blend
d_1 = density of Component 1
d_2 = density of Component 2
D = specified or required density of blend

Solving the equation for X gives the grams of Component 1 in 100 g of the blend. This is the same as wt %. Substracting X from 100 gives the wt % of Component 2. Equation 12–1 is derived as follows:

1. Assume that the total weight of the blend will be 100 g. Then, let X = g of Component 1 in the blend. Therefore, $100 - X$ is the weight of Component 2 in the blend.
2. The total number of cc of Component 1 in 100 g of the blend is equal to the weight of Component 1 in the blend divided by density (d_1).

$$\text{Volume in cc of Component 1} = \frac{X}{d_1}$$

3. The total number of cc of Component 2 in the blend is equal to the weight of Component 2 in the blend divided by its density (d_2).

$$\text{Volume in cc of Component 2} = \frac{100 - X}{d_2}$$

4. The total volume in cc of 100 g of the blend, therefore, is equal to the sum of the cc of Component 1 and the cc of Component 2.

$$\text{Total volume in cc of 100 g of blend} =$$
$$\frac{X}{d_1} + \frac{100 - X}{d_2}$$

5. The density of the blend is equal to the weight divided by the volume.

Since the weight is 100 g and the total volume is

$$\frac{X}{d_1} + \frac{100 - X}{d_2} \text{ cc,}$$

the density is as follows:

$$\text{Density of blend} = \frac{100}{\dfrac{X}{d_1} + \dfrac{100 - X}{d_2}}$$

6. Since the density desired is D g/cc, the equation then becomes,

$$\frac{100}{\dfrac{X}{d_1} + \dfrac{100 - X}{d_2}} = D$$

7. Solving for X gives Equation 12–1.

As an example, assume that the composition of a "Freon" 12–isobutane blend with a density of 1.0 g/cc at 70°F is desired. The composition of the blend is obtained as follows:

1. The densities of "Freon" 12 and isobutane at 70°F are:

Density of "Freon" 12 = 1.32 g/cc
Density of isobutane = 0.56 g/cc

2. Substituting these values in Equation 12–1, along with the value of 1.0 g/cc for D, gives

$$X = \frac{100 \, (1.32 \times 0.56 - 1.32 \times 1.0)}{1.0 \, (0.56 - 1.32)}$$

3. Solving the equation for X gives a value of 76.4 g of "Freon" 12. The number of grams of isobutane in the blend is $100 - 76.4$, or 23.6 g. The composition of the blend with a density of 1.0 g/cc, therefore, is 76.4% "Freon" 12 and 23.6% isobutane.

Three-Component Blends

There are many combinations of three components that have the same density. The simplest approach is to calculate the compositions of several of these combinations and plot the data on a triangular coordinate chart. The two points at each end of the density plot fall on the base lines for two-component blends. The compositions of these two blends are calculated using Equation 12–1 and plotted on the triangular coordinate chart. A straight line is drawn between the two points. All possible combinations of the three components that have the specified density fall on this line. In view of the number of different combinations that have the same den-

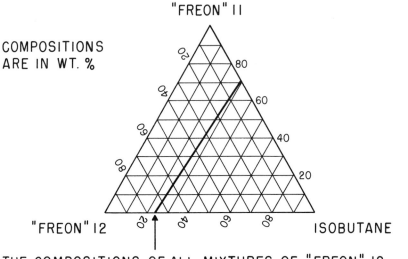

THE COMPOSITIONS OF ALL MIXTURES OF "FREON" 12, "FREON" 11, AND ISOBUTANE WITH A DENSITY OF 1.0 g/cc FALL ON THIS LINE.

Figure 12-1 Mixtures of "Freon" 12, "Freon" 11, and isobutane with a density of 1.0 g/cc at 70°F.

sity, the selection of the final blend will have to be based upon some other factor in addition to the density.

The procedure for preparing a triangular coordinate plot of compositions with a particular density can be illustrated as follows: assume that the compositions of "Freon" 12, "Freon" 11 and isobutane blends with a density of 1.0 g/cc are required. The "Freon" 12–isobutane blend with a density of 1.0 g/cc was shown in the previous section to have a composition of 76.4 wt % "Freon" 12 and 23.4 wt % isobutane. Similarly, the composition of the "Freon" 11–isobutane blend with a density of 1.0 g/cc can be calculated using Equation 12-1. The composition of this blend is 70.6% "Freon" 11 and 29.4% isobutane. These two compositions are then plotted on a triangular coordinate chart and a straight line drawn between the points as shown in Figure 12-1. All combinations of the three components that have a density of 1.0 g/cc fall on this line.

EXPERIMENTAL DETERMINATION OF DENSITY

In some instances, it is desirable to determine the density of a propellant blend experimentally. If precise results are not required, the density may be obtained by adding the propellant blend to a volumetrically calibrated pressure tube and thermostating the tube to the desired temperature. The weight of propellant added is obtained by weighing the glass tube and the volume of the propellant is read directly from the tube. If more accurate results are desired, the densities can be obtained with a hydrometer. The procedure and apparatus for this method are described in Reference 2.

The densities of complete aerosol formulations can be obtained by using the same procedures as those described for the propellant blends. If the relative proportions of the propellant and concentrate are known, and the densities of each are also known, the density of the product can be calculated. In many cases, the concentrate consists of a complex mixture of ingredients and it is difficult to calculate the density of the concentrate itself. In these cases, it is usually easier to obtain an approximate density of the concentrate experimentally by weighing a known volume of the concentrate. The density of the aerosol product is then calculated using the experimentally determined concentrate density.

EFFECT OF TEMPERATURE

The DOT regulations specify that an aerosol package must not become liquid full at 130°F. In order to determine the volume of any given weight of a product at 130°F, it is necessary to know the density. An indication of whether or not the container will be liquid full at 130°F can be obtained by assuming that the entire contents of the aerosol package consist only of propellant. Generally, the propellants expand more with a rise in temperature than any aerosol product. The volume fill of the container at 130°F can be calculated by using the propellant density data given in the literature, such as that for the "Freon" propellants in Reference 1. If the container will not be liquid full with propellant alone, it is unlikely that it will become liquid full with the aerosol product itself.

The change in density with temperature can also be obtained by measuring the densities at several temperatures. A curve illustrating the change in density with temperature can then be drawn. Densities at temperatures either higher or lower than those determined experimentally can be approximated by extrapolation.

REFERENCES

1. "Freon" Aerosol Report FA-22, "Vapor Pressure and Liquid Density of 'Freon' Propellants."
2. "Tentative Hydrometric Determination of Aerosol Liquid Densities," CSMA Aerosol Guide, June, 1957, p. 33.

13

STABILITY

There are many properties that an aerosol package should have in order to be successful. One that is an absolute necessity is adequate shelf life or storage stability. If an aerosol leaks as a result of container corrosion or the product itself deteriorates because of instability, the aerosol is not satisfactory regardless of its other properties. The premature marketing of an aerosol product with poor shelf life can be very costly if the product has to be withdrawn from the market. It is important, therefore, to take every precaution to be certain that an aerosol product has sufficient storage stability.

Poor shelf life can result for a number of reasons. Leakage of the product may occur because of pinhole corrosion even though the product appears to be stable. The odor may change without any noticeable container corrosion, and the product itself may deteriorate and lose its effectiveness or activity. A valve may fail to operate after a period of time either because the formulation has affected the valve gasket or because physical or chemical changes in the product result in the formation of particles that clog the valve orifice. Since the causes for poor storage stability are varied, an aerosol product must be subjected to a number of tests in order to insure that it has satisfactory shelf life.

In many cases adequate information for making reasonable predictions already exists and storage stability tests can be minimized or avoided. Aerosol loaders and suppliers, such as the propellant, valve and container manufacturers, have accumulated extensive data on the storage stability of hundreds of products; also, conclusive evidence on the stability of aerosol propellants under numerous conditions is readily available. In addition, a comprehensive understanding of the systems and materials that cause corrosion can often be helpful in avoiding potentially unstable products.

118

CORROSION

Corrosion in metal containers is one of the major reasons for the poor shelf life of aerosol products. It can be slight as far as the effect on the materials of the container is concerned but still be sufficient to have an adverse effect upon the aerosol components, such as perfumes. In extreme cases corrosion can result in perforation of the aerosol containers with loss of product. Most corrosion is the result of electrochemical processes, and some of the fundamental concepts of electrochemical corrosion are discussed in the folowing section. The electrochemical aspects of corrosion have been summarized in an excellent booklet by LaQue, May, and Uhlig.[1]

Fundamentals of Corrosion

In order for electrochemical reactions to take place, certain conditions are necessary. There must be present an electrolyte for conducting electric current (assumed to move in the direction of positive charges), two electrodes, an anode and a cathode with a difference in potential, and a path for electrons to flow from the anode to the cathode. An electrolyte is a liquid containing ions that conduct the current (positive ions). Water is an example of an electrolyte that contains hydrogen and hydroxyl ions. The two electrodes may consist of two different metals placed in an electrolyte solution. The electrochemical process with different metals is called galvanic action. The electrodes can also form at two different locations on the same piece of metal. A potential difference can arise from differences in the environmental conditions, differences in structure of the metal, concentrations of impurities, etc. The electrochemical processes are called local action in this case.

The path for the electrons can consist of a piece of wire or if the two electrodes are located on the same piece of metal, the electrons can flow directly from the anode to the cathode through the metal. The complete system is called a corrosion cell and is illustrated in Figure 13–1a.

The potential difference between the two electrodes is the driving force that actuates the electrochemical process. The degree to which the potentials of the electrodes differ depends upon many factors, such as the types of metals, their environment, etc.[1] However, once the conditions exist so that the electrochemical process can occur, a number of reactions take place simultaneously at both the anode and the cathode. At the anode, positively charged metal ions leave the anode and enter the electrolyte solution. This loss of metal from the anode, if allowed to continue, will ultimately result in destruction of the anode. Most of the corrosion that occurs in electrochemical processes takes place at the anode for this reason. Meanwhile, the free electrons that are left in the anode as a result of

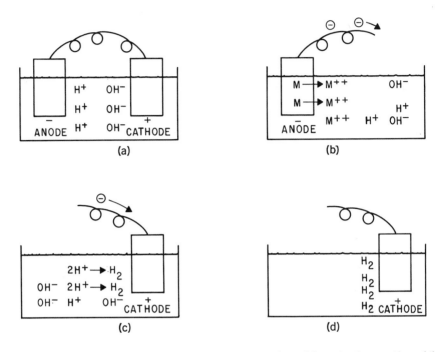

Figure 13–1   (a) Corrosion cell; (b) Anodic reaction; (c) cathodic reaction; (d) Cathodic polarization.

the loss of the positively charged metal ions, flow from the anode to the cathode through the metallic path. This anodic electrochemical process is illustrated in Figure 13–1b.

Meanwhile, a different series of reactions takes place at the cathode. Positively charged hydrogen ions migrate through the electrolyte solution to the cathode. Here the hydrogen ions meet the electrons that travel from the anode to the cathode and are neutralized to form hydrogen gas. This cathodic process is illustrated in Figure 13–1c.

Polarization and Depolarization

The hydrogen that is formed at the cathode can either escape as hydrogen gas or remain and accumulate at the cathode. When sufficient hydrogen molecules collect at the anode, they form a barrier and interfere with the neutralization of other hydrogen ions by the electrons in the cathode. This slows down the electrochemical process and reduces the rate of corrosion. This effect of hydrogen molecules on the cathode is called polarization and the extent to which an electrode can be polarized depends upon such factors as the environmental conditions present and the type of metal forming the electrode. Polarization is illustrated in Figure 13–1d.

Oxygen can have a pronounced effect upon polarization. If oxygen is dissolved in the electrolyte, it can react with the hydrogen on the cathode to form water or hydrogen peroxide and hydroxyl ions.[1] In any case, these reactions remove the hydrogen molecules from the cathode and allow the neutralization of hydrogen ions with electrons, and thus corrosion, to continue. The reaction of oxygen with hydrogen is called depolarization and oxygen is termed a cathodic depolarizer.

Relative Areas of the Anode and Cathode

The relationship of the area of the anode to that of the cathode is one of the factors that can have a marked effect upon the rate of corrosion. When the area of the cathode is large compared to that of the anode, corrosion is favored. This is because the polarizing hydrogen that is formed at the cathode is spread out over a comparatively large area. Under these conditions the hydrogen is easily accessible to attack by oxygen which removes it from the cathode and allows corrosion to continue. When the area of the cathode is small compared to that of the anode, the rate of corrosion is much lower because the polarizing hydrogen molecules are able to concentrate on a smaller area and consequently are more difficult to remove. In order to minimize corrosion, therefore, it is desirable to have the area of the cathode small relative to that of the anode.

The area relationship of the anode and cathode and its effect upon corrosion become very important when applied to aerosol tinplate containers. In these containers, the area of the tin coating is very large compared to that of the iron, which is usually exposed only through minute holes in the tin coating. If, in the tinplate container, iron is the anode and tin the cathode, then corrosion will be rapid because of the unfavorable relationship of the large area of the cathode and the small area of the anode. There is an additional reason why this particular situation is undesirable in aerosol tinplate containers. Since corrosion occurs mostly at the anode, it will be concentrated on the relatively small areas of iron available through the holes in the tin coating and pinholing can result in a short time. The most desirable condition is obtained when the polarity is reversed and the tin is anodic and the iron is cathodic. In this case, if corrosion does occur, the tin will be attacked initially instead of the iron. Generalized detinning would occur instead of pinholing and the rate of corrosion would be much lower because of the favorable relationship of the large anode area and the small cathode area.

Effect of Oxygen upon Polarity of Tinplate Containers

Tin and iron are similar to each other in their electrochemical properties[2] and comparatively small changes in the environment or the product may determine which metal will be the cathode and which the anode. One of the environmental factors that influences the polarity of the metals in the tinplate system is the concentration of oxygen. In the presence of low concentrations of oxygen, tin is anodic and iron is cathodic. This produces a comparatively low rate of corrosion and is one reason why vacuum packed food products are stable in tinplate cans.[2] At higher concentrations of oxygen, the polarity of the system is reversed with tin cathodic and iron anodic. This condition promotes corrosion.

The effect of oxygen upon corrosion in tinplate containers can be summarized as follows: low concentrations of oxygen retard corrosion because there is little depolarization of the cathode and because tin is anodic and the iron cathodic. High concentrations of oxygen promote corrosion as a result of the depolarizing activity of oxygen and because tin is cathodic and the iron is anodic.

STABILITY OF AEROSOL PROPELLANTS

There are very pronounced differences in the stability of the various propellants, depending upon the conditions to which the propellants are exposed. A knowledge of these differences in stability is necessary in order to avoid the use of propellants that are unsuitable for certain aerosol systems. The stabilities of various propellants may be summarized as follows:

"Freon" 22

"Freon" 22 propellant is stable in practically all nonaqueous systems and in aqueous systems which are neutral or acidic. "Freon" 22 decomposes in alkaline systems but the reactions involved have not been determined.[6]

"Freon" 12

"Freon" 12 propellant is stable in both nonaqueous and aqueous aerosol systems (except under extremely alkaline conditions) and, therefore, is suitable for use with almost any aerosol product.

"Freon" 114

"Freon" 114 propellant also is very stable in both nonaqueous and aqueous systems and may be used with any aerosol product.

"Freon" 21

"Freon" 21 fluorocarbon, like "Freon" 22, is stable in aqueous systems that are either neutral or acidic. It decomposes rapidly under alkaline conditions and is not suitable for use in such systems. As far as is known, "Freon" 21 is stable in nonaqueous systems. However, "Freon" 21 has been used to such a limited extent in aerosols that there is not much information available on its stability in aerosols.[7]

"Freon" 11

"Freon" 11 propellant is one of the most commonly used "Freon" fluorocarbons. As a result, the stability of "Freon" 11 has been the subject of a number of investigations.[6,8-10] The stability of "Freon" 11 depends upon the particular type of aerosol system that is used. In some systems "Freon" 11 is perfectly stable while in others, its use must be avoided.

"Freon" 11 and Aqueous Systems. The "Freon" propellants are not miscible with water. Therefore, aqueous systems containing the "Freon" propellants must be formulated as oil-in-water emulsions, water-in-oil emulsions, three-phase systems in which the water is layered over the propellant, or as three-component systems in which the third component is a cosolvent for water and the "Freon" propellant.

In oil-in-water emulsions, or in three-phase systems with metal present, "Freon" 11 propellant usually is unstable and should not be used. Metal appears to act as a catalyst for the reaction of "Freon" 11 with water, and reaction products are formed which cause corrosion of the metal. Aerosols containing water and "Freon" 11 possibly could be packaged in glass if the product did not come into contact with metal.

The decomposition of "Freon" 11 by water in the presence of metals is not a simple hydrolysis reaction. "Freon" 21 and "Freon" 112 were identified among the reaction products and are considered to be formed by the following reactions, where M is metal.[10]

$$2 \ CCl_3F + M \longrightarrow MCl_2 + CCl_2FCCl_2F \qquad (13\text{–}1)$$
$$2 \ CCl_3F + 2H_2O + 2M \longrightarrow 2 \ CHCl_2F + MCl_2 + M(OH)_2 \quad (13\text{–}2)$$

Acid is also formed, which indicates that some acid forming reaction such as the following may also take place. The specific reaction responsible for the acid formation has not been identified.

$$CCl_3F + H_2O \longrightarrow \underset{Cl}{\overset{F}{C}} = O + 2HCl \qquad (13\text{–}3)$$
$$\underset{(H_2O)}{\longrightarrow} CO_2 + HCl + HF$$

As a result of the instability of "Freon" 11 in the presence of water and metals, "Freon" 11 is not suitable for use with oil-in-water emulsion products such as shaving lathers, starch sprays, and window cleaners. The use of lacquer-lined containers and lacquer-coated valve cups does not solve the problem of instability because there are always enough pinholes or other imperfections in the lacquer lining or coating on the valve cup so that there is some contact of the formulation with metal.

Nitromethane is a fairly effective inhibitor for retarding the decomposition of "Freon" 11 by water in metal containers. The evidence indicates that nitromethane functions as an inhibitor in this system by deactivating the metal surface so that it no longer acts as a catalyst for the "Freon" 11 propellant–water reaction.[10] The use of nitromethane in combination with other inhibitors for water-based systems has been suggested.[11] Nitromethane has not been tested sufficiently in aqueous-based aerosol products as yet, however, to determine how effective an inhibitor it is for such systems.

In contrast to its instability in oil-in-water emulsions, "Freon" 11 can be quite stable in water-in-oil emulsions.[12,13] In water-in-oil emulsions "Freon" 11 is part of the outer phase and the water is the dispersed phase. The dispersed water droplets also are surrounded by the oil soluble emulsifying agent which has concentrated at the interface between the water and the "Freon" propellants. This prevents the water from coming in contact with the metal, which is the catalyst for the decomposition.

"Freon" 11 and Nonaqueous Systems. "Freon" 11 propellant is stable in most nonaqueous systems and has been widely used in products such as aerosol insecticides, room deodorants, hair sprays, etc. However, under certain conditions, "Freon" 11 will react with ethyl alcohol via a free radical reaction mechanism to give acetaldehyde, hydrogen chloride, and "Freon" 21.[9]

$$CCl_3F + CH_3CH_2OH \longrightarrow CH_3CHO + HCl + CHCl_2F$$
$$(13\text{--}4)$$

Secondary reactions then follow to give acetal and ethyl chloride.

$$CH_3CHO + 2CH_3CH_2OH \longrightarrow CH_3CH(OC_2H_5)_2 + H_2O$$
$$(13\text{--}5)$$

$$CH_3CH_2OH + HCl \longrightarrow CH_3CH_2Cl + H_2O \qquad (13\text{--}6)$$

These reactions were of particular interest because of the corrosion observed on various occasions with hair sprays formulated with ethyl alcohol and "Freon" 12–"Freon" 11 propellant mixtures. The free radical reaction between "Freon" 11 propellant and ethyl alcohol is catalyzed by low concentrations of oxygen and corrosion in hair spray containers was sometimes observed in products loaded in the summer, where the warm conditions promoted effective purging of air during the loading of the samples. The products, therefore, had relatively low concentrations of oxygen.

High concentrations of oxygen inhibit the reaction between "Freon" 11 propellant and ethyl alcohol and the products loaded in winter tended to have higher concentrations of oxygen. The property of oxygen of functioning as a catalyst for free radical reactions at low concentrations and acting as an inhibitor at higher concentrations has been observed for other free radical reactions.[9]

There are many compounds that are known to act as inhibitors for free radical reactions, and an intensive investigation of potential inhibitors for the free radical reaction between "Freon" 11 and ethyl alcohol was carried out. The compound which was selected as the most suitable and effective inhibitor for this reaction in aerosol products was nitromethane. "Freon" 11 S, which contains nitromethane as a stabilizer for the reaction, is available commercially for use with aerosol products containing alcohols.[9]

Aqueous Ethyl Alcohol Mixtures.   The use of nitromethane as a stabilizer either for aqueous systems with "Freon" 11 propellant or for alcoholic systems with "Freon" 11 has been discussed. On the basis of the stabilizing effect of nitromethane in these two systems, it might be expected that nitromethane would be equally effective in aqueous ethyl alcohol mixtures. Unfortunately, this does not seem to be the case. Nitromethane is fairly effective as a stabilizer in 95% ethyl alcohol (by volume)–"Freon" 11 mixtures but not in mixtures containing higher percentages of water.

Unstabilized mixtures of 95% ethyl alcohol and "Freon" 11 cause quite rapid corrosion in metal containers. The addition of nitromethane stabilizes the mixtures so that the stability is almost equal to that of absolute ethyl alcohol and unstabilized "Freon" 11 propellant. Many of the components of aerosol products, such as hair sprays, may also act either as stabilizers or catalysts for the free radical reaction between "Freon" 11 and ethyl alcohol. Therefore, it is difficult to predict the storage stability of any product formulated with 95% ethyl alcohol and "Freon" 11 as two of the components. If the other ingredients also have a stabilizing effect, then the product may have sufficient storage stability for marketing. If the other components are catalysts for the free radical reaction, the storage stability may be too poor for marketing.

Nitromethane was ineffective in stabilizing mixtures of 90% ethyl alcohol and 80% ethyl alcohol with "Freon" 11. These mixtures caused rapid corrosion of metal containers regardless of whether nitromethane was present or not.

EFFECT OF OTHER AEROSOL COMPONENTS

Acids, Alkalies, and Salts

One component necessary for electrochemical reactions is an electrolyte solution. Since inorganic acids, alkalies, and salts are electrolytes, it is not surprising that these materials have a long history of corrosion. The hydrogen ion is one of the main factors involved in corrosion since it is involved in the cathodic reaction with electrons which forms free hydrogen.[3] The rate of corrosion depends upon the acidity of the solution, with a dividing line occurring in the neighborhood of pH 4.5. The rate of corrosion is fairly low above a pH of 4.5 but becomes quite rapid at lower pH values.

Alkalies are usually considered to be less corrosive than acids but they can produce anodic attack at localized areas which results in a pitting type of corrosion.[4]

Salts are good electrolytes and their solutions are highly conducting. As a result they can be quite corrosive. The corrosiveness of a salt solution will depend both upon the concentration and type of salt present. The acid aluminum salts that are used in antiperspirant products are quite corrosive and must be packaged in glass.

Chlorinated Solvents

Chlorinated solvents, such as methylene chloride, methyl chloroform, and trichloroethylene are used in aerosol formulations as solvents for active ingredients, as vapor pressure depressants, or just as low cost substitutes for "Freon" 11 propellant. In some products, they have a more specific function, such as dry cleaning solvents, etc.

The chlorinated solvents usually are somewhat unstable in the presence of metals, heat, moisture, and oxygen and may decompose to some extent with the formation of acidic decomposition products. If this occurs during shipment in steel drums or in solvent applications involving the cleaning of metal components at elevated temperatures, corrosion problems may arise. As a result, many compounds have been tested as inhibitors for increasing the stability of the chlorinated solvents. A number of patents have been issued which disclose many additives that are reported to be effective stabilizers.[14] Most of the common commercial chlorinated solvents now contain stabilizers.

Methylene chloride is quite stable in most aerosol formulations and has been used in both oil-in-water and water-in-oil emulsion systems. Methyl chloroform appears to be less stable in aqueous systems. Any aqueous formulation containing a chlorinated solvent should be checked thoroughly for shelf life before marketing is considered.

Surface-Active Agents

There are many products on the market containing surface-active agents that have a perfectly satisfactory shelf life, such as shaving lathers, upholstery cleaners, window cleaners, etc. However, some surface-active agents are known to be very corrosive in some metal containers. One of the best known classes of the surfactants in this category are the sodium alkyl sulfates. A number of early attempts were made to package aerosol shampoos based upon detergents of the sodium alkyl sulfate type but the products generally were too corrosive in metal containers of that period to be marketed.[15] Some of these products appeared to have satisfactory storage stability judging by accelerated storage stability tests at elevated temperatures but subsequently were found to cause perforation of containers in a relatively short time at room temperature.

There are a number of reasons why the sodium alkyl sulfates are corrosive. They ionize in solution and also contain inorganic salts as impurities. Therefore, their aqueous solutions are good electrolytes and conductors of electricity. Another important factor, as shown by Root,[16] is that in these systems the tin is cathodic and the iron anodic. Therefore, corrosion is localized at the minute areas where iron is accessible, and pinholing results. This is an excellent example of the rapid corrosion that can occur when the system contains a cathode with a much larger area than that of the anode. Root also examined several other formulations where the tin was found to be anodic and the iron cathodic. He reported that these formulations appeared to have much better storage stability.

In contrast to the sodium alkyl sulfates, the anionic triethanolamine salts of the fatty acids have been used extensively in aerosol foam products and the rate of corrosion has been found to be acceptably low. There are several reasons why these surfactants are less corrosive than the sodium alkyl sulfates. Since the triethanolamine fatty acid soaps are salts of a weak acid and a weak base, they tend to hydrolyze in solution rather than ionize. Also, they do not contain inorganic salts as impurities. Therefore, the solutions of the triethanolamine salts are much weaker electrolytes than solutions of the sodium alkyl sulfates. West[17] has suggested that the fatty acid soaps should be considered as corrosion inhibitors. Another factor may be the relative solubilities of the tin and iron salts of the fatty acids and the lauryl sulfates. LaQue and Copson[4] have indicated that corrosion is reduced when the corrosion products are insoluble. The tin and iron salts of the lauryl sulfates would be expected to be more soluble in the aqueous phase than the corresponding salts of the fatty acids.

Nonionic surfactants in general appear to cause much less corrosion than anionic agents. Since these compounds do not ionize their solutions are less conductive and this is certainly one of the factors responsible for the lower rate of corrosion.

Cationic agents, such as the quaternary ammonium compounds, have been used in a number of products as germicidal agents where the concentration is very low. It is likely that at much higher concentrations they could cause corrosion because their solutions are fairly good electrolytes.

Water

Water alone can cause considerable corrosion in metal aerosol containers. The corrosive effects of water are due largely to the presence of dissolved solids and gases.[4] The corrosive action of water, therefore, will vary considerably, depending upon the source of the water. Distilled water is not very corrosive but sea water, with its relatively high concentration of salts, is quite corrosive. Water containing appreciable concentrations of carbon dioxide will cause significant corrosion of steel but will not have much effect upon tin.

In view of the difference in corrosive activity observed with water from various sources, it is important in storage stability tests to use the same water that will be used in production. The stability of a test pack prepared in the laboratory with distilled water may be considerably different from that of samples prepared with deionized or raw water. Nitromethane is a fairly good inhibitor for retarding the corrosion of tinplate containers by water.[10]

ALUMINUM CONTAINERS

Aluminum as a material of construction for aerosol containers has a number of advantages. It is a good example of a metal highly resistant to corrosion because of a protective oxide film. The oxide film forms easily in the presence of oxidizing agents or oxygen and is reported to be stable under most conditions in the pH range from 4.5–8.5.[5]

Storage stability studies have shown that certain combinations of the propellants with alcohols can corrode aluminum containers. For example, mixtures of anhydrous ethyl alcohol with Propellant 114 were reported to cause perforation of aluminum containers during tests at elevated temperature. The corrosion was atttributed to the formation of the aluminum alcoholate and hydrogen.[16] The increase in pressure resulting from the formation of hydrogen was also noted in other tests carried out with alcohols and propellants in the presence of aluminum strips.[8] The presence of 1.5–2.0% water was observed to have an inhibiting effect upon the formation of the alcoholate, which evidently is favored by anhydrous conditions.[16,18]

Combinations of Propellant 11 and ethyl alcohol are known to cause severe corrosion of aluminum containers manufactured in the United States. The addition of water does not sufficiently inhibit the corrosion inducing processes. Since Propellant 11 can react with ethyl alcohol or water to form acidic reaction products, it is assumed that these products attack the protective oxide film on aluminum and cause corrosion. However, many hair sprays formulated with combinations of Propellant 12, Propellant 11, and ethyl alcohol are packaged in aluminum containers in Europe and apparently have adequate storage stability. The reason for the difference has not been determined.

STORAGE STABILITY TESTS FOR AEROSOL PRODUCTS

There are no standard storage stability tests for aerosol products. Since a wide variety of products, containers, valves, and propellants exists, it is difficult to devise a series of tests that would adequately cover all variables. Theoretically, an ideal test for an aerosol package would be one in which the product was tested under all the conditions to which it would be exposed during consumer use. This, of course, would be difficult to determine. A suggested product check that is helpful in avoiding premature marketing of an aerosol product has been published.[19] A tentative method for storage tests of aerosol insecticides, which is valuable as a guide in setting up storage stability tests, is also available.[20]

Most aerosol products are used at room temperature and it would be most desirable to test the storage stability at room temperature. The disadvantage of this is that it requires long-term testing, which sometimes extends into years. Many marketers are reluctant to wait for the results of room temperatures tests because of the extended periods required. As a result, tests at elevated temperatures are generally used in conjunction with room temperature tests. Tests at elevated temperatures have been used in the rubber and petroleum industries for many years and it is well known in these industries that in many cases, the storage stability of a product at elevated temperature does not necessarily correlate with that at room temperature.

The same situation exists with respect to accelerated storage stability tests with aerosol products. In some cases, the lack of correlation between the results obtained from room temperature tests and those from tests at higher temperatures was striking. One well-known example is that of aerosol shampoos formulated with surfactants, such as sodium lauryl sulfate. When these products were packaged in metal containers, the products appeared to be essentially noncorrosive when tested at 130°F for two or three months. However, the containers perforated within a few weeks at room temperature.[15] According to Foresman, no typical perforation corrosion of the type caused by shampoos has ever been observed at elevated temperatures. Foresman recommended a temperature of 100°F for accelerated tests because he considered it close enough to room temperature so that it was not unrealistic and yet the slight increase in temperature increased the rate of aging. He observed that in some cases storage for three months at 100°F indicated the stability of a formulation for one year at room temperature.

Glessner studied the storage stability of water-based insecticides and reported that with some products an accelerated factor of 2–3.5 to 1 was observed with tests at 100°F.[11] He also pointed out that the results of tests at 100°F or 130°F could be misleading, however, and might cause the rejection of a product with adequate stability at room temperature.

Therefore, the results of tests at elevated temperatures must be viewed with a certain amount of caution. Many formulators carry out storage stability tests at both room temperature and at elevated temperatures. If the short term tests at elevated temperatures produce results that look sufficiently promising, limited marketing of a product may be considered. If the results of tests at the elevated temperature indicate that the storage stability of the product may be doubtful, marketing may be delayed until some of the longer term tests at room temperature have been completed.

Although there are no standard accelerated storage stability tests, there are some conditions that are used by a considerable number of formulators for testing the stability of their products. These are as follows:

Products in Drawn Two-Piece Containers

Storage tests are carried out for one and two months at 130°F. Most formulators do not extend the tests at 130°F beyond two months. Very short term tests at 130°F, such as a few days, are in a sense, not really accelerated tests, since the product may reach this temperature for a short period of time during transport or during storage.

Products in Three-Piece Sideseam Containers

Storage tests in three-piece sideseam containers for three, six, and twelve months at temperatures from 100–110°F are common. In some cases, shorter term tests at 120°F are used.

One effect that has been observed with some sideseam containers is that the sideseam solder fatigues during extended periods at elevated temperatures such as 130°F.[15] Under these conditions, what often is being tested is the resistance of the soldered side-seam to fatigue at 130°F rather than the resistance of the container to corrosion by the product. This is not a problem with 2P side-seam containers, however.

In some cases, cycling temperature tests are used in which the containers are stored for short periods of time at 120°F, for example, then stored successively at temperatures of 98°F, 70°F, and 40°F. These tests are intended to subject the product to the extremes of temperature to which it might be exposed during shipment and storage before it reaches the consumer.

Glass Bottle Products

Glass bottle tests may be carried out at temperatures of 100°F or slightly higher.

Ovens and Sample Storage

Many times a single thermometer or thermocouple is placed at one location in the oven. If the temperature at this location reads 130°F, it is sometimes assumed that the temperature in all other spots in the oven also is 130°F. It is advisable to check the temperature at the different locations in an oven to be certain that there are no areas where the differences in temperature may be significant. If samples of one product are stored in an oven at a location where the temperature is 130°F and others are stored in the same oven but at a location where the temperature is 135°F, comparisons of the two sets of samples after the same aging period may lead to false conclusions about the relative storage stabilities of the two sets of samples.

A convenient and effective way to store samples is to use boxes constructed of perforated iron sheets, equipped with a hinged lid. The boxes can be made to hold any desired number of samples and they can be stacked in an oven, thus utilizing the space available in the most effective way. If any aerosol containers explode during storage, the metal boxes usually will retain the pieces. Of greatest importance is the fact that metal boxes provide protection for the technician who is removing the samples after storage.

The number of samples tested depends upon a variety of factors, such as the desired reliability, available storage space, the number of formulation variables involved, the different types of containers and valves, etc. Some companies store samples in inverted, upright, and sidewise positions in order to simulate the conditions to which the product might be exposed during transportation and storage. Obviously, storing one or two samples of a product at a higher temperature is not going to produce a very significant result but there is also a limit to the number of samples that can be tested.

Evaluation of Products after Storage

There are two main considerations in planning storage stability tests; the stability of the product itself and the stability of the complete aerosol package, which includes the container and the valve. As far as the product is concerned, it is the use for which the product is designed that determines the types of storage stability tests. If chemical reactions occur during storage that reduce the effectiveness of the product, this must be determined. If the product is a disinfectant, for example, then measurement of disinfectant activity before and after aging is one criterion of stability. Aerosol perfumes are checked for odor stability and aerosol foam products for foam stability after storage. The criterion of product stability is: Does the product perform its function adequately after storage.

In determining the suitability of a container and valve for a product. there are many factors to consider. Corrosion can lead to off-odors, discoloration of the product, and leakage. Corrosion may be due to decomposition of the product, or the product itself may be corrosive. The extent of corrosion after a storage stability test can be determined by opening the container and removing the product. This is done by cooling the sample sufficiently so that there is no pressure in the container. The bottom is then removed with a can opener. The interior of the container, the can bottom, the dip tube, and the valve cup are inspected for indications of attack, such as detinning, rusting, etc. In many cases it is sufficient merely to record the extent of corrosion by describing the condition of the container in words. In other cases it is simpler to use a corrosion rating system, such as the following:

CORROSION RATING SYSTEM

Corrosion Rating	Description
0	No corrosion.
1	Generalized corrosion either liquid phase or vapor phase—spotty detinning.
2	Complete detinning either liquid phase or vapor phase.
3	Complete detinning both liquid phase and vapor phase.
4	Complete detinning both liquid phase and vapor phase with spotty rusting.
5	Complete rusting throughout container or leakage.

The corrosion rating system can be made as simple or as detailed as desired and can be devised to fit the particular product that is being tested. The effect of the product upon the valve during storage is of the utmost importance. Whether or not the valve functions satisfactorily after aging may be determined merely by discharging the sample. If the valve fails to operate properly, the reason for the failure can often be determined by taking the valve apart and examining the valve components. It is often informative to compare the valve components from the test sample with those from an unaged valve. This will usually show if excessive swelling or deterioration of the valve gaskets or plastic components has occurred. Sometimes, valve failure can be eliminated by changing valves but more often it is necessary to modify the formulation.

Quite often the first signs of corrosion in an aerosol container appear on the valve cup, which is usually uncoated. Coated valve cups may be obtained and may help in retarding corrosion. However, crimping the valve may cause strains in the coating so that the uniformity of the coating on the valve is impaired.

The frozen concentrate which has been removed from the container is placed in a beaker and allowed to warm up to room temperature. Inspection of the concentrate will determine if discoloration or some other indication of decomposition, such as the formation of insoluble reaction products, has occurred. The measurement of properties, such as pH, can be carried out to determine if acidic reaction products have formed.

The subject of corrosion and storage stability is very complicated and the literature on corrosion and the causes of corrosion is very extensive. Additional sources of information regarding corrosion of various aerosol products are given in References 21 through 28.

REFERENCES

1. F. L. LaQue, T. P. May and H. H. Uhlig, "Corrosion in Action," The International Nickel Company, Inc., 1955.
2. F. L. LaQue and H. R. Copson, "Corrosion Resistance of Metals and Alloys," 2nd ed., Chap. 13, Reinhold Publishing Corporation, New York, N.Y., 1963.
3. Reference 2, Chap. 3.
4. Reference 2, Chap. 2.
5. Reference 2, Chap. 8.
6. "Freon" Technical Bulletin B-2, "Properties and Applications of the 'Freon' Fluorinated Hydrocarbons."
7. "Freon" Aerosol Report FA-28, 'Freon' 21, Aerosol, Solvent, and Propellant."
8. H. M. Parmelee and R. C. Downing, *Soap and Sanitary Chemicals* **26** (1950) CSMA 2 ("Freon" Aerosol Report A-19).
9. P. A. Sanders, *Soap Chem. Specialties* **36**, 95 (1960) ("Freon" Aerosol Report A-51), U.S. Patent 3,085,116.

10. P. A. Sanders, *Soap Chem. Specialties* **41**, 117 (1965) ("Freon" Aerosol Report A-64).
11. A. S. Glessner, *Aerosol Age* **9**, 98 (1964).
12. P. A. Sanders, *J. Soc. Cosmetic Chemists* **9**, No. 5, (1958) ("Freon" Aerosol Report A-49).
13. P. A. Sanders, *Soap Chem. Specialties* **42**, 74, 135 (1966) ("Freon" Aerosol Report A-66).
14. B. P. 773,187, U. S. Patent 2,185,238, U. S. Patent 3,391,689, U. S. Patent 2,567,621, U. S. Patent 2,923,747, and U. S. Patent 3,159, 582.
15. R. A. Foresman *Aerosol Age* **1**, 34 (1956).
17. C. W. West, *Aerosol Age* **6**, 20 (1961).
18. E. D. Giggard *Aerosol Age* **6**, 20 (1961).
19. Proc. 38th Mid-Year Meeting, CSMA, 53 (May 1952).
20. CSMA Aerosol Guide, 65 (June 1957).
21. R. A. Foresman, Jr., Proc. 39th Ann. Meeting, CSMA, 32, (December 1952).
22. M. J. Root, "Factors in Formulation Design," in H. R. Shepherd, "Aerosols: Science and Technology," Interscience Publishers, Inc., New York, N.Y., 1961.
23. M. J. Root, *Aerosol Age* **4**, 29 (1959).
24. C. W. West, *Aerosol Age* **6**, 20 (1961).
25. M. F. Johnson, *Aerosol Age* **7**, 20 (1962).
26. L. M. Garton, *Aerosol Age* **7**, 108 (1962).
27. W. J. Pickett, Proc. 40th Ann. Meeting, CSMA, 43 (December 1953).
28. M. J. Root, Proc. 42nd Ann. Meeting, CSMA, 41 (December 1955).

14

THE FLAMMABILITY OF AEROSOLS AND AEROSOL PROPELLANTS

The flammability of an aerosol product is one of its most important properties If an aerosol is flammable, it may be potentially hazardous when shipped, stored in warehouses, or used in the home. The flammability of aerosol propellants is equally important, not only because of the influence the propellants have upon the properties of aerosol products themselves, but also because of the possible hazards involved in the manufacture, transportation, storage, and handling of the propellants. The flammability properties of both aerosols and propellants have to be known because of the regulations that apply to flammable products.

There are a number of tests that evaluate and differentiate various aspects of flammability. Therefore, it is not sufficient merely to report that a product or propellant is flammable: the particular test or tests used to determine the flammability must also be specified. As far as aerosol products are concerned, the tests of primary importance are the flame extension, flash back, drum and flash point tests. The most important criteria for the flammability of propellants or propellant blends are whether they will form explosive mixtures with air when completely vaporized, flammable fractions during evaporation, or have a flash point.

THE FLAMMABILITY OF AEROSOL PRODUCTS

Flammability Tests

The flammability tests for aerosol products were developed by the Bureau of explosives and are listed in Agent T. C. George's Tariff No. 19, Paragraph 73.300, effective September 5, 1966. These tests are described below.

Flame Extension Test. This test indicates how far the spray from an aerosol will extend the flame when the spray passes through the flame. The test equipment consists of a base 4-in. wide, 2-ft long, marked in 6-in. intervals. A rule 30 in. long and marked in inches is supported horizontally on the side of the base and about 6 in. above it. A plumber's candle of such height that the top third of the flame is at the height of the horizontal rule, is placed at the zero point in the base.

The aerosol container is placed 6 in. in back of the candle and sprayed so that the spray passes through the top third of the flame and at right angles to it. The height of the flame should be about 2 in. Two observers are required for the test. One observer notes how far the flame is extended while the other operates the container. The flame will bend about 2 in. regardless of the flammability of the product and this is part of the recorded extension. The tests should be carried out in an area with good ventilation since the decomposition products of the aerosol may be irritating and toxic. The test is positive, if the flame projects more than 18 in. beyond the ignition source with the valve fully opened, or if the flame flashes back and burns at the valve with any degree of valve opening.

Drum Tests. These tests were developed to indicate what hazard might result if aerosol containers were sprayed or ruptured and leaked into a closed area where a flame or some other source of ignition was present. This situation could result if a fire occurred in warehouses or during transit in trucks or boxcars.

The apparatus consists of a 55-gal. open-head drum fitted with a hinged cover over the open end. The cover does not have to be airtight but should open with a pressure of about 5 lbs. The closed end of the drum is equipped with three shuttered openings, top, side, and bottom, for introduction of the spray. The openings have a 1-in. diameter. A 6-in. square opening is cut in the closed end and a piece of safety glass attached, so that the inside of the drum can be observed during the test. A plumber's candle is placed an equal distance from the ends on the bottom of the drum.

OPEN-DRUM TEST   The open-drum test is carried out with the hinged end in a completely open position and with all three shutters closed. The spray from the aerosol with the valve fully opened is directed into the upper half of the open end and above the ignition source for one minute. Any significant propagation of flame through the vapor–air mixture away from the ignition source is considered a positive result. Any minor or unsustained burning is not a positive result.

CLOSED-DRUM TEST   The closed-drum test is conducted with the hinged cover dropped against the open end of the drum so that it makes a reasonably secure closure. The top shutter only is opened and the aerosol is sprayed into the drum with the valve fully opened for one minute. After the drum atmosphere has been cleared from any previous tests, the test is repeated through the side and bottom vents. Any explosion or rapid burning of the vapor–air mixture is considered a positive result.

There are several precautions that should be observed in carrying out the flame extension and closed-drum tests. For the flame extension apparatus, some laboratories use an asbestos backboard marked with vertical lines at 1-in. invervals instead of a ruler. In many cases it is easier to read the flame extension with this equipment than with the ruler. However, in carrying out the test, the candle and aerosol container should be placed far enough out from the asbestos board so that the flame does not come in contact with the board. Otherwise, the extended flame may creep along the board. This increases the flame extension and the results may be in error.

In carrying out the closed-drum test, the candle should not be allowed to burn alone in the drum for any extended period before the aerosol is tested. When the candle is burning, the supply of oxygen in the drum is being depleted with the formation of noncombustible products. Therefore, it is possible that an erroneous result could be obtained because of the limited supply of oxygen in the drum.

The drum should be cleared completely of any decomposition products from previous tests before carrying out a determination. The noncombustible products and the depleted supply of oxygen resulting from the previous tests could affect the tests results.*

FLASH POINT   The aerosol unit is cooled to about 50°F below zero and the cold aerosol product is transferred to the test apparatus. The aerosol is allowed to increase in temperature at a rate of about 2°F per minute and the test flame taper passed across the sample at intervals of 2°F until the sample flashes, reaches a temperature of 20°F, or has evaporated completely. The level of the liquid does not remain constant during the determination.

The method of determining the flash point of aerosol products is described in the CSMA Aerosol Guide. The ASTM designation for the flash point test is D 1310–55T "Flash Point of Volatile Flammable Materials by Tag Open-Cup Apparatus."

---

* A description of the Flame Projection Apparatus, Open-Drum Apparatus, and the Closed-Drum Apparatus, and the methods of testing aerosol products may be obtained either from the Bureau of Explosives, 63 Vesey Street, New York, New York, or the Chemical Specialties Manufacturers Association, 50 East 41st Street, New York, N.Y.

THE DOT REGULATIONS

At the present time flammability tests are not required for any aerosol product packaged in a nonrefillable metal container with a capacity of 35 in.^{3} (19.3oz) or less. For aerosols packaged in containers with a capacity greater than 35 in.^{3} (19.3 oz) evaluation by the flame extension, flash back and drum tests is required and if the product is found to be flammable by any one of these tests, specification packaging and marking, etc., required by the DOT Regulations must be followed.

Until recently, all aerosol products which were classified as compressed gases according to the DOT definition (see Chapter 5) had to be tested by the flame extension, flash back and drum tests. If any one of the tests was positive, the flash point of the aerosol had to be determined. A product that did not flash below 20°F was exempt from specification packaging and labeling assuming that it met the other conditions specified in the exemptions (see Paragraph 73.306 in Agent T. C. George's Tariff No. 19). If the product flashed below 20°F, it either had to be shipped in metal cylinders specified for flammable compressed gases or else a special permit had to be obtained.

The flash point provision (subparagraph 73.306 (a)(3)(iv) has now been eliminated from the DOT regulations (see the Federal Register of Thursday, June 26, 1969, page 9869). Since the flash point was the only flammability restriction in the exemptions, its removal also eliminated the need for carrying out the flame extension, flash back and drum tests because the purpose of these tests was to determine if a flash point test was necessary. However, it should be emphasized that this applies only to products packaged in metal containers with a capacity of 35 in.^{3} (19.3 fl oz) or less. (Note 1–subparagraph 73.306). Flame extension, flash back and drum tests must still be carried out for products packaged in containers with a capacity over 35 in.^{3}

In addition to the DOT regulations, there are many other Federal, state, and local regulations covering aerosol products. Several of these are considered in the next section.

THE FEDERAL HAZARDOUS SUBSTANCES LABELING ACT

The law known as the Federal Hazardous Substances Labeling Act became effective in 1960 and is administered and enforced by the Food and Drug Administration. Precautionary labeling is required on any hazardous substance intended or suitable for household use. According to the act,* the contents of self-pressurized containers are *extremely flammable* if a

---

*CSMA Publication "Laws, Regulations and Agencies of Interest to the Aerosol Industry," Fifth Edition, July 1966.

flashback (a flame extending back to the dispenser) is obtained at any degree of valve opening and the flashpoint is less than 20°F. Contents of self-pressurized containers are *flammable* if a flame projection exceeding 18 in. is obtained at full valve opening or a flashback is obtained at any degree of valve opening.

THE NEW YORK CITY FIRE DEPARTMENT REGULATIONS

The use of flammable hydrocarbons as propellants or as components of propellant blends has caused a considerable amount of controversy regarding the potential hazard of aerosol products formulated with the propellants. The New York City Fire Department has issued a series of regulations pertaining to an aerosol product and/or its propellant for products that are to be sold in New York City. The flammability tests must be carried out on the propellants alone as well as the completed aerosol package. These regulations describe the tests that are to be carried out on aerosol products, or their propellants, the type of labeling for the aerosols which is determined by the flammability tests, and the restrictions covering the storage of certain aerosol products.

The following exerpts from the regulations have been taken from the CSMA publication on the laws and regulations of interest to the aerosol industry.*

A. A product and/or its propellant shall be classified as *Extremely Flammable,* if
  (a) A flame flashes and burns toward the valve located at a distance of six (6) inches at any degree of the valve opening in normal, inverted, or side positions when tested with the flame projection apparatus similar to that used by the Bureau of Explosives,
  (b) When the flash point is below 20°F when tested by the Bureau of Explosives modified Tagliabue Open Cup test method, after the contents in the container have been chilled to 5°F below the boiling point of the propellant, or to minus 50°F, whichever is lower, and the contents thoroughly mixed before placing in test cup.
B. A product and/or its propellant shall be classified as *Flammable* if,
  (a) A flame flashes and burns at a length exceeding eighteen (18) in. at full valve opening in normal, inverted or side po-

*CSMA Publication "Laws, Regulations and Agencies of Interest to the Aerosol Industry," Fifth Edition, July 1966.

sitions when tested with the flame projection apparatus similar to that used by the Bureau of Explosives,

— OR —

(b) When the flashpoint is between 20°F and 100°F when tested with the Bureau of Explosives modified open cup test method.

The New York City Fire Department Regulations are still being contested and the final outcome has not yet been resolved.

THE FLAMMABILITY OF PROPELLANTS

The most important basis for judging the flammability of a propellant is whether it will form an explosive mixture with air when completely vaporized. If a propellant is flammable and forms explosive mixtures with air, special precautions have to be observed in manufacturing, handling, shipping and storing the propellant in plants. The "Freon" propellants are all nonflammable and, therefore, no hazards are involved. Problems do arise when the hydrocarbon propellants, isobutane, n-butane, and propane, and dimethyl ether and vinyl chloride are introduced. Such propellants are extremely flammable and form explosive mixtures with air.

The concentration of the flammable propellant in the blend determines whether a mixture of nonflammable and flammable propellants can form an explosive mixture. The maximum concentration of a flammable propellant that can be added to a nonflammable propellant without the blend becoming flammable in air when vaporized is referred to as the flammability limit of the flammable component in the nonflammable propellant and is reported in weight percent in the liquefied propellant blend.

Flammability Limits in Air

A gas is considered to be flammable if it will form explosive mixtures with air or if a flame is self propagating in the mixture of gas and air.[1] Flammable mixtures liberate enough energy on combustion of any one layer to ignite the neighboring and each successive layer of unburned gas throughout the mixture.

There is only a certain range of concentrations in air where any flammable gas is capable of forming explosive mixtures with air. The lowest concentration of gas that will form a flammable mixture in air is termed the lower flammability limit of the gas and the highest concentration the upper flammability limit. A gas will form flammable mixtures at all concentrations in between the lower and upper flammability limits. The reasons for these limits are as follows. A lower flammability limit exists because there has to be a sufficient concentration of the gas in air so that when the gas ignites, enough energy is liberated to ignite the neighboring layers of unburned gas. An upper flammability limit occurs because as the concentration of gas in air is increased, it replaces that amount of air corresponding to the volume of added gas. This, of course, decreases the supply of available oxygen. When the amount of added gas reaches a certain limit, the amount of oxygen present in the mixture is insufficient to burn the gas and no explosion will occur with the mixture. The flammability limits of gases in air are usually reported in volume percent. For example, the flammability limits of a typical gas might be stated to be 2.0–8.5 vol %.

Mixtures of nonflammable and flammable propellants are tested for flammability by discharging the liquefied propellants into a test container, allowing the propellants to vaporize completely, and then checking the resulting mixture of the vaporized propellants and air with a suitable ignition source. For example, a liquefied propellant blend consisting of 75% "Freon" 12 propellant and 25% vinyl chloride by weight—"Freon" 12–vinyl chloride (75/25)—did not form an explosive mixture at any concentration and is nonflammable in air. However, a "Freon" 12–vinyl chloride (74/26) blend did form explosive mixtures with air. The flammability limit of vinyl chloride in "Freon" 12 propellant, therefore, is 25% by weight. The "Freon" 12–vinyl chloride (74/26) mixture is flammable since it forms explosive mixtures with air. The lower and upper concentrations of the vaporized "Freon" 12–vinyl chloride (74/26) blend in air that are explosive are the flammability limits of this particular gas mixture. The maximum concentration of vinyl chloride (25%) that can be added to "Freon" 12 before the blend becomes flammable in air is the flammability limit of vinyl chloride in "Freon" 12 and is reported in weight percent in the liquefied propellant blend.

Methods of Determining Flammability Limits. A standard procedure for determining the flammability limits of gases in air involves the use of an explosion burette or eudiometer tube, as described by the Bureau of Mines.[1] The eudiometer tube is a glass tube 5 cm diameter and 150 cm in height, and is equipped at the bottom with an ignition source, such as an electric arc or a match that can be ignited by a hot wire. Suitable inlets and mixing devices are provided so that mixtures of the flammable gas and air with known compositions can be prepared in the tube. Gas–air mixtures are generally considered to be flammable if, when the heat source is ignited, flame propagation in the tube occurs from bottom to top. Thus, the flammability limits of any flammable gas in air can be determined by preparing mixtures with increasing concentrations of the gas and noting whether or not the mixture propagates a flame.

The determination of the flammability limits of gases in air is not as simple as it might seem, as Coward and Jones, Bureau of Mines, have noted.[1] The data in Table 14–1 show the different flammability limits for the same gas that have been reported. These differences may result because the type of apparatus used to determine the flammability limits was different. There is no doubt that in some cases, the flammability limits obtained with the eudiometer tube do not represent the flammability limits in larger spaces. For example, Haenni et al.,[2] reported that mixtures of 84% "Freon" 12 and 16% ethylene oxide were nonflammable when tested in a conventional eudiometer tube. However, when the tests were carried out in a 13.1-ft³ autoclave, it was necesary to reduce the concentration of ethylene oxide to 12% in order to obtain a nonflammable mixture.

TABLE 14-1 FLAMMABILITY LIMITS OF VARIOUS PROPELLANTS IN AIR

| Propellant | Flammability Limits in Air (vol %) | | Reference |
	Lower Limit	Upper Limit	
n-Butane	1.9	8.5	2, 8
	1.9	9.0	3
Dimethyl ether	3.4	18.0	3, 8
Isobutane	1.8	8.4	2, 3, 4, 8
Propane	2.2	9.5	2, 3, 8
Propellant 142b	9.1	14.8	6, 8
	6.0	15.0	5
Propellant 152a	5.1	17.1	6
	3.7	18.0	8
Vinyl chloride	4.0	22.0	3, 7, 8

The determination of flammability limits at the "Freon" Products Laboratory was carried out using a modified 55-gal stainless steel drum. This apparatus was used in order to minimize any error that might result from the use of the small eudiometer tube. The open end of the drum was covered with polyethylene film so that as the ignition sources were activated the inside of the drum could be observed. A copper tube was led through the top of the drum about equidistant from the front and back and extended 4 in. below the top. A glass pressure burette was connected to the copper tube so that known quantities of the liquefied propellant mixture could be discharged into the drum. Stirring was provided by a fan blade, which was attached to a rod that led to the outside through a rubber stopper in the drum. The unit was rotated by a variable speed motor.

Both match and electric arc ignition were used in the drum. For match ignition, the source was located at the midbottom of the drum. A nichrome wire spiral holding two kitchen matches was constructed, and the assembly was supported by a small frame. The wires were led through a rubber stopper in the drum. Electric arc ignition was obtained with two ¼ in. diameter copper rods extended through the sides of the drum. One was insulated with a nylon bushing, while the other was part of the drum. It was secured to the drum with brass bushing. The length of the arc could be adjusted with the rods. A 15,000-V ignition transformer was used to supply the current; a small momentary push button was connected into the circuit. The drum was grounded permanently to avoid shock.

Another reason that different flammability limits have been obtained for the same gas is that different ignition sources were used. At the "Freon" Products Laboratory, data indicated that, in some cases, the electric arc gave higher flammability limits than match ignition. Coward and Jones[1] have reported that if the ignition source were not sufficiently strong, some flammable mixtures could not be ignited, particularly if the composition of the mixtures was such that they were near the flammability limit. They emphasized that the flammability of the gas mixture and not the capacity of the ignition source to initiate flame was of primary significance in testing.

Flammability of Two Component Propellant Blends. Mixtures of nonflammable and flammable propellants have been of considerable interest to the aerosol industry for many years. The flammable hydrocarbons are low in cost and the blends are less expensive than the nonflammable fluorinated propellants alone. As a general rule, the mixtures have been formulated in an attempt to achieve a maximum cost savings with as little increase in the flammability of the propellant as possible.

A hydrocarbon gas, such as propane, will form explosive mixtures in air when the concentration in air falls within a specific range as defined by the lower and upper flammability limits. When a nonflammable propellant like "Freon" 12 is added to air, the flammability range of propane in air is decreased and the propane is less flammable in the "Freon" 12–air mixture. As the concentration of "Freon" 12 in air increases, the flammability range of propane continues to decrease until ultimately it reaches zero and the propane will not form an explosive mixture with the "Freon" 12–air combination regardless of how much propane is added. This effect is shown in Table 14–2.

TABLE 14-2   EFFECT OF "FREON" 12 UPON THE FLAMMABILITY OF
             PROPANE IN AIR

"Freon" 12 in Air (vol %)	Flammability Limits of Propane (vol %)	
	Lower Limit	Upper Limit
0	2.2	9.5
6.0	3.0	7.0
12.0	3.7	4.5
13.0	Nonflammable	

The data in Table 14–2 are illustrated graphically in Figure 14–1. This is a typical flammability limit curve and shows which compositions of propane, "Freon" 12 propellant, and air are flammable. The curve also shows how increasing the concentration of "Freon" 12 in air decreases the flammability range of propane.

The flammability curve in Figure 14–1 is plotted in volume percent, but it can be used to determine if any blend of "Freon" 12 and propane

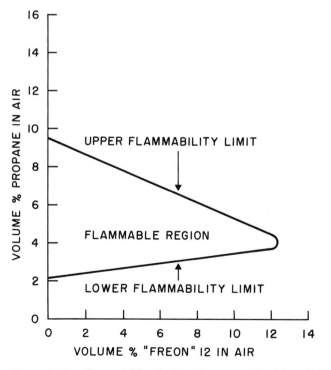

Figure 14–1   Flammability limits of propane in "Freon" 12-air mixtures.

(whose composition is expressed in weight percent in the liquid phase) will form an explosive mixture in air when vaporized. In order to do this, it is necessary to convert the weight percent ratios of the propellants in the liquid phase to volume ratios as gases after vaporization. Gas volume ratios (which are the same as mole ratios), are obtained by dividing the weight percents by the molecular weights of the propellants. For example, assume that it is necessary to know if a "Freon" 12–propane (76.7/23.3) weight percent propellant blend will form an explosive mixture with air when vaporized. The ratio of the gas volumes of the propellants after vaporization are obtained as follows:

$$\text{Moles of "Freon" 12 in 76.7 g of "Freon" 12} = \frac{76.7}{120.9} = 0.63$$

$$\text{Moles of propane in 23.3 g of propane} = \frac{23.3}{44.0} = 0.53$$

Therefore, the gas volume ratio of "Freon" 12–propane is

$$\frac{0.63}{0.53} = \frac{6}{5}.$$

The relationship between other weight ratios in the liquid phase and the volume ratios of the resultant gases when the liquefied propellants are vaporized is shown for various mixtures of "Freon" 12 and propane in Table 14–3. The data in Table 14–3 can be plotted on the flammability limit curve for "Freon" 12–propane mixtures in air as shown in Figure 14–2.

TABLE 14-3   RELATIONSHIP BETWEEN LIQUID PHASE AND VAPORIZED
COMPOSITION FOR VARIOUS "FREON" 12 PROPELLANT/
PROPANE BLENDS

Liquid Phase Composition (wt %)		Volume Percent Ratio of the Gases after Vaporization of the Liquefied Propellants		
"Freon" 12	Propane	"Freon" 12	Propane	Line*
91.6	8.4	6.0	1.5	A
89.1	10.9	6.0	2.0	B
84.5	15.5	6.0	3.0	C
76.7	23.3	6.0	5.0	D
70.1	29.9	6.0	7.0	E

*Lines in Figure 14-2.

The lines A to E in Figure 14–2 show how the volume concentrations of "Freon" 12 propellant and propane change in air as increasing amounts

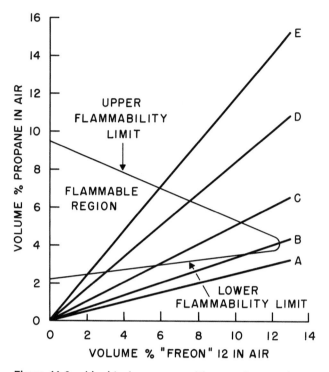

Figure 14–2   Liquid phase compositions and gas volume ratios for "Freon" 12-propane blends

of any one of the liquid mixtures is vaporized. The resulting "Freon 12–propane–air mixtures do not become flammable until the concentration of propane in air corresponding to the lower flammability limit is reached. This is the concentration where the volume ratio line intersects the lower flammability limit curve. The "Freon" 12–propane–air mixtures remain flammable with increasing concentrations of "Freon" 12 and propane until the concentrations exceed the upper flammability limit. The total weight of any of the mixtures of the liquefied propellants required to produce a concentration of gases falling within the flammable region will depend entirely upon the volume of air into which the propellants are vaporized.

Line *A* shows that a "Freon" 12–propane (91.6/8.4) wt % blend will not form an explosive mixture with air at any concentration of the propellants since line *A* does not cross the flammable region. The rest of the mixtures, corresponding to lines *B* to *E,* will form explosive mixtures with air at certain concentrations in air.

In order to determine experimentally if a given mixture of nonflammable and flammable propellants will form an explosive mixture with air, increasing amounts of the test mixture are added liquid phase to the eudiometer tube or modified drum and are allowed to vaporize. The mixture of the vaporized propellants in air is then tested with a suitable ignition source after each addition to determine if an explosion occurs. Thus, assume that it was desirable to know if a Freon 12–propane (76.7/23.3) wt % blend would form an explosive mixture with air at any concentration. The blend is prepared in a suitable container and a small amount added to the drum and allowed to vaporize. If the "Freon" 12–propane–air mixture did not explode when tested with an ignition source, the drum would be aired and a larger quantity of the blend added and tested. This procedure would be continued until either an explosion occurred at some concentration or it was reasonably certain that further additions of the blend would not produce an explosion. The latter point can usually be determined by observing the effect of increasing additions of the blend upon the ignition source. When the blend has a composition that is close to being flammable, increasing additions of the blend to the drum will cause the flame to increase in size up to a point. Further additions of the blend will either extinguish the flame or cause it to decrease in size. If there is any doubt about the flammability, the region in which the flame reached maximum size is rechecked with smaller increments of the propellant blend in the drum. With the particular "Freon" 12–propane (76.7/23.3) blend under discussion, an explosion would occur when the concentration of propane in the drum exceeded the lower flammability limit of about 2.6 vol % propane (see line D, Figure 14–2).

The above procedure indicates if a given mixture will form an explosive mixture with air, but it will not show what the lower and upper flammability limits of the propellant blend are in air. However, the flammability limits may be obtained by taking a sample of the propellant blend–air mixture in the drum immediately before each test and analyzing the mixture for composition. Compositions of mixtures that are nonflammable and flammable may be obtained by this procedure and a flammability curve can be drawn if enough mixtures are tested. In any of these tests, the incremental additions of the propellant blends to the drum should be small enough so that one addition does not produce a composition with air that falls below the lower flammability curve while the next addition produces a composition above the upper flammability limit. This is most likely to occur where the flammable range in air is very small.

Flammability of Three-Component Propellant Blends.   The flammability properties of three-component blends, such as mixtures of "Freon" 12 propellant, "Freon" 11 propellant and vinyl chloride, may be illustrated in a number of ways. A curve showing the limit of flammability in air for any specific mixture of "Freon" 12 and "Freon" 11, such as "Freon" 12–"Freon" 11 (50/50), with vinyl chloride can be constructed by obtaining the necessary data. A curve of this type, similar to that in Figure 14–1 for "Freon" 12–propane mixtures, will show not only the compositions of mixtures of "Freon" 12–"Freon" 11 (50/50) and vinyl chloride that are flammable in air, but also the range of concentrations in air that are flammable. However, the data required for the construction of these curves are quite extensive and the experimental work is time consuming. Furthermore, the curves show only the flammability properties of one specific "Freon" 12–"Freon" 11 mixture with the flammable component.

For most purposes, it is generally sufficient to have a triangular coordinate chart that shows the compositions of all combinations of the three components that form explosive mixtures with air. A chart of this type does not show the flammability limits in air for any mixture nor the range of concentrations in air which will produce flammable mixtures but it does show if any particular blend of the three components will form an explosive mixture with air.

As an example, assume that a triangular coordinate chart showing the flammability properties of "Freon" 12, "Freon" 11, vinyl chloride blends is required. A fairly accurate chart can be drawn by obtaining the flammability limit of vinyl chloride in "Freon" 12, "Freon" 12–"Freon" 11 (50/50), and "Freon" 11 and plotting the compositions of these mixtures on the triangular coordinate chart. The flammability limit of vinyl chloride in the mixtures can be obtained with an explosion burette or eudiometer tube, or a larger piece of equipment such as the modified closed drum.

The flammability limit of vinyl chloride in "Freon" 12 was obtained at the "Freon" Products Laboratory by preparing various mixtures of "Freon" 12 and vinyl chloride and testing them in the modified closed drum to determine if they would form an explosive mixture with air. It was found that a "Freon" 12–vinyl chloride (75/25) wt % mixture when vaporized did not form an explosive mixture in air at any concentration. However, when the concentration of vinyl chloride in "Freon" 12 was increased by 1% to give a liquid blend with a composition of 74 wt % "Freon" 12 and 26% vinyl chloride, the mixture did explode in air. Therefore, the flammability limit of vinyl chloride in "Freon" 12 is 25% by weight. The flammability limits of vinyl chloride in "Freon" 12–"Freon" 11 (50/50) and "Freon" 11 were obtained by using the same procedure. The results are given in Table 14–4.

The data for the nonflammable mixtures in Table 14–4 can be transferred to a triangular coordinate chart as shown Figure 14–3. Thus, by using the triangular coordinate chart it is possible to determine whether any particular "Freon" 12–"Freon" 11–vinyl chloride composition will form an explosive mixture with air.

TABLE 14-4   FLAMMABILITY OF "FREON" 12–"FREON" 11-VINYL
             CHLORIDE MIXTURES

Mixtures	Wt % Ratios of "Freon" Propellant– Vinyl Chloride in the Liquid Mixtures	
	Nonflammable Mixtures	Flammable Mixtures
"Freon" 12–Vinyl Chloride	75/25	74/26
"Freon" 12–"Freon" 11 50/50)–Vinyl Chloride	77/23	76/24
"Freon" 11–Vinyl Chloride	79/21	78/22

Unpublished "Freon" Products Laboratory Data.

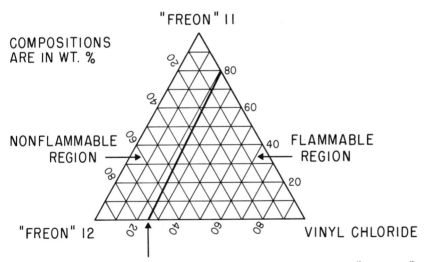

COMPOSITIONS
ARE IN WT. %

"FREON" 11

NONFLAMMABLE
REGION →

FLAMMABLE
REGION ←

"FREON" 12

VINYL CHLORIDE

LIMIT OF FLAMMABILITY OF VINYL CHLORIDE IN "FREON" 12
/ "FREON" 11 MIXTURES – COMPOSITIONS TO THE LEFT OF
THIS LINE DO NOT FORM EXPLOSIVE MIXTURES IN AIR
WHEN COMPLETELY VAPORIZED.

Figure 14–3   Flammability of "Freon" 12–"Freon" 11-vinyl chloride blends. (Unpublished "Freon" Products Laboratory Data)

At times, the possibility of determining the approximate flammability of propellant blends by spraying the mixtures into the closed drum with the candle as the ignition source has been proposed. The test would be carried out by spraying the blend into the drum until either an explosion occurred or the candle was extinguished. In a sense, this would be a modification of the DOT closed drum test, except that spraying would be continued beyond the one minute limitation of the DOT. Since the volume of the drum and the amount of blend sprayed into the drum could be determined, the composition of the mixture in the drum could be approximated by calculation.

Unfortunately, the method gives erroneous results. The reason for this can be seen by considering what takes place in the drum during the test. As the blend is sprayed, continuous combustion of the propellants occurs. The combustion of the blend not only diminishes the concentration of oxygen in the drum, but also forms nonflammable combustion products, such as carbon dioxide and water. In some cases where the candle is extinguished during the test, the gas mixture that extinguishes the candle has an entirely different composition than the initial mixture of air and propellant blend. As a result, it is possible for a propellant blend that normally forms an explosive mixture with air when tested in a eudiometer tube or modified closed drum to extinguish the candle when sprayed into a closed drum.

The Flammability of Evaporating Mixtures

The flammability limits of vinyl chloride with the nonflammable "Freon" propellants are listed in Table 14–4. These data provide an indication of the potential hazard that might occur if any of the propellant blends vaporized completely in air. A situation of this type might result from a liquid phase leak from a cylinder, tank car, or storage tank.

However, another possibility that could lead to a potentially hazardous situation must be considered. The composition of the vapor above any mixture of propellants with different boiling points is different from that of the liquid phase. For example, the vapor over "Freon" 12–"Freon" 11 (50/50) propellant has a composition of approximately 86 wt % "Freon" 12 and 14 wt % "Freon" 11.[11] Therefore, whenever any mixture of propellants with different boiling points is allowed to evaporate, fractionation occurs and the composition of the vapor changes continually during the evaporation. Under these conditions, it is entirely possible that a mixture of nonflammable and flammable propellants, which itself is nonflammable in air when completely vaporized, will form flammable fractions at some point during evaporation. This situation could occur as a result of a spill or a vapor phase leak from a storage container. If a storage tank or some other propellant container ruptured, flash vaporization of the propellant would occur until the liquid had cooled down to its boiling point at atmospheric pressure. Evaporation at atmospheric pressure would then take place.

If a propellant blend that is nonflammable under evaporating conditions is desired, and if the blend contains a flammable propellant as one of the components, it is necessary to determine if the blend will form flammable fractions during evaporation. In order to obtain the necessary information, the change in composition of the vapor phase of the blend during evaporation is first determined by taking samples of the vapor phase at intervals during the evaporation and analyzing the samples for composition by the gas chromatograph.

If the flammable propellant is either the lowest or the highest boiling component in the blend, usually the blend will form flammable fractions during evaporation. If the flammable propellant is the lowest boiling component, the initial vapor phase fraction will contain the highest proportion of the flammable propellant and will be the most flammable (assuming that only one flammable component is present). Actually, this is the only fraction that needs to be analyzed or tested. If the flammable component is the highest boiling of the propellants, the last fraction will contain the highest proportion of the flammable propellant and will be the most flammable. This is the only fraction that needs to be analyzed and tested. If there are only two components in the blend, the flammable component will, of necessity, be either the lowest or highest boiling component (except for azeotrope formation).

When the blend consists of two nonflammable propellants and a third flammable propellant, then if the flammable propellant has a boiling point between those of the two nonflammable propellants, one of the intermediate fractions will contain the highest proportion of the flammable propellant and will usually be the most flammable. As an example, the change in composition of the vapor phase during evaporation of a propellant blend consisting of 75 wt % "Freon" 12–"Freon" 11 (50/50) and 25% vinyl chloride is illustrated in Figure 14–4. The data show that initially the vapor phase is richest in "Freon" 12 propellant and poorest in "Freon" 11 propellant. During evaporation, the concentration of "Freon" 12 constantly decreases while that of "Freon" 11 increases. Vinyl chloride has a boiling point between that of "Freon" 12 and "Freon" 11. As evaporation proceeds, the concentration of vinyl chloride in the vapor phase increases until it reachts a maximum of 36% after 60–64% of the mixture has evaporated. Beyond this point, the concentration of vinyl chloride

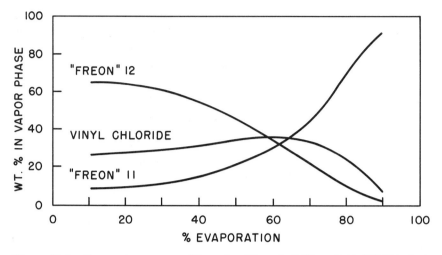

Figure 14–4   Vapor phase composition of a "Freon" 12-"Freon" 11 (50/50)/ vinyl chloride (75/25) mixture during evaporation. (Unpublished "Freon" Products Laboratory Data)

in the vapor phase decreases as evaporation continues. The fraction with the maximum concentration of vinyl chloride had a composition of 34% "Freon" 12, 30% "Freon" 11, and 36% vinyl chloride.

If various three-component combinations that form explosive mixtures with air when completely vaporized have already been determined and data such as that illustrated in Figure 14–3 on the triangular coordinate chart are available, then the flammability or nonflammability of the vapor-phase fraction can be determined after analysis of the fraction by locating the vapor phase fraction on the triangular coordinate chart and noting whether it falls in the flammable or nonflammable region. Thus, by referring to Figure 14–3, it can be seen that a gas with a composition of 34% "Freon" 12, 30% "Freon" 11, and 36% vinyl chloride does fall in the flammable region and, therefore, will form explosive mixtures with air.

If the limits of flammability have not been determined and data of the type illustrated in Figure 14–3 are not available, a propellant blend having the composition corresponding to that of the vapor-phase sample could be prepared and tested to determine if it formed an explosive mixture with air. In the preceding example, if data for the limits of flammability of vinyl chloride in "Freon" 12-"Freon" 11 propellant mixtures were not available, it would be necessary to prepare a propellant blend with a composition of 34% "Freon" 12, 30% "Freon" 11, and 36% vinyl chloride and test this blend to determine if it formed explosive mixtures with air.

When working with three-component mixtures, one of the most effective ways of characterizing the flammability properties of the evaporating mixtures is to obtain enough data so that a triangular coordinate chart can be prepared which shows the mixtures that give flammable fractions during evaporation and those that do not. For example, assume that a chart showing the flammability during evaporation of all the different mixtures of "Freon" 12, "Freon" 11, and vinyl chloride is needed. All that is required in order to construct a fairly accurate curve on a triangular chart is to determine the maximum concentration of vinyl chloride that can be added to "Freon" 12 alone. "Freon" 11 alone, and "Freon" 12–"Freon" 11 (50/50) propellant without producing a propellant blend that gives flammable fractions during evaporation. These data can be obtained by preparing mixtures of vinyl chloride with the three propellants, allowing the mixtures to evaporate, analyzing the vapor-phase fractions obtained during evaporation, and determining whether the fractions will form explosive mixtures with air. Actual experimental data are given in Table 14–5.

TABLE 14-5 LIMITING CONCENTRATION OF VINYL CHLORIDE IN "FREON" 12–"FREON" 11 PROPELLANT MIXTURES FOR NONFLAMMABLE EVAPORATION

Propellant	Limiting Concentration of Vinyl Chloride in Propellant (wt %)
"Freon" 12	9
"Freon" 12–"Freon" 11 (50/50)	15
"Freon" 11	8

Unpublished "Freon" Products Laboratory Data.

The data in Table 14–5 are then transferred to a triangular coordinate chart as shown in Figure 14–5. All compositions to the left of curve A give nonflammable fractions during evaporation while those to the right of curve A give flammable fractions. It is convenient to illustrate on the same chart the data for the maximum concentration of vinyl chloride for nonflammable mixtures in air when the blend is completely vaporized (Figure 14–3). The triangular chart in Figure 14–5 gives the compositions of all mixtures which are nonflammable when completely vaporized and also the compositions of mixtures which do not form flammable fractions during evaporation.

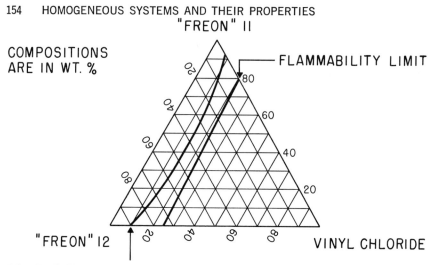

Figure 14–5   Flammability of "Freon" 12/"Freon" 11-vinyl chloride blends. (Unpublished "Freon" Products Laboratory Data)

The DOT Regulations

If a propellant or propellant blend is packaged in a container with a capacity no greater than 35 in.³ (19.3 oz), flammability tests are not required regardless of whether the propellant is classified as a compressed gas according to the definition in Paragraph 73.300 (a) of Agent T. C. George's Tariff No. 19. If, however, the capacity of the container is greater than 35 in.³ the standard flame extension, flash back, and drum tests must be carried out in order to establish the flammability properties of the propellant. If any one of these tests is positive, the propellant is classified as a flammable compressed gas and must be packaged and labeled, etc. as specified by the DOT regulations. No additional flammability tests are necessary. If the container has a capacity no greater than 50 in.³ (27.7 oz) standard aerosol containers are permitted but for containers with a capacity greater than 50 in.³, cylinders with the proper specifications must be used.

If the preceding flammability tests are negative, additional tests must be carried out in order to establish if the propellant will form an explosive mixture with air. According to the definition in Paragraph 73.300 (a) of Agent T. C. George's Tariff No. 19., a compressed gas will not only be classified as a flammable compressed gas if it gives a positive result with the flame extension flash back, or drum tests, but also: "if either a mixture of 13% or less by volume with air forms a flammable mixture or the flammability range with air is greater than 12% regardless of the lower limit. These limits shall be determined at atmospheric pressure and temperature. The method of sampling and the test procedure must be acceptable to the Bureau of Explosives." The flammability range is defined as the difference between the mininum and maximum percentage by volume of the material in mixture with air that forms a flammable mixture.

If any propellant or propellant blend is nonflammable in air when completely vaporized, it is not classified as a flammable compressed gas. However, if it does form explosive mixtures with air, the lower and upper flammability limits of the propellant or propellant blend in air must be determined. If the complete flammability curve for mixtures of the blend with air is available, it is possible to determine from this curve if any propellant or propellant blend will be classified as a flammable compressed gas.

This is illustrated in Figure 14–6 using "Freon" 12 propellant–propane mixtures as the example. The flammability limit curve for "Freon" 12–propane mixtures in air is shown and a line indicating all compositions of "Freon" 12 and propane with a total concentration of 13 vol% in air has been drawn.

The 13 vol% line crosses the flammability curve and any mixture of "Freon" 12 and propane that forms an explosive mixture in air at a concentration of 13% or less will be classified as a flammable compressed gas. Also shown in Figure 14–6 are the volume concentration lines of three "Freon" 12–propane blends. The weight ratios of the blends in the liquefied state are shown and the lines A, B, and C, indicate the volume concentrations of "Freon" 12 propellant and propane after the liquids have been vaporized.

A "Freon" 12–propane (90/10) blend is nonflammable in air, and this is illustrated by the fact that line A, which shows the volume ratios of the two components at various concentrations in air, does not cross the flammability limit curve. The "Freon" 12–propane (89/11) blend is an interesting example of a mixture that does form an explosive mixture with air but does not fall under the DOT definition of a flammable compressed gas. The "Freon" 12–propane (89/11) blend forms explosive mixtures with air since line B crosses the flammability limit curve, although the range of concentrations where flammability occurs is very small. However, the concentrations of the blend in air where explosive mixtures are formed are all greater than 13 vol%. Also, the total range of flammability of the mixtures in air is not greater than 12 vol%. Therefore, this mixture cannot be classified as a flammable compressed gas. The "Freon" 12–propane (88/12) blend forms a flammable mixture with air at a concentration in air of slightly less than 13 vol%. This mixture would, therefore, be classi-

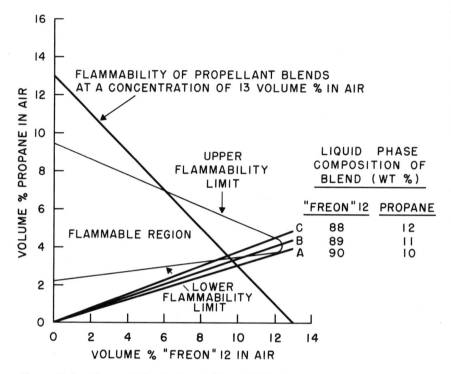

Figure 14-6   Flammability limits of "Freon" 12-propane blends in air and the DOT regulations.

fied as a flammable compressed gas and would have to be shipped in accordance with the specifications indicated by the DOT regulations.

The Flash Points of Propellants and Propellant Blends

The flash point of propellants and propellant blends is often used as an indication of the flammability and potential hazard of the propellants. For certain regulations, such as those of the New York Fire Department, the flash point of the propellants must be determined. For single propellants, the flash point might be considered an indication of whether or not the propellant will form an explosive mixture with air. For blends of nonflammable and flammable propellants, the flash point determination might be considered as a rough indication of whether or not the propellant blend will form a flammable fraction during evaporation.

The flash points of a number of common flammable propellants and propellant blends are listed in Table 14–6.

TABLE 14-6 FLASH POINTS OF PROPELLANTS AND PROPELLANT BLENDS

Propellant or Propellant Blend	Flash Point (°F)	Reference
n-Butane	−101	2
Dimethyl Ether	−42	3
"Freon" 12–Propane (91/9)	−67, −72	9
"Freon" 12–Vinyl Chloride (65/35)	−2	2
Isobutane	−117	2
Propane	−156	2
Propellant A ("Freon" 12– "Freon" 11–Isobutane (45/45/10)	No flash to dryness	9
Propellant 142b	No flash to dryness	9
Propellant 152a	> −58	9

REFERENCES

1. H. F. Coward and C. W. Jones, Bulletin 503, Bureau of Mines, "Limits of Flammability of Gases and Vapors" (1952).
2. Phillips Hydrocarbon Aerosol Propellants, Technical Bulletin 519, 1961. Phillips Petroleum Company.
3. "Isotron" Technical Information File, "The Isotron Blends," 2, No. 1, Pennsalt Chemicals Corporation.
4. Technical Service Report, "Ucon" Hydrocarbon Propellants, Union Carbide Chemicals Company, 1961.
5. "Freon" Technical Bulletin D-60, "Properties of K-1452b, 1-Chloro-1,1-Difluoro-ethane, CH_3CClF_2."
6. "Genetron" Aerosol Propellants, General Chemical Division, Allied Chemical Corporation.
7. Vinyl Chloride Monomer, Technical Bulletin, The Dow Chemical Company, 1954.
8. Matheson Gas Data Book, The Matheson Company, Inc., 1965.
9. "Freon" Products Laboratory Data, unpublished.
10. E. O. Haenni, R. A. Fulton, and A. H. Yeomans, Ind. Eng. Chem. 51, (May 1959).
11. "Freon" Aerosol Report FA-22, "Vapor Pressure and Liquid Density of Freon Propellants."

EMULSIONS, FOAMS AND SUSPENSIONS

15

SURFACES
AND
INTERFACES

Water possesses many properties that make it a desirable component for use in aerosol systems. It is an excellent solvent, nonflammable, readily available, inexpensive, and low in toxicity. However, its immiscibility with the liquefied gas propellants is an obvious disadvantage. Fortunately, there are several ways in which an immiscible state can be overcome. One way is to use the so-called three-phase system in which the water is layered over the denser fluorocarbon propellants or below the lighter hydrocarbon propellants. Water can also be combined with the propellants to a very limited extent by the addition of a cosolvent, such as ethyl alcohol, which is miscible with both water and the propellants. Finally, water can be mixed with the propellants in the presence of a suitable surface-active agent to give an emulsion system. The product can be discharged either as a foam or a spray. The latter procedure, i.e., the formation of an emulsion system with the propellants, is the predominant method that is used at the present time.

The chemistry of emulsions and foams is one of surfaces and is intimately involved with the realm of attractive and repulsive forces between molecules. These forces are responsible for surface and interfacial tension, the strength of interfacial films, the solubility properties of compounds, the formation of molecular complexes, and the stability of emulsions and foams.

THE EIGHT COLLOIDAL SYSTEMS

Emulsions and foams belong to the class of substances designated as colloidal systems. A colloidal system consists of a dispersion of one material in another; most definitions of colloidal systems have been based upon the particle size of the dispersed material. In the present discussion, a colloidal system is considered to be one in which the particle size of the dispersed phase is larger than molecular but small enough so that interfacial forces are a significant factor in determining the properties of the system. This is similar to the definition proposed by Sennett and Oliver[1] and, as they state, it excludes true solutions and boulders in a lake, but includes about every dispersion in between.

Since colloidal systems consist of a dispersion of one substance in another, they are heterogeneous in nature and are extremely commonplace.

There are eight colloidal systems, some examples of which are listed below.[2]

Solid-in-Solid

An example of this colloidal system can be observed in many churches. The ruby glass of the stained glass windows is a dispersion of metallic gold in glass. Some minerals are solid-in-solid dispersions.

Solid-in-Liquid

Suspensions are solid-in-liquid systems. Aerosol powders, which consist of powders suspended in the aerosol propellants, are examples of this system as long as they are confined in the aerosol container. Once they have been sprayed and the propellant evaporates, then the result is a solid-in-gas, or powder-in-air system.

Solid-in-Gas

Dust and smokes are solid-in-gas systems. The haze of cigarette smoke or of a forest fire is due to small particles of carbon dispersed in the air. The particle size of the carbon particles is often very small and may be as low as 0.01μ. Aerosol powders after discharge produce a solid-in-gas system because the liquefied gas propellant vaporizes and leaves the solid powder particles suspended in the air.

Liquid-in-Solid

This colloidal system which includes many precious gems, is of particular interest to women. The opal is a dispersion of water in silicon dioxide, and the pearl is a dispersion of water in calcium carbonate. Gortner[2] has pointed out that valuable pearls have been destroyed by storage in a safety deposit box where the humidity is low.

Liquid-in-Liquid

Emulsions, which are dispersions of one immiscible liquid in another, fall in this class. Milk, which consists of a dispersion of fat droplets in water, is one of the most familiar examples of an emulsion. There are many examples of emulsions in the aerosol field, but these are generally emulsions only as long as they are in the container under pressure. After discharge, most of these products change to foams or sprays.

Liquid-in-Gas

This colloidal system has caused considerable inconvenience to travelers. Fogs and mists, dispersions of water droplets in air, are examples of this system. In the aerosol field, water-based insecticides and deodorants produce water-in-air, or liquid-in-gas dispersions after they have been discharged.

Gas-in-Solid

The absorption of gases by materials such as carbon black may result in gas-in-solid systems. Some minerals are reported to contain finely dispersed gas bubbles.

Gas-in-Liquid

The familiar everyday foams belong in this class. Foams are dispersions of gases in liquids, such as air bubbles in water. Aerosol foams, i.e., shaving lathers and shampoos, are gas-in-liquid dispersions after discharge and consist of vaporized propellant dispersed in an aqueous phase.

INTERFACIAL FORCES

Interfacial forces govern the properties of interfacial regions. These regions are present in all colloidal systems and are the boundaries between two immiscible liquids, a liquid and a gas, or between a solid and a liquid, etc. Interfacial regions possess very special properties and these properties distinguish the colloidal systems from noncolloidal systems, such as true solutions.

Three of the forces that are active in the interfacial regions are the same attractive forces that exist between gas molecules and are known collectively as the van der Waals forces of attraction. These forces[3-5,9] can result in weak bonds between molecules. These bonds in turn are involved in the affinity of one compound for another and thus in the formation of molecular complexes, etc.

Dipolar Molecules

Dipoles in molecules are the basis for the van der Waals forces. Molecules possessing dipoles may be electrically neutral as a whole, but nevertheless possess localized charges that have resulted from a separation of the negative and positive charges in the molecule. Because of this separation, the

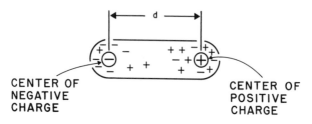

CENTER OF   CENTER OF
NEGATIVE   POSITIVE
CHARGE   CHARGE

Figure 15–1   A dipolar molecule

center of negative charge does not coincide with the center of positive charge (Figure 15–1).

The separation of charges occurs for several reasons. In asymmetrical molecules, such as ethyl alcohol and ethyl chloride, it occurs because the various atoms that make up the molecule do not all have the same affinity for electrons. Thus, in the two examples cited above, the chloride and oxygen atoms have a greater affinity for electrons than the other atoms in the molecule and tend to draw electrons towards themselves. They thus become the centers of negative charge in these molecules. This results in a deficiency of electrons in some other part of the molecule which causes a local positive charge to develop. Because they have an affinity for electrons, chlorine and oxygen atoms are referred to as electronegative.

The magnitude of the charge separation in dipolar molecules is indicated by the dipole moment of the molecule. The dipole moment, u, is equal to the product of the distance, d, between the centers of negative and positive charge, and the charge, e, itself. Thus, the dipole moment of the molecule ilustrated in Figure 15–1 is $u = de$, where d is the distance between the charges and e is the magnitude of the charge.

In asymmetrical molecules the charge separation is permanent and the molecules have a permanent dipole. Molecules that are symmetrical do not have a permanent dipole. However, a temporary dipole can arise in these molcules when the molecules are subjected to an electric field. This causes a displacement of the molecular electrons in the structure and an induced dipole results. One source of an electric field that can cause an induced dipole is a molecule with a permanent dipole.

Van der Waals Forces of Attraction

Attractive forces between molecules were postulated long before the time of van der Waals (see Reference 4 for an excellent article on the historical development in this field). Initially, the existence of attractive forces was suggested to explain surface tension. Later, in 1743, Clairault pointed out that these forces had to be applicable to all molecules. However, it remained for van der Waals in 1881 to lay the groundwork for the subsequent investigations which have led to the modern conception of these forces.

The well-known equation of state derived by van der Waals to account for the deviation of gases from the ideal gas law is

$$\left(P + \frac{a}{V^2}\right)(V - b) = RT. \qquad (15–1)$$

In this equation the constant a expresses the attractive force between molecules. Van der Waals did not specify or attempt to define the nature of these forces. Subsequent investigations have shown that the original van der Waals force actually contains three components and that all three of these forces are electrostatic in nature. These component forces have been given the names of the men who are most identified with them and are known respectively, as Keesom, Debye, and London forces The nature of these attractive force.s will be considered in the following sections.

Keesom Forces (Dipole–Dipole Interactions). Keesom, in an attempt to explain van der Waals forces in fundamental terms, proposed that in any group of dipolar molecules (molecules with permanent dipoles), there would be a certain number of collisions between the molecules[6] As a result of their dipoles, some of the molecules would be oriented with respect to an adjacent molecule so that their dipoles would be aligned as shown in Figure 15–2.

Overall there is a net attractive force which results from the interactions between the opposite charged ends of the dipoles. This is also referred to as the orientation effect since it is the orientation of the dipoles that produces the attractive force. An examination of the equation which Keesom developed[3] to account for this effect shows that the dipole–dipole interactions are important for molecules with permanent dipoles at low or moderate temperatures, where the kinetic energy associated with the temperature is less than the energy of the dipole-dipole interaction.

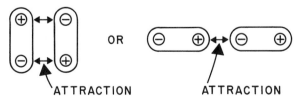

Figure 15–2   Dipole–dipole interaction.

Debye Forces (Dipole–Induced Dipole Interactions).   Although the orientation effect advanced by Keesom was a considerable step forward, it failed as a general explanation of the van der Waals forces for two reasons. Many molecules that did not have dipole moments were known to possess attractive forces. Also, the orientation effect would be expected to disappear at high temperatures because the increased kinetic energy of the molecules would make orientation unlikely. However, it was known that the attractive forces did not disappear at higher temperatures.

Debye accounted for the presence of van der Waals forces at high temperature by introducing another concept which did not depend upon temperature. Debye pointed out that the electron structure of a molecule is not fixed or rigid but consists of mobile clouds of electrons. When a molecule is approached by a dipolar molecule, the electric field associated with the dipolar molecule interacts with the mobile electron cloud of the first molecule and causes a separation of the charges. This effect is called polarization, and when it occurs, the polarized molecule temporarily becomes a dipole. The orientation of the two molecules, the dipolar molecule and the polarized molecule with an induced dipole, as they approach each other is such that an attractive forces exists between them.[7] The attractive force which results is called a dipole-induced dipole interaction because the dipolar molecule induces a dipole in the original molecule. It is illustrated in Figure 15–3.

The Keesom orientation effect helped explain the attraction between the molecules of dipolar gases at moderate temperature and the Debye induction effect explained the persistance of the van der Waals forces at high temperatures, but neither of these effects explained the attractive force between molecules that do not possess dipoles, such as the rare gases.

London—van der Waals Forces (Instantaneous Dipole-Induced Dipole Interactions).   It became apparent at this point that there must be a third force present which was still unknown. It remained for London to clarify the nature of this force.[8,9] London suggested that even molecules without permanent dipoles, such as the rare gases, have *instantaneous dipoles.* He pointed out that in any molecule, although the charge distribution may be perfectly symmetrical when averaged over a period of time, at any one

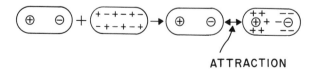

ATTRACTION

Figure 15–3   Dipole-induced dipole interaction.

instant there will be a region where the concentration of electrons is higher than average, thus producing a local negative charge. Therefore, there has to be a corresponding region where there is a less than average concentration of electrons and this produces a local positive charge.

Thus, each molecule is momentarily a dipole, and these instantaneous dipoles appear and disappear with great rapidity because of the high orbital speed of the electrons. These instantaneous dipoles act on the mobile electron clouds of adjacent molecules and induce in them a corresponding dipole. London showed that this always results in a net attractive force between the two molecules.

The London–van der Waals force (also called the London dispersion force) is the universal attractive force between molecules and in most cases accounts for a far greater portion of the attractive forces between molecules than the Keesom and Debye forces, even with dipolar gases. The London dispersion forces are additive for a group of molecules and this is why the adhesion forces between macroscopic particles are often attributed to the London–van der Waals attractive forces.[10]

Hydrogen Bonds

Hydrogen bonds between molecules result from dipole–dipole interactions but they have special properties and are usually treated separately.[3,5] They are mentioned at this point because they play an important role in determining the properties of interfacial areas. In some cases they contribute more to the properties than the London–van der Waals forces.

As previously mentioned, an electronegative atom is one that has an affinity for electrons. Chlorine and oxygen are examples of atoms with high electronegativity. Hydrogen has a relatively low electronegativity and when it is attached to an atom with a higher electronegativity, such as oxygen, then the electrons are displaced towards the oxygen and the hydrogen becomes positive. The molecule then is dipolar and undergoes the electrostatic interactions which are characteristics of a dipolar molecule.

The formation of a dipole in an alcohol, where R = alkyl, is illustrated in Figure 15–4a. Since the hydrogen is positive, it will be attracted to a second electronegative atom. The hydrogen atom therefore becomes a bridge between two electronegative atoms and hydrogen bonded compounds are often referred to as hydrogen bridge compounds. The attractive force between the positive hydrogen atom and the second electronegative atom is usually indicated by a dotted line. Alcohols are associated in the liquid state as a result of hydrogen bonding. A polymeric structure of an alcohol that results from hydrogen bonding is illustrated in Figure 15–4b.[11]

The hydrogen bond forms readily because hydrogen has a small radius, which allows it to approach closely to the second electronegative atom.

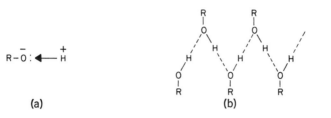

(a)   (b)

Figure 15–4   Hydrogen bonding in an alcohol. (a) Dipole formation. (b) Hydrogen bridge and polymer formation. From "Mechanism and Structure in Organic Chemistry" by Edwin S. Gould. Copyright © 1959 by Holt, Rinehart and Winston. Reproduced by permission of Holt, Rinehart and Winston.

The water molecule itself is dipolar, $\overset{+}{H_2} - \overset{-}{O}$, and water molecules have, therefore, a considerable attraction for each other. As a result, water molecules become linked to each other through hydrogen bonds and water exists in a polymeric state. A number of investigations on the structure of water have been carried out and are reported in References 12–14. Ferguson[5] has illustrated the structure of water with a planar representation as shown in Figure 15–5

A considerable amount of energy is required to break hydrogen bonds and this is why water, in spite of its apparently small molecular size, has such a high boiling point. The hydrogen bonds must be broken before the individual molecules can be vaporized. This is also why ethyl alcohol has a much higher boiling point (78°C.) than the isomeric dimethyl ether (– 25°C.). The latter does not form hydrogen bonds.

Hydrogen bonding is very important in surface and interfacial chemistry. The major part of the force causing the surface tension of water is due to hydrogen bond formation. Hydrogen bonding undoubtedly has a considerable effect upon the properties of most interfacial films. The affinity of many compounds for water is the result of hydrogen bonding.

Ion–Dipole Interactions

There is another attractive force that has not been mentioned up to this point but is quite important. This is the interaction between an ion and a

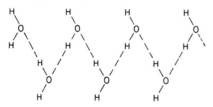

Figure 15–5   Polymeric structure of water. Lloyd N. Ferguson, "Electron Structures of Organic Molecules," © 1952, Prentice-Hall, Inc.

dipole, such as water. Ferguson[5] states that it appears likely that all ions are hydrated in aqueous solution. This interaction is of considerable importance in surface chemistry because many investigators believe that hydration accounts for much of the strength of interfacial films formed by surface active agents. Ion–dipole interactions are also assumed to play a part in the formation of molecular complexes formed from combinations of anionic surface-active agents and long-chain polar alcohols or acids.

SURFACE TENSION

It has been known for many years that the surface of a liquid behaves as if it were covered with a skin that has properties different from those of the bulk of the liquid underneath. Thus, if a small ring is lowered beneath the surface of the liquid and slowly raised, it is found that considerably more force is required to pull the ring through the surface and into the air than was necessary to move the ring up through the bulk of the liquid.

This effect (called surface tension) is due to certain attractive forces between the molecules. A molecule in the bulk of the liquid is attracted by the other molecules that surround it. Since the attractive forces are exerted on all sides of the molecule the forces tend to cancel each other out. However, the molecules in the surface of the liquid are in a completely different environment. Here they are strongly attracted by the molecules underneath them in the bulk of the liquid. However, there are essentially no molecules in the vapor phase above to provide a counter attractive force. Therefore, the attractive forces acting upon the molecules in the surface are unbalanced and tend to draw the molecules from the surface into the body of the liquid.

This force pulling on the surface molecules reduces the total number of molecules in the surface region, and thus increases the intermolecular distance between those that remain. As Fowkes[15] has pointed out, the molecules on the surface resist this separation from each other because of the attractive forces between them. He pictures the molecules as being attached to each other by springs which represent the attractive forces and it is then easy to visualize that it requires work to pull the molecules further apart. As a result, the surface molecules exist in a state of constant tension and their natural inclination is to move closer to each other to reduce this tension. Therefore, surface tension causes the surface to contract and decrease the total surface area.

The molecular structure of the liquid surface can, therefore, be considered to be disordered or disoriented compared to that of the bulk phase. This effect is not confined to the surface molecules only but extends some

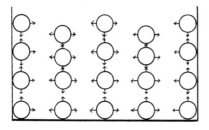

Figure 15–6   Molecular illustration of surface tension.

distance into the bulk of the liquid itself. There is some disagreement among various surface chemists as to the distance to which the disorder extends into the bulk phase but the minimum distance is reported to be five to ten molecular layers.[16]

A simplified version of the molecular explanation of surface tension is illustrated in Figure 15–6. The attractive forces are shown as acting only in a vertical or horizontal direction.

Since it was necessary to expend energy in order to separate the molecules in the surface, this energy must still be present in the surface region and can be released when the tension between the surface molecules is reduced by decreasing the total surface area. This is the meaning of surface free energy and the reason for its existence in the surface region.

Becher[17] has shown that surface tension can be defined on a physical basis as the work in ergs necessary to generate one square centimeter of surface. This is illustrated with a small rectangular wire frame containing a liquid film upon which work can be done. Assume that the small rectangular wire frame *ABCD* in Figure 15–7a is suspended vertically and

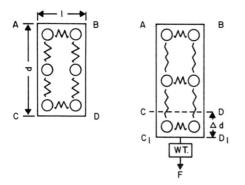

Figure 15–7 Physical interpretation of surface tension.

contains a film of liquid as shown. The film of liquid consists of molecules held together by attractive forces which are indicated by the springs suggested by Fowkes. The lower end of the frame, CD, is moveable.

The surface area of the film, considering that it has two sides, is $2ld$. Now assume that a force F is exerted on the end of the frame CD by attaching some weights to the frame. This moves the end of the frame CD downwards a distance Δd to a new position C_1D_1. This is shown in Figure 15–7b.

The work done to bring the film to the new position is

$$W = F\Delta d. \tag{15–2}$$

The force F exerted by the weights is opposed by the resistance of the forces of attraction between the molecules (as indicated by the springs). This counter opposing force exists along the length CD. If γ is the opposing force in dynes per centimeter resulting from the forces of attraction then the force along the length CD is γl. Since there are two sides to the film, the force that opposes the expansion of the film is $2\gamma l$. Substituting this expression for the force F in Equation 15–2 gives

$$W = 2\gamma l \Delta d. \tag{15–3}$$

The increase in surface, ΔS, that occurred when the film was expanded is $2l\Delta d$. Therefore, substituting ΔS for this expression in Equation 15–3 gives

$$W = \gamma \Delta S, \quad \text{or } \gamma = \frac{W}{\Delta S}. \tag{15–4}$$

Therefore, the surface tension γ is the work in ergs required to form one square centimeter of surface. Becher points out that since the surface tension is considered as the force acting along a 1-cm length and since an erg is equal to a dyne cm, it is appropriate to define surface tension in dynes cm. This is the usual unit for surface tension.

Thus far very little has been said about the specific intermolecular forces that are responsible for surface tension. In a symmetrical, nonpolar hydrocarbon liquid, such as *n-hexane,* the attractive forces are due to the London dispersion forces since the liquid is nonpolar and does not have a permanent dipole moment. The London dispersion forces can be represented by the symbol γ_{HC}^d. Therefore, the surface tension of the hydrocarbon, γ_{HC}, is due entirely to the London–van der Waals forces of attraction between the molecules.

This relationship can be shown as follows:

$$\gamma_{HC} = \gamma_{HC}^d. \tag{15–5}$$

In water, a polar liquid, the major intermolecular forces are hydrogen bonding (dipole–dipole interaction), which is symbolized by $\gamma_{H_2O}^H$, and the London–van der Waals forces, $\gamma_{H_2O}^d$. Therefore, the surface tension of water, γ_{H_2O}, is equal to the sum of the hydrogen bonding forces and the London–van der Waals forces

$$\gamma_{H_2O} = \gamma_{H_2O}^H + \gamma_{H_2O}^d. \tag{15–6}$$

Water has a surface tension of 72.8 dynes/cm at 20°C. In this case, the hydrogen bonding contributes more to the surface tension (51.0 ± 0.7 dynes/cm) than the London dispersion forces (21.8 ± 0.7 dynes/cm).[17]

INTERFACIAL TENSION

In the previous section it was mentioned that the surface of a liquid acts as if a skin or membrane were present as a result of the surface tension. The same effect occurs at the interface between two immiscible liquids for essentially the same reason, i.e., the unbalanced attractive forces exerted on the molecules in the interfacial region. Fowkes[15] indicated that at the interface between two liquids, such as water and a hydrocarbon, there are actually two interfacial regions. One is the interfacial region associated with the aqueous phase and the other is the interfacial region of the hydrocarbon phase (Figure 15–8).

At first glance it might appear that the interfacial tension, $\gamma_{H_2O/HC}$, between the water and hydrocarbon, might be equal to the sum of the two individual surface tensions of water, γ_{H_2O} and hydrocarbon γ_{HC}, because the molecules of each liquid in their own interfacial regions are pulled towards their own bulk phases by the same intermolecular forces that cause the surface tension. However, there is now an additional force at the interface between the two liquids that opposes the forces causing the surface tension of water and the surface tension of the hydrocarbon. This

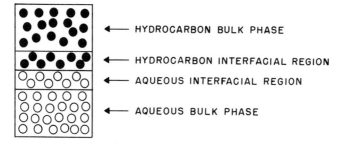

HYDROCARBON BULK PHASE

HYDROCARBON INTERFACIAL REGION

AQUEOUS INTERFACIAL REGION

AQUEOUS BULK PHASE

Figure 15–8   The two interfacial regions at a water-hydrocarbon interface.

additional force is the London–van der Waals attractive force between the water molecules and the hydrocarbon molecules. Thus, in the aqueous interfacial regions, the intermolecular forces exerted by the water molecules in the bulk phase tend to pull the interfacial water molecules into the bulk phase. However, this is resisted by the London dispersion forces between the water molecules in the interfacial region and the adjacent hydrocarbon molecules which tend to pull the interfacial water molecules in the opposite direction towards the hydrocarbon region. Therefore the tension on the aqueous side is equal to the surface tension of water minus the attraction of the hydrocarbon molecules for the water molecules.

The same situation exists in the hydrocarbon interfacial region. The tension in the hydrocarbon interfacial region is equal to the surface tension of the hydrocarbon minus the attraction of the water molecules for the hydrocarbon molecules in the interfacial regions.

It was shown previously that the surface tension of water arose from the two intermolecular forces, hydrogen bonding and the London attractive forces between the water molecules (Equation 15–6). The portion contributed by the London forces is called the dispersion component of the surface tension. The same condition applies to the surface tension of other liquids, although in the case of hydrocarbons, the intermolecular forces are practically all due to the London dispersion forces since there is no hydrogen bonding.

Fowkes[15] has indicated that at an interface between two liquids, such as water and a hydrocarbon, the magnitude of the London interaction between the water molecules and the hydrocarbon molecules which tends to pull the molecules in an opposite direction from that of the surface tension forces, can be determined by the expression

$$(\gamma_{H_2O}^{d}\ \gamma_{HC}^{d})^{1/2},$$

where $\gamma_{H_2O}^{d}$ and γ_{HC}^{d} are the dispersion components of the surface tension of water and hydrocarbon, respectively. Thus, the tension in the interfacial region of water would be

$$\gamma_{H_2O} - (\gamma_{H_2O}^{d}\ \gamma_{HC}^{d})^{1/2}$$

and that in the interfacial region of the hydrocarbon would be

$$\gamma_{HC} - (\gamma_{H_2O}^{d}\ \gamma_{HC}^{d})^{1/2}.$$

Since the interfacial tension at the water–hydrocarbon interface is the sum of the tensions in the aqueous and hydrocarbon interfacial regions, the interfacial tension can be expressed as follows:

$$\gamma_{H_2O/HC} = \gamma_{H_2O} + \gamma_{HC} - 2\ (\gamma_{H_2O}^{d}\ \gamma_{HC})^{1/2}. \qquad (15\text{–}7)$$

The values of the surface tension for water (72.8 dynes/cm at 20°C) and various hydrocarbons are known and therefore Fowkes was able to calculate the magnitude of the dispersion component for the surface tension of water. He reported that the dispersion component contributed 21.8 ± 0.7 dynes/cm at 20°C to the surface tension of water and consequently, hydrogen bonding contributes 72.8 − 21.8 or about 51/dynes/cm. Hydrogen bonding, therefore, makes the larger contribution to the surface tension of water.

One of the practical aspects of interfacial tension is that it is an indication of the ease with which an emulsion can be prepared. The formation of an emulsion involves a considerable increase in the interfacial area of the dispersed liquid, and, as in the case of surface tension, the increase in the interfacial area is opposed by the interfacial tension. Therefore, if the extension is high, emulsion formation is difficult and may require the expenditure of considerable mechanical energy. However, if the interfacial tension is low, emulsions may be formed with relative ease, and if the tension is sufficiently low, emulsions may form spontaneously.[18,19] It should be emphasized that a low value of interfacial tension indicates only the relative ease of emulsification and not the stability of the emulsion once it is formed.

There are a number of ways of measuring the surface and interfacial tension of conventional liquids but special equipment is required for the liquified gas propellants because they are under pressure. Recently, Kanig and Shin described a pressure tensiometer constructed from a compatability tube which utilizes the capillary rise principle[20] The experimental values they reported compare very well with the theoretical calculated values available previously.[21]

The surface tensions of the liquified gas propellants are low, as is expected of compounds in which the attractive forces between the molecules are comparatively low. The surface tensions of four fluorocarbon compounds and their interfacial tensions against water are listed in Table 15–1.

TABLE 15-1   SURFACE AND INTERFACIAL TENSIONS OF
            FLUOROCARBON PROPELLANTS

Compound	Surface Tension (Dynes/cm at 25°C)	Interfacial Tension (Dynes/cm at 25°C)
Propellant 12	8.4	56.7
Propellant 114	11.7	54.6
Propellant 11	18.9	49.7
Propellant 113	18.4	49.1

From Kanig and Shin, Proceedings of the Scientific Section of The Toilet Goods Association, Inc.[20]

Figure 15–9   Structure of a typical surface-active compound.

SURFACE-ACTIVE AGENTS

A surface-active agent is a compound attracted to and adsorbed at an interface. The interface can be the surface of a liquid, or the boundaries between two immiscible liquids, a solid and a liquid, or two solids. The term surface-active agent is a general designation for all materials which are surface-active but most agents have certain properties that make them desirable for specific uses. Thus, they may be classified primarily as emulsifying agents, wetting agents, foaming agents, suspending agents, detergents, etc., depending upon their particular function at an interface. A detailed listing of synthetic surface-active agents along with their compositions, properties, and source is given in Reference 22.

In order to possess the property of surface activity, a material must have a specific type of molecular structure. As a general rule, surface-active agents have two different parts, a polar part and a nonpolar part. The polar section is designed to have an affinity for water and is termed the hydrophilic (water loving) portion. The other part of the molecule has an aversion for water and is called the lipophilic (oil loving) portion. Thus, surface-active agents are constructed with two, almost opposite, parts (Figure 15–9).

There are many polar structures which confer affinity for water upon the molecule. These include ionizable groups, such as sulfate, carboxylate, sulfonate and quaternary ammonium groups, and nonionic structures, such as the polyoxethylene chains that have an affinity for water as a result of hydrogen bonding forces. Lipophilic properties are usually obtained with hydrocarbon chains.

It can be seen, therefore, that a molecule with this type of structure will seek a region, such as an oil–water interface, where it can satisfy its solubility tendencies by placing its lipophilic tail in the oil phase and its hydrophilic polar head in the aqueous phase. (Figure 15–10).

At an aqueous surface, the surface-active agent can orient itself by immersing its polar head in the aqueous phase and allowing its nonpolar, lipophilic tail to extend into the air. This is shown in Figure 15–11.

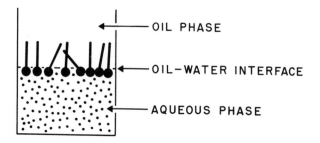

Figure 15-10   Orientation of a surface-active compound at an oil-water interface.

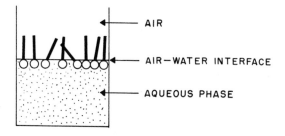

Figure 15-11   Orientation of a surface-active agent at an aqueous surface.

Types of Surface-Active Agents

Surface-active agents are divided into three broad groups, depending upon their structure, and are classified as anionic, nonionic, and cationic agents. For a detailed description of the various types of agents within any particular group, see Reference 23.

Anionic Surface-Active Agents.   Typical examples of compounds that fall into the anionic class are the salts of the fatty acids, such as sodium stearate and triethanolamine palmitate, and the various detergents, such as sodium lauryl sulfate. These agents ionize in solution and the negative ion that results is adsorbed at the interface and imparts a negative charge to the interface.

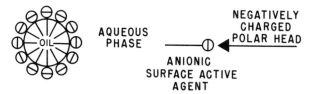

Figure 15-12   Orientation of an anionic surface-active agent at an oil-water interface.

In these compounds, the carboxylate or sulfate ionic head is the polar part that confers affinity for water upon the molecule and the hydrocarbon chain is the part that has an affinity for oil. The anionic agents therefore orient themselves at an oil–water interface with the hydrocarbon chain extending into the oil phase and the ionized polar head in the aqueous phase. The orientation of an anionic agent at the interface of an oil droplet is illustrated in Figure 15–12.

Nonionic Surface-Active Agents. Nonionic surface-active agents do not ionize in solution. Esters of the fatty acids, such as glycerol stearate, diethanolamides of the fatty acids, and the ether alcohols produced by condensation of ethylene oxide with the fatty alcohols are typical nonionic surface-active agents. An example of the latter is polyoxyethylene (4) lauryl ether, where the number in parenthesis indicates the number of moles of ethylene oxide that were condensed with one mole of lauryl alcohol. The structure of this ether alcohol is $CH_3(CH_2)_{11}-O-(CH_2CH_2O-)_4H$.

In a molecule of this type the lipophilic properties result from the presence of the hydrocarbon chain and the hydrophilic properties from the oxygen atoms in the ether linkages and the hydroxyl group. The affinity for water results from the hydrogen bonding of the oxygen atoms with water molecules as shown in Figure 15–13.

Cationic Surface-Active Agents. Cationic surface-active agents ionize in solution and impart a positive charge to the interface at which they are adsorbed. By far, these agents constitute the smallest of the three classes of surface-active agents. The quaternary ammonium compounds are typical of a cationic agent and the structure of one of these, benzyl dimethyl ammonium chloride, is

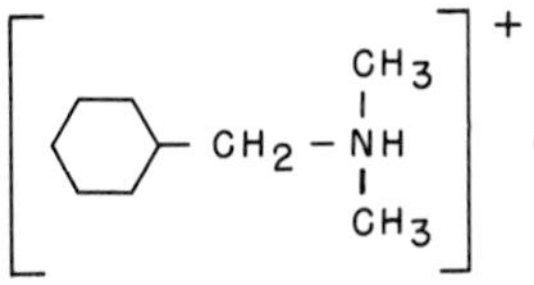

$$\left[\bigcirc\!\!\!\!- CH_2 - \overset{\overset{\displaystyle CH_3}{\displaystyle |}}{\underset{\underset{\displaystyle CH_3}{\displaystyle |}}{N}H} \right]^{+} \quad Cl-$$

Ampholytic Surface-Active Agents. Ampholytic surface-active agents are a special class of compounds that contain both an acidic and a basic group in the same molecule. The molecule can be either negatively or positively charged in solution, depending upon the acidity of the medium. Thus, they

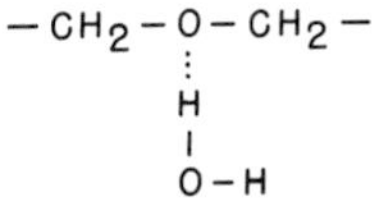

$$-CH_2-O-CH_2-$$
$$\vdots$$
$$H$$
$$|$$
$$O-H$$

Figure 15–13   Hydrogen bonding with the polyoxyethylene fatty ethers.

can be considered to be either anionic or cationic, depending upon the conditions. Thus far, they have not been used to any great extent in industry and only relatively few ampholytic surface-active agents are available commercially. Dodecyl-beta-alanine[23] is a typical ampholytic surface-active agent and has the structure $C_{12}H_{25}NHC_2H_4COOH$.

Selection of a Surface-Active Agent

Unfortunately, the selection of a surface-active agent or blend of surface-active agents for preparing a specific emulsion is still, to a considerable extent, a black art. The extraordinary large number of surface-active agents that are commercially available make an empirical trial and error method of selection a costly and time consuming process. There are, however, several general rules which are helpful. One is Bancroft's rule that the phase in which the surface-active agent is the most soluble will be the continuous phase.[25] Thus, a water soluble agent tends to produce oil-in-water emulsions and an oil soluble agent tends to give water-in-oil emulsions. There are other factors involved that determine whether or not a system will be oil-in-water or water-in-oil so that Bancroft's rule is not always applicable.

In most cases, a blend of several emulsifiers is more effective than a single emulsifier by itself.[26] Such blends usually contain both lipophilic and hydrophilic agents. In selecting blends, it should be realized that mixtures of cationic and anionic surface-active agents are incompatible as a result of their opposite charges and will react with each other. However, nonionic surface-active agents can be combined with either anionic or cationic agents.

One of the most useful tools for selecting an emulsifier or blend of emulsifiers is the HLB system devised by Griffin.[17,26] The letters HLB stand for hydrophile–lipophile balance, and the HLB number of an emulsifier shows the relationship between the hydrophilic and the lipophilic portions of the emulsifier. A low HLB number indicates that the emulsifier is predominantly lipophilic and, therefore, would tend to form water-in-oil emulsions while a high HLB number indicates that the emulsifier is predominantly hydrophilic and will tend to form oil-in-water emulsions. Thus, emulsifiers with HLB numbers in the range of 3–6 are useful for preparing water-in-oil emulsions, and those with HLB numbers of 8–18 are used for oil-in-water emulsions.

As Griffin has pointed out, the HLB number indicates the water or oil solubility of the emulsifier; and the HLB system is, in a sense, a refinement of Bancroft's rule that water soluble emulsifiers should be chosen for oil-in-water emulsions and vice versa.

A simple method of diminishing the list of potential surface-active agents is to ask that various manufacturers of surface-active agents recommend products for the particular application being considered.

Effect of Surface-Active Agents upon Surface and Interfacial Tension

Most surface-active agents lower the surface tension of water at relatively low concentrations, as illustrated in Figure 15–14.[17]

The reason that surface-active agents lower the surface tension of water is that after being adsorbed at a surface they form a new surface in which the intermolecular forces of attraction are different from those in the original surface. The surface is now predominantly lipophilic since it is composed essentially of the lipophilic tails of the surface-active agent. The lower surface tension of this lipophilic surface is due to the fact it has a lower surface energy because the forces between the lipophilic tails are the lower magnitude London–van der Waals dispersion forces.

The effect of surface-active agents upon interfacial tension can be explained in the same manner. The orientation of a surface-active agent at an oil–water interface results in the formation of a new interface composed primarily of the surface active agent. The hydrophilic portion of the surface-active agent has an affinity for the water molecules in the aqueous phase and the lipophilic hydrocarbon chain is attracted by the molecules in the oil phase. Since these forces oppose each other, the agent concentrates at the interface rather than dissolving in either the oil or aqueous phase. Interfacial tension now decreases because of the lowered interfacial energy.

A high interfacial tension deters efforts of emulsifying one liquid in another, and the opposite state is true for a lowered interfacial tension. A main function of an emulsifying agent is, therefore, to lower the interfacial tension at the oil–water interface so that the system can be emulsified with relative ease. It must be emphasized that a low interfacial tension does not

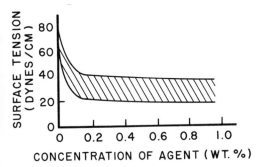

Figure 15–14   General effect of surface-active agents upon surface tension. After Becher.[17]

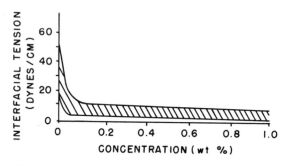

Figure 15–15   The effect of most surface-active agents upon interfacial tension. After Becher.[17]

increase the stability of the emulsion but only makes it easier for emulsification to take place.

The range of interfacial tensions expected with solutions of most surface-active agents is illustrated in Figure 15–15.[17]

The interfacial tensions in water–propellant systems containing surface-active agents were determined by Kanig and Shin.[20] The data for systems containing 0.1% of a nonionic surface-active agent ("Igepal" CO–530) are shown in Table 15–2. "Igepal" CO–530 is a nonylphenoxypolyoxyethylene compound.

TABLE 15-2   INTERFACIAL TENSIONS IN WATER–PROPELLANT SYSTEMS

Propellant	Interfacial Tension (Dynes/cm at 25°C)
Propellant 11	12.3
Propellant 113	6.2
Propellant 12	3.5
Propellant 114	2.2

From Kanig and Shin, Proceedings of the Scientific Section of the Toilet Goods Association. Inc.[20]

The interfacial tensions in Table 15–2 are similar to those illustrated in Figure 15–15 for a surface-active agent concentration of 0.1%.

General Properties of Surface-Active Agents in the Bulk Phase

Micelle Formation.   If the concentration in water of an ionizable surface-active agent, such as sodium lauryl sulfate, is gradually increased and the electrical conductivity of the solution measured during the addition of the agent, it will be found that initially there is a straight line relationship be-

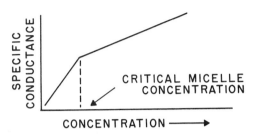

Figure 15-16   Effect of micelle formation upon electrical conductivity.

tween the concentration of the surface-active agent and the conductivity. Ultimately, a definite change in the slope of the concentration versus conductivity plot takes place. The concentration of the agent at which this change occurs is referred to as the critical micelle concentration (CMC) of the agent and the change in slope of the plot is due to the formation of aggregates of the surface-active agent called micelles. An idealized illustration of the change in slope of the conducivity–concentration plot for an anionic surface–active agent is shown in Figure 15–16.

Not only does a change occur in the slope of the electrical conductivity curve, but similar changes have been reported for other properties of the solution, such as surface tension, osmotic pressure, refractive index, vapor pressure, and density.[17]

The structure of the micelles that are formed at the critical micelle concentration has been the subject of intense and continuing investigation. X-ray studies have indicated that some micelles have a laminated structure with definite spacing between the layers as shown in Figure 15–17a. In this structure, the hydrocarbon tails of the surface-active agent are oriented toward the interior of the micelle where they are probably held together by London–van der Waals dispersion forces. The hydrophilic polar heads extend into the aqueous phase.

One difficulty with this structure, as Becher[17] has indicated, is that there

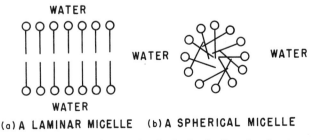

(a) A LAMINAR MICELLE   (b) A SPHERICAL MICELLE

Figure 15-17   Laminar and spherical micellar structures.

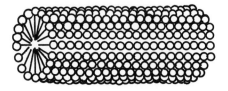

Figure 15–18   Sausage-shaped micellar structure. After Rosevear.[30]

is no simple way for the laminar micelle to end, and it might be expected to continue indefinitely from thermodynamic considerations. For this reason, Hartley[27] suggested a spherical structure for micelles as shown in figure 15–17b. The spherical micelle is much more disordered than the laminar structure.

Other workers have suggested disc and sausage-shaped micelles to account for some of the properties that have been observed.[28–30] A typical sausage-shaped micelle is illustrated in Figure 15–18. Debye considered that the ends of the micelle were capped so that the entire micelle had a hydrophilic surface.

Becher and Arai[31] have proposed a *log boom* rather than a radial sausage structure for the micelles in solutions of three polyoxyethylated lauryl alcohols. Their conclusion was based upon the results of light-scattering and hydrodynamic measurements. The log boom structure of the miscelles is illustrated in Figure 15–19.

Micelles also form in nonaqueous solutions, and in this case the structure is inverted, compared to that in aqueous solutions. In a spherical micelle, for example, the polar heads of the molecules are oriented towards the interior of the micelle with the nonpolar hydrocarbon tails extending in the oil phase[32] (Figure 15–20).

Surface Tension Lowering versus Micelle Formation.   A plot of surface tension versus concentration for a typical surface-active agent shows that the initial addition of the agent causes a marked decrease in surface tension which continues until the critical micelle concentration is reached. At this point, the surface tension tends to level off with continued addition of the surface-active agent (Figure 15–21).

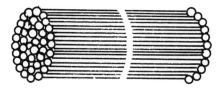

Figure 15–19   Log boom micellar structure. From Becher and Arai, *J. Colloid Science,* © Academic Press.

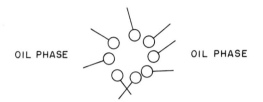

OIL PHASE   OIL PHASE

Figure 15–20   Spherical micellar structure in nonaqueous solutions.

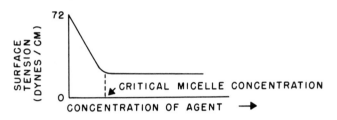

Figure 15–21   Effect of concentration upon surface tension.

Figure 15–21 can be interpreted as follows: Initially, increasing concentrations of the surface-active agent causes a continued decrease in the surface tension because the molecules first orient themselves at the air–water interface. This continues until the critical micelle concentration is reached, at which point further addition of the surface-active agent results in the formation of more micelles rather than a continued adsorption at the air–water interface.[32]

Addition of Hydrocarbon to an Aqueous Micellar System.   When a hydrocarbon (or oil) is gradually added to an aqueous solution containing

WATER

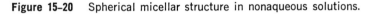

WATER

Figure 15–22   Solubilization of a hydrocarbon by micelles.

micelles of a surface-active agent, initially the hydrocarbon dissolves in the interior of the micelles because of the attraction between the hydrocarbon chains of the surface-active agent and the hydrocarbon molecules (illustrated in Figure 15–22 with a laminar micelle). This process is called solubilization, and the system exhibits the properties of a true solution because the vapor pressure of the hydrocarbon decreases in proportion to its mole fraction in solution in accordance with Raoult's law.

The micelle continues to swell and increase in size as more hydrocarbon is added until ultimately it ruptures. According to Fowkes and Ross[32] this point is reached when the diameter of the micelle is about 500 Å.

REFERENCES

1. P. Sennett and J. P. Oliver, "Colloidal Dispersions," in S. Ross, "Chemistry and Physics of Interfaces," Chap. 7, American Chemical Society Publications, Washiugton, D. C., 1965.
2. R. A. Gortner, "Outlines of Biochemistry," 3rd ed., Chap. 1, John Wiley & Sons, Inc., New York N.Y., 1949.
3. D. K. Sebera, "Electronic Structure and Chemical Bonding," Chap. 11, Blaisdell Publishing Company, Waltham, Mass., 1964.
4. H. Margenau, *Rev. Mod. Phys.* **11**, 1 (1939).
5. L. N. Ferguson, "Electron Structures of Organic Molecules," Chap. 2, Prentice Hall, Inc., New York, 1952.
6. W. H. Keesom, *Physik Z.* **22**, 129 (1921).
7. P. Debye, *Physik Z.* **21**, 178 (1920).
8. F. London, *Physik Z.* **60**, 491; **63**, 245 (1930).
9. F. London, *Trans. Faraday Soc.* **33**, 8 (1937).
10. C. N. Davies, "Aerosol Science," Chap. 11, Academic Press, Inc., New York, N.Y., 1966.
11. E. S. Gould, "Mechanism and Structure in Organic Chemistry," Holt, Rinehart, and Winston, Inc. New York, N.Y., 1959.
12. J. Bernal and R. H. Fowler, *J. Chem. Phys.* **1**, 520 (1933).
13. G. S. Forbes, *J. Chem. Educ.* **18**, (1941).
14. N. D. Coggeshall, *J. Chem. Phys.* **18**, 978 (1950).
15. F. M. Fowkes, "Attractive Forces at Interfaces," in S. Ross "Chemistry and Physics of Interfaces," Chap. 1, American Chemical Society Publications, Washington, D. C., 1965.
16. W. Drost-Hansen, "Aqueous Interfaces," in S. Ross "Chemistry and Physics of Interfaces," Chap. 3, American Chemical Society Publications, Washington, D. C., 1965.
17. P. Becher, "Emulsions: Theory and Practice," 2nd ed., Chap. 2, Reinhold Publishing Corp., New York, N.Y., 1965.
18. A. E. Alexander and J. H. Schulman *Trans. Faraday Soc.* **36**, 960 (1940).
19. E. G. Cockbain and J. H. Schulman, *Trans. Fadaray Soc.* **36**, 651 (1940).
20. J. L. Kanig and C. T. Shin, *Proc. Sci. Sec. Toilet Goods Assoc.* **38**, 55 (1962).
21. "Properties and Applications of the 'Freon' Fluorocarbons," Freon Technical Bulletin B-2

22. "Detergents and Emulsifiers," 1968 Annual, John W. McCutcheon, Inc., Morristown, N. J.
23. A. M. Schwartz, J. W. Perry, and J. Berch, "Surface Active Agents and Detergents," Vol. 2, Interscience Publishers Inc., New York, N.Y., 1958.
24. J. A. Kitchener and P. R. Mussellwhite, "The Theory of Stability of Emulsions," in P. Sherman, "Emulsion Science," Chap. 2, Academic Press, Inc., New York, N.Y., (1968).
25. W. D. Bancroft, *J. Phys. Chem.* **17,** 591 (1913); **19,** 275, (1915).
26. W. C. Griffin, "Emulsions," Encyclopedia of Chemical Technology, Volume 8, 2nd ed., pp. 117–154, John Wiley & Sons, Inc., 1965.
27. G. S. Hartley, "Paraffin-Chain Salts," p. 45, Paris, Hermann et Cie, 1936.
28. P. Debye, *J. Phys. Chem.* **53,** 1 (1949).
29. P. Debye, and W. E. W. Anacker, *J. Phys. Colloid Chem.* **51,** 18 (1947).
30. F. B. Rosevear, *J. Soc. Cosmetic Chemists* **19,** 581 (1968).
31. P. Becher, and H. Arai, *J. Colloid Sci.* **27,** 634 (1968).
32. S. Ross and F. M. Fowkes, "Emulsions and Dispersions." Course Booklet for ACS Short Course, 1968.

16

GENERAL PROPERTIES OF EMULSIONS

This chapter in the main is concerned with conventional nonaerosol emulsions. Aerosol emulsions and foams by themselves constitute only a very small part of the general field of surface chemistry, and there is relatively little basic information available about these aerosol systems. However, the properties of the aerosol systems are governed by the same factors that determine the properties of the conventional nonaerosol emulsions and foams. Therefore, in order to learn about aerosol emulsions and foams it is necessary to understand the principles that have been found to apply to conventional systems. For this reason, some general properties of emulsions and foams are considered before discussing the more specific and narrower field of aerosol systems.

There are several reasons for the lack of fundamental information about aerosol systems. One is that such systems have appeared only recently, and there has not been sufficient time to develop much basic information. Also, much of the equipment needed for basic studies of systems under pressure has yet to be designed. Even at the present time, there is no suitable microscopic equipment available for viewing emulsions under pressure. Finally, although aerosol emulsions are extremely useful and necessary for formulating many products, the emulsions themselves are only a means to an and and are not the final form in which the product is applied or used. Thus, an aerosol may be present in the container as an emulsion but after discharge it will be either a foam or spray. This has tended to focus attention on the end product rather than on the intermediate emulsion.

DEFINITION OF AN EMULSION

Becher[1] has listed nine emulsion definitions suggested by various workers. Some of these include a condition of stability while others impose a size

limit upon the dispersed droplets. For the purposes of the present discussion, an emulsion is defined merely as a dispersion of one immiscible
liquid in another. This is essentially the definition of a colloidal system
used previously and thus includes aerosol emulsions where there is relatively little information about the droplet size of the dispersed phase.

TYPES OF EMULSIONS

In most emulsion systems, water is one of the immiscible liquids. The
other liquid is usually organic in nature and is generally referred to as the
oil phase, regardless of its structure.

There are two general types of emulsions. In one type, the oil is dispersed throughout the water in the form of small droplets. This is called
an oil-in-water emulsion and is abbreviated o/w. The oil is referred to as
the dispersed or internal phase and the water as the continuous or external
phase. In the second type of emulsion, the situation is reversed and the
water is dispersed throughout the oil phase. This is called a water-in-oil
emulsion and is abbreviated w/o. In this system the oil is the continuous
phase and the water the dispersed phase.

This terminology applies to most aerosol emulsions where the liquified
gas propellant is considered to be the oil component of the emulsion. A
dispersion of the propellant in an aqueous phase is called an oil-in-water
emulsion, and a dispersion of water droplets in liquefied gas propellant is
called a water-in-oil emulsion.

DETERMINATION OF EMULSION TYPE

A number of different methods are available to determine whether a conventional emulsion is oil-in-water or water-in-oil.[1,2] If an aerosol is formulated with a concentrate which itself is an emulsion system, then all of
these methods are applicable to the concentrate since it is not under
pressure. Some of the procedures can also be used for an aerosol which
contains propellant.

Drop Dilution
This method is useful for testing aerosol concentrates that are themselves
an emulsion. In carrying out the test, a drop of the concentrate is allowed
to fall upon a water surface. If the drop disappears or disperses throughout
the water, then the concentrate is an oil-in-water emulsion since adding the
concentrate to water merely dilutes the emulsion. If, however, the drop
remains intact on the water surface, then the oil phase must be the external
phase and the emulsion is a water-in-oil type.

If the concentrate is a water-in-oil emulsion, then the aerosol prepared from the concentrate will also be a water-in-oil emulsion since the addition of propellant to the concentrate will merely extend the oil phase. However, if the concentrate is an oil-in-water emulsion, it does not necessarily follow that the aerosol will also be an oil-in-water emulsion. In this case, it is possible that addition of propellant to the oil-in-water concentrate might cause phase inversion to a water-in-oil emulsion. Generally, it is possible to judge if this might occur by other factors, such as the proportion of the oil phase in the original concentrate, the quantity of propellant added, the type of emulsifying agent present, etc.

Because of technical difficulties drop dilution is not a very satisfactory method for aerosol emulsions. Although it is possible to invert a glass bottle containing an aerosol emulsion (without dip tube) over a second glass bottle containing only water and transfer the emulsion to the second bottle by connecting the valve stems on the two bottles, the emulsion foams as it is being transferred, thus making any conclusions of doubtful significance. Cooling the bottles prior to transfer reduces the pressure and in turn the tendency to foam.

Electrical Conductivity

The conductivity of an emulsion depends essentially upon that of the continuous phase. Oil has relatively low conductivity while water has a high conductivity. Therefore, an oil-in-water emulsion will have a high conductivity since water is the continuous phase. On the other hand, a water-in-oil emulsion will have low conductivity since oil is the continuous phase.

A simple apparatus for judging the conductivity of emulsions has been described by Griffin.[3] The apparatus contains a neon bulb that lights brightly if the emulsion is the oil-in-water type and only dimly or not at all for water-in-oil emulsions. The equipment was adapted for use with aerosol products by modifying a compatibility tube so that the conductivity of the aerosol could be determined in the tube.[4]

Creaming

Creaming occurs when the dispersed phase of an emulsion either settles to the bottom as a result of gravitation or rises to the top because it is less dense than the continuous phase. Becher[1] apparently was the first to point out that the manner in which an emulsion creams can be used to determine the emulsion type—an excellent method for aerosol emulsions because there is a fairly large difference in density between the liquefied gas propellants and water. The "Freon" propellants are heavier than water and when they are dispersed in an aqueous phase to give an oil-in-water emulsion,

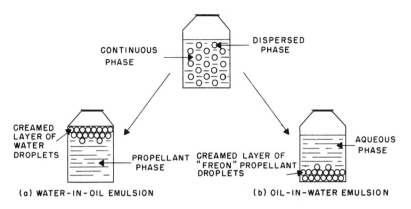

CONTINUOUS
PHASE

DISPERSED
PHASE

CREAMED
LAYER OF
WATER
DROPLETS

PROPELLANT
PHASE

CREAMED LAYER OF
"FREON" PROPELLANT
DROPLETS

AQUEOUS
PHASE

(a) WATER-IN-OIL EMULSION   (b) OIL-IN-WATER EMULSION

Figure 16-1   Creaming of aerosol emulsions with the "Freon" propellants.

the propellant droplets will settle in time and form a milky, dense, creamed layer at the bottom of the glass container. If the emulsion is the water-in-oil type, then the creamed layer will appear at the top since the dispersed water droplets, being less dense than the propellant, will rise to the top (Figure 16–1).

The hydrocarbon propellants, such as isobutane, are less dense than water and, therefore, creaming occurs in a reverse manner from that of the "Freon" propellants. When an oil-in-water emulsion containing a dispersed hydrocarbon propellant creams, the emulsified hydrocarbon propellant droplets will rise to the top since they are less dense than water, and the creamed layer, therefore, appears at the top. In a water-in-oil emulsion, the dispersed water droplets settle to the bottom since they are heavier than the continuous propellant phase.

If the composition of the propellant phase has been adjusted to have a density close to that of the aqueous phase (by use of propellant blends or propellant–solvent combinations), then it is more difficult to determine emulsion type by creaming because the densities of the two phases may change slightly after being mixed. Also, if the two phases have approximately the same density, creaming may not occur for a considerable period of time. From a product stability standpoint this is desirable, of course. However, when creaming does occur, it is usually possible to determine emulsion type by noting that the proportion of the emulsion that creamed. For example, if an emulsion is prepared with 20% propellant and 80% aqueous phase, and after creaming, it can be seen that the milky creamed layer constitutes roughly 20% of the emulsion, then the creamed layer must consist of propellant droplets and the emulsion must be an oil-in-water regardless of whether the creamed layer appears at the top or bottom. If the creamed layer constitutes about 80% of the emulsion, then it must be a water-in-oil emulsion.

General Observations of the Aerosols

Usually there is little difficulty in determining whether an aerosol emulsion is oil-in-water or water-in-oil merely from a knowledge of the emulsifying agents that are present and the concentration of propellant. Most aerosol foam products are formulated with water soluble or dispersible surface-active agents and contain 10% propellant or less. Practically all of these emulsions have been shown to be the oil-in-water type and will foam when discharged unless significant quantities of a foam depressant, such as isopropyl alcohol, are present.

Water-based systems designed to produce a fairly fine spray upon discharge such as room deodorants and space insecticides, are generally formulated with an oil soluble, nonionic surface-active agent, such as polyglycerol oleate. These products normally contain a fairly high percentage of propellant, i.e., 30% or higher, and discharge as nonfoaming fine sprays. They have been shown to be the water-in-oil type.

Bancroft's rule appears to be generally applicable to aerosol emulsions in that the water soluble surface-active agents tend to produce oil-in-water emulsions and the oil soluble agents water-in-oil emulsions. This seems to be a more important factor than the concentration of propellant, although the latter undoubtedly plays a part in determining the type of emulsion. For example, an aerosol formulated with 30% of an aqueous shaving lather concentrate containing water soluble and dispersible surface-active agents and 70% of a "Freon" propellant was an oil-in-water emulsion and discharged as a stream that foamed vigorously upon contact. On the other hand, a product formulated with 30% water and 70% of a "Freon" propellant containing the oil soluble, nonionic polyglycerol oleate, was a water-in-oil emulsion and discharged as a fine, nonfoaming spray.

Occasionally an aerosol formulated as a water-in-oil emulsion will invert to an oil-in-water emulsion during storage. When this occurs, the spray changes from the initial fine, nonfoaming spray to a coarse, streamy, foamy spray. The change is usually unmistakable.

EMULSION STABILITY

Emulsion stability is generally considered to consist of two aspects, creaming, which leads to phase separation, and coalescence, which results in the breaking of an emulsion.[3,5] Coalescence results when two droplets come into contact and combine to form a larger droplet. Coalescence may be preceded by the intermediate step of flocculation where the droplets are held together in an aggregate or cluster but still maintain their own individuality.

Some workers prefer to consider emulsion stability only from the viewpoint of coalescence which leads to a change in the number of droplets and their size. They divorce creaming from emulsion stability on the basis that creaming merely separates the initial emulsion into two emulsions, a dilute emulsion and a concentrated emulsion with the number of dispersed droplets remaining the same.[2,6] However, Becher[1] has pointed out that although creaming does not represent actual breaking of an emulsion, it is an indication of a situation that may lead to emulsion breaking. Creaming is favored by large droplet size which may be an indication of coalescence. The closer packing of the droplets in the creamed layer also increases the tendency for coalescence. Schulman[7] considered phase separation (creaming) to be sufficiently valid for the comparison of relative emulsion stabilities when all the samples were prepared in the same way.

Creaming is quite important in aerosol products, because a product that has creamed must be shaken before use. If significant creaming occurs while the product is being used, then the discharge characteristics will change noticeably during use, and the consumer will end with a product that is discharging either propellant or concentrate, depending upon whether the propellant is denser or lighter than the aqueous phase. For this reason, there is additional justification in the aerosol field for considering creaming as one aspect of emulsion stability.

The term *phase separation* is used in the present text to indicate any visible separation of the two phases of an emulsion, whether it occurs as a result of creaming or coalescence of the emulsified droplets to form a discrete separate phase. Complete phase separation is the term used to indicate that the emulsion has broken.

Creaming

Three factors have a major influence upon the rate of creaming of an emulsion. They are the relative densities of the dispersed phase (d_1) and the continuous phase (d_2), the droplet radius (r), and the viscosity of the continuous phase (η). This follows from Stoke's law[8] which gives the sedimentation rate (u) of a spherical particle in the liquid. In Equation 16–1, g is the acceleration of gravity.

$$u = \frac{2gr^2 \, (d_1 - d_2)}{9\eta} \qquad (16-1)$$

If d_1 is larger than d_2, then the dispersed phase will settle to the bottom. If d_1 is less than d_2, the rate of sedimentation, u, will have a minus sign indicating that the dispersed phase will rise to the top (creaming).

Equation 16–1 shows that the rate of creaming is reduced by decreasing the radius of the droplets, decreasing the density difference between the dispersed and continuous phases, and increasing the viscosity of the continuous phase.

Little can be done to reduce the droplet size of an aerosol after the propellant has been added except mechanical agitation of the samples. Homogenization would be difficult because it would have to be carried out either in a closed system under pressure or at a sufficiently low temperature so that propellant losses would be minimized. The latter would be impractical with most propellants. If the concentrate is an emulsion itself, then homogenization of the concentrate prior to addition of the propellant might result in an aerosol emulsion in which the ultimate droplet size was smaller.

One of the most effective methods for minimizing creaming is to adjust the composition of the propellant phase so that it has about approximately the same density as that of the aqueous phase. This can be done in several ways. If the aerosol contains very little organic material besides the propellant, then a propellant that has approximately the same density as that of the aqueous phase can be used. Methods for calculating the compositions of propellant blends with a specific density were described in Chapter 12. For oil-in-water emulsions, blends of "Freon" 12, "Freon" 114 and isobutane are generally satisfactory. The only problem with these blends is that they are usually flammable because of the quantity of hydrocarbon propellant that needs to be added to bring the density of the propellant blend down to approximately that of water.

Sometimes the properties required for a particular aerosol product either permit or actually necessitate the presence of a higher boiling hydrocarbon solvent in the concentrate, such as odorless mineral spirits, VM & P naphtha, etc. In this case it is possible to adjust the proportions of the "Freon" propellant and hydrocarbon solvent so that the density of the resulting organic phase is about equal to that of water. This avoids the necessity for using a flammable propellant to obtain the desired density. In these cases "Freon" 12 is normally used by itself as the propellant because of the vapor depressant effect of the hydrocarbon solvent.

The viscosity of the aqueous phase may be increased by the addition of suitable thickening agents.[5] Combinations of surface-active agents with polar compounds, such as the fatty alcohols, often produce highly viscous external aqueous phases in which the rate of creaming is very low.[9,10]

Flocculation

Flocculation takes place when the droplets in an emulsion approach sufficiently close to each other so that they form clusters or aggregates. In

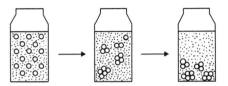

Figure 16–2   Flocculation of droplets in an emulsion.

these clusters the droplets may remain as individual droplets, they may coalesce into a larger droplet, or they may redisperse as single droplets when the emulsion is agitated or shaken. These clusters will cream more rapidly than individual droplets because the radius of the cluster is larger than that of the droplets (Figure 16–2).

Coalescence

Coalescence occurs when emulsified droplets come into contact with each other and unite to form a larger droplet. This leads to phase separation and complete breaking of the emulsion. If this occurred to any significant degree while a product was being used, the spray or foam characteristics would change. A product of this type would be of little value for the consumer market. Coalescence and complete phase separation are illustrated in Figure 16–3.

The reason that emulsions coalesce is that they are thermodynamically unstable. Energy is required to form the emulsion initially (by shaking, homogenization, etc.,) and this energy is concentrated in the interfacial regions around each droplet where it is manifested as interfacial tension. Therefore, the larger the interfacial area of an emulsion, i.e., the smaller the droplet size and the greater the number of droplets, the greater is the interfacial energy. Processes tend to proceed in a direction which will decrease the total energy of the system and the dispersed droplets can

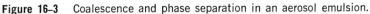

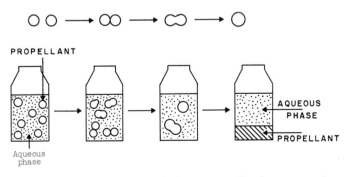

Figure 16–3   Coalescence and phase separation in an aerosol emulsion.

decrease the interfacial energy by coalescing to form larger droplets with an over-all decrease in interfacial area. Therefore, coalescence is a spontaneous process.

In an unstabilized aerosol system consisting of propellant and water, creaming, coalescence, and complete phase separation occur within a matter of seconds after the bottle has been shaken. Even if the density of the propellant is adjusted to match that of the aqueous phase so that creaming is negligible, coalescence occurs so rapidly that complete phase separation again results in a few seconds. An emulsion that is to be marketed as a consumer product must have the droplets protected against coalescence while the consumer is using the product. The fact that an aerosol product creams during storage is not too important as long as a simple shake of the container redisperses the creamed layer and as long as creaming does not occur while the product is being used. If, however, the emulsion breaks readily, the container may have to be shaken vigorously in order to reemulsify the propellant. Most consumers do not shake this hard.

Coalescence of the dispered droplets can be delayed by the addition of suitable surface-active materials to the system. Since the mechanism by which this stabilization is achieved depends to a considerable extent upon whether the emulsion is oil-in-water or water-in-oil, the stabilization of each of these two types of emulsions by surface-active agents is considered separately.

STABILIZATION OF OIL-IN-WATER EMULSIONS

Two major theories have been advanced to account for the stability of oil-in-water emulsions. One involves the presence of an electric charge on the droplets so that there is a mutual repulsion between the droplets when they approach each other. The second involves the formation of a protective interfacial film of surface-active agent around the droplets which prevents them from coalescing should they come into contact. There has been and still is a certain amount of controversy about the relative importance of these two mechanisms but the general view at the present time appears to be that since each contributes to the stability of emulsions in a different way they are both important. The extent to which any one mechanism influences the stability of an emulsion system depends upon such factors as the type and concentration of emulsifying agent, the particle size, etc.

The way in which the two mechanisms operate in stabilizing emulsions can be seen by considering a simplified picture of the processes that lead to the breaking of an emulsion. In the first place, the dispersed droplets must come into contact, either as a result of a collision, or by flocculation due to an attraction between the droplets. After contact, the droplets can either separate and redisperse again, remain flocculated but maintain their own individuality, or coalesce. If the rate of coalescence is sufficiently high, complete phase separation will occur and the emulsion will break. If the droplets possess an electric charge, this can decrease the rate and intensity of collisions and it also will determine the extent to which the emulsion droplets flocculate. The electric charge, therefore, is the primary stabilizing factor. If, however, the droplets come into contact, then the strength and nature of the interfacial film determine if coalescence will take place. The interfacial film is the second and final stabilizing factor.

The degree to which these two mechanisms operate in an emulsion depends to a considerable extent upon the type of emulsifying agent present. In emulsions stabilized with either anionic or cationic agents, the droplets are charged and both the electric charge and the strength of the interfacial film are quite important. The droplets in emulsions stabilized with nonionic emulsifying agents also can be charged but the electric charge on the droplets is much smaller and the predominant stabilizing factor is the strength of the interfacial film. Emulsions can also be stabilized with gums and in these systems the strength of the interfacial film is the only important factor.

Stabilization by Electric Charge

The theory of the stabilization of dispersed particles by electric charge is based upon the fact that when two droplets with the same type of charge approach each other there is always a repulsion between them (Figure 16–4). This repulsive force decreases the possibility of the particles approaching close enough so that flocculation or coalescence can occur.

The fact that the stability of certain colloidal dispersions was due to the electric charge on the particles was recognized around the turn of the century as a result of investigations of colloidal sols. These colloidal sols were dilute dispersions of submicroscopic solid particles. The small particle size of the solids was obtained by mechanical means and there were no surface-active agents present. Schulze, as early as 1882[11] reported that the addition of electrolytes to colloidal dispersions caused them to become

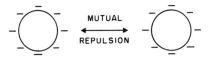

Figure 16–4   Mutual repulsion between two particles with the same charge.

unstable and flocculate. Multivalent ions were found to have a much greater effect than monovalent ions. A few years later Linder and Picton[12] discovered that the particles of a colloidal sol migrated under the influence of an electric field. This indicated that the particles were charged and from the direction in which the particles moved they were able to determine the sign of the charge on the particles. This phenomenon is called electrophoresis and the velocity of migration was a function of the charge on the dispersed particles. Since the sign of the charge on the dispersed particles could be determined, it soon became apparent that the multivalent ions, which Schulze had found to have a disproportionate effect upon the flocculation of the particles, had an electric charge opposite in sign to that on the dispersed particles. The effect of electrolytes upon dispersions has since become known as the Schulze–Hardy rule and can be stated as follows: "the precipitating power of an electrolyte depends upon the valence of the ion whose charge is opposite to that on the colloidal particles."[13] Subsequently Hardy[14] and Burton[15] demonstrated that the stability of the dispersions was related to their movement in an electrical field. Since the movement was a function of the degree of charge on the particles, this showed a relationship between the charge on the particles and the stability of the dispersion.

The Electric Double Layer.   An electric charge can arise on a particle by ionization, absorption of other ions, i.e., hydroxyl ions from solution, or by friction between the interfacial area of the dispersed particles and the surrounding medium. In the case of emulsions, the charge can arise as a result of the orientation of an ionized surface-active agent, such as sodium stearate, at the oil–water interface. The molecule can be pictured with its hydrocarbon tail anchored in the oil droplet and the negatively charged polar head extending in to the aqueous phase. In this case the droplets would be negatively charged.

The actual structure and distribution of charges at the interface of a dispersed particle or an oil droplet has been the subject of many investigations and the picture has not been completely clarified as yet. Helmholtz proposed in 1879[16] that the interfacial area consisted of a double layer of oppositely charged particles. According to this view there is a layer of ions absorbed at the surface of the particle with another layer of opposite charges located in the continuous phase adjacent to the surface charges so that a double layer of electrical charges is formed (Figure 16–5a). Helmholtz referred to the potential drop across the layer of charges as the electrokinetic potential.

The Heimholtz model of the double layer had the disadvantage that it required a sharp potential drop at the interface between the two series of charges, and it seemed unlikely, considering the general mobility of ions, that the outer layer of ions would remain oriented as required by the theory.[1] In order to avoid some of these difficulties, Gouy[17] assumed that the double layer was diffuse and that the outer layer decreased in electrical density at increasing distances from the interface. Although this was an improvement there were still some deficiencies in the theory and Stern[18] suggested a compromise between the two views which assumes that the double layer consists of two parts, one of which is approximately a single ion in thickness and remains attached to the surface of the particle while the other (the Gouy layer) is diffuse and extends into the surrounding liquid. The potential drop from the first ionic layer is sharp while that in the second diffuse layer gradually decreases as the distance from the interface increases. The modified Stern version of the double layer is illustrated in Figure 16–5b.

The Potential Energy Barrier (The D.L.V.O. Theory).   When electrically charged particles or droplets approach each other, there are two different forces which affect the behavior of the particles.[19] These forces are: (1) The London–van der Waals forces of attraction, and (2) the electrostatic repulsive forces resulting from the interaction of the electrical double layers on the particles.

The theory concerned with the behavior of these charged particles is called the D.L.V.O. theory from the initials of the scientists Derjaguin and Landau[20] and Verwey and Overbeek[21] who authored it.

If the London–van der Waals forces of attraction between two particles is greater than the repulsive forces regardless of the distance that separates the particles, the particles continue to approach until they come into contact. Flocculation then occurs with the formation of agglomerates. Under other conditions the repulsive forces between the particles are larger than

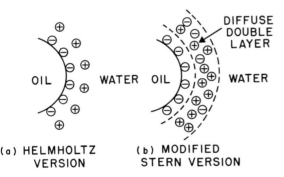

(a) HELMHOLTZ
     VERSION

(b) MODIFIED
     STERN VERSION

Figure 16–5   Simplified version of the electric double layer.

the London–van der Waals attractive force and this tends to prevent contact between the particles or droplets, thus decreasing the possibility of flocculation and coalecence. The repulsive force (potential energy barrier) between the particles increases in intensity as the droplets approach each other until it reaches a maximum. If the kinetic energy of the droplets is sufficiently large so that they can overcome this energy barrier, the London —van der Waals attractive forces take over and bring the particles or droplets together.

The London–van der Waals attractive forces and the electrostatic repulsive forces between the particles oppose each other and since they operate independently the net interaction between the two particles is determined by the additive effect of the forces. Therefore, the curve that shows how the force between the particles varies with distance is obtained by superimposing the curve for the London–van der Waals attractive force upon the curve that indicates how the repulsion varies with distance.

Many factors, including the radius of the particles, the electric charge, the electrolyte concentration, etc., affect the shape of potential energy curve for two approaching electrical particles. A typical potential energy curve for two electrically charged particles is ilustrated in Figure 16–6.

If the potential energy barrier X is much larger than the kinetic energy of the particles, then when the droplets collide they would be expected to bounce off each other without contact.[19] Under these conditions, the emulsions would be stable to flocculation.

The curve in Figure 16–6 shows a secondary minimum at a distance of about 100 Å. This is predicted by the D.L.V.O. theory and occurs because the London–van der Waals attractive force falls off more slowly with distance than the repulsion. This minimum becomes significant for large particles, such as those in emulsions, and it seems probable that weak aggregates form at this distance in which the droplets are separated by a liquid film. Because the forces that hold the droplets together in the aggre-

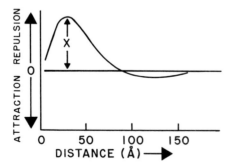

Figure 16–6   A typical energy diagram for two charged particles. After Kitchener and Mussellwhite.[19]

gate are very weak, the droplets would be expected to redisperse under any slight mechanical shearing force. In highly concentrated emulsions the droplets are crowded together and the high viscosity of these emulsions is considered to be due to the formation of aggregates in the region of the secondary minimum.

It should be emphasized that the D.L.V.O. theory was developed to account for the stability of dispersions of submicroscopic inorganic solid particles. Attempts to extend the theory to include emulsions where the oil droplets are stabilized by interfacial films of an emulsifying agent should be viewed with caution.[19] The D.L.V.O. theory is useful in interpreting the effect of electric charge in stabilizing emulsions against flocculation but it does not apply to coalescence, which depends upon the disruption of the interfacial film of emulsifying agent.

Some evidence indicating that electrically charged emulsion droplets behave similarly to colloidal inorganic dispersions as far as flocculation is concerned, is based upon the observation that the effect of added electrolytes upon the flocculation of emulsions follows the Schulze–Hardy rule. This was reported for emulsions of xylene in aqueous sodium oleate by Martin and Hermann[22] and by van den Tempel[23] for emulsions stabilized with the sodium salts of surface-active acids.

The Interfacial Film

There is also considerable evidence which shows that the stability of emulsions to coalescence results from the presence of the interfacial film of emulsifying agent and not the electric charge. Thus, Limburg[24] found that the addition of electrolytes to dilute oil-in-water emulsions without emulsifying agents present decreased the electrophoretic velocity (thus the charge) and the emulsion stability, as expected. However, when an emulsifying agent (saponin) was added to the emulsion, the addition of electrolyte to the emulsion caused the same decrease in electric charge as in the preceding case but had no effect upon emulsion stability. Limburg concluded that the protective film of emulsifying agent was a more important factor in emulsion stability than electric charge.

King and Wrzesinski[25] carried out a similar but more extensive series of experiments in which they determined the effect of electrolytes upon the stability and electrokinetic potential of oil-in-water emulsions stabilized with a number of surface-active agents including sodium oleate and saponin. They came to essentially the same conclusion as Limburg and stated that there was no relationship between the electrokinetic potential on the droplets and the stability of the emulsions as long as sufficient emulsifying agent was present to cover the droplets with a monomolecular film of the emulsifying agent. King believed that the coherent protective film provided by the emulsifying agent was the major source of stability.

Recently, Kremnev[26] measured the electric potential in emulsions of benzene in water stabilized with sodium oleate and reported that although the addition of small amounts of sodium chloride caused the initially high electric potential to decrease sharply, there was no corresponding decrease in the stability of the emulsion. Kremnev stated that the coalescence of the emulsified droplets was prevented by a factor other than the repulsive force in the electric double layer.

Additional evidence was provided by Derjaguin and Titievskaya[27] who found that when small amounts of sodium chloride were added to an aqueous sodium oleate solution in which two bubbles were pressed together, the salt caused the thickness of the water layer separating the bubbles to decrease until the electric double layers surrounding the bubbles overlapped. The bubbles did not coalesce, however.

Nature of the Interfacial Film. The investigations discussed in the previous section indicated that once the droplets were in contact, some factor other than electric charge was responsible for the stability of emulsions to coalescence. This factor is generally considered to be the nature and strength of the interfacial film of emulsifying agent that surrounds the droplet. The importance of the strength of the interfacial film can be seen when it is realized that in order for coalescence to occur, the molecules of the surface-active agent oriented at the interfaces of the two droplets have to be displaced and pushed aside so that the molecules of the oil droplets can come into contact. Therefore, resistance of the interfacial film to rupture and displacement is a necessary condition for stability of an emulsion to coalescence.

Besides the work discussed previously, there is a considerable amount of additional evidence for the theory that the stability of emulsions to coalescence is due to the strength of the interfacial film. Much of this evidence has resulted from investigations of molecular complexes that are formed between surface-active-agents and long-chain polar compounds at interfacial regions. These complexes have been shown to have a considerable effect upon emulsion and foam properties, both in the aerosol and non-aerosol fields. This effect is attributed to the strong interfacial film formed by these complexes.

An additional factor involved in the strength of the interfacial film is hydration of the film. There seems to be little doubt that hydration may be a very important factor in the strength of many interfacial films but its role remains to be clarified.

Molecular Complexes. Molecular complexes, also known as association complexes or molecular addition compounds, have been postulated since

1823 (see Reference 28 for a list of the early papers in this field). The complexes formed between free fatty acids and their salts (the so-called acid soaps) were the first reported but their existence was disputed until McBain and his co-workers demonstrated in 1933 that the acid soap consisting of equal molecular proportions of lauric acid and potassium laurate was a true chemical individual.[28-30] Many other acid soap complexes have been reported, and the pearlescence of some of the early lotions containing stearic acid was attributed to the formation of the acid soap complex.[31] The sodium stearate–stearic acid complex subsequently was investigated by Ryer[32] and more recently by Goddard and Kung[33] who found that the ratio of stearic acid to sodium stearate in the complex was 1:1.

The real stimulus for the investigation of complex formation, however, was provided by the work of Schulman and his co-workers who showed that molecular complex formation was not confined to the acid soaps but was a much broader phenomenon than had been suspected. Schulman also demonstrated that complex formation could have a considerable effect upon the properties of interfacial films and the stability of emulsions and thus revealed the importance of complex formation in the field of surface chemistry.

Schulman and Rideal reported in 1937 that when a monomolecular film of a water insoluble compound, such as cetyl alcohol or cholesterol, was spread over the surface of an aqueous solution containing sodium cetyl sulfate, the molecules of the surface-active agent penetrated the surface film of the cetyl alcohol and formed a complex with the alcohol.[34] Evidence for the existence of the complex was indicated by the fact that it was much more resistant to disruption by pressure than the film from either of the components by themselves. This is one reason why a complexed interfacial film around oil droplets is better able to retard coalescence than the film from the surface-active agent alone.

Schulman then extended his investigation to a study of the effect of complex formation upon the stability of oil-in-water emulsions prepared with mineral oil.[7] He observed that emulsions prepared with a surfactant alone, such as sodium cetyl sulfate, were unstable. However, the addition of a long-chain polar compound, such as cetyl alcohol or cholesterol, to the system resulted in the formation of excellent, stable emulsions. This effect of the alcohols upon emulsion stability was attributed to complex formation between the sodium cetyl sulfate and the cetyl alcohol at the droplet interface. Thus, Schulman's results at the oil-water interface paralleled those at the air-water interface.

It was also noted that oleyl alcohol was relatively ineffective in promoting complex formation. This was considered to be due to the steric hindrance resulting from the kinked structure of the double bond in oleyl alcohol which prevented close association between the oleyl alcohol molecules and the sodium cetyl sulfate.

The existence of the complexes formed between the alkyl sulfates and the long-chain alcohols was subsequently confirmed by Epstein[35] and Goddard and Kung[33] by analysis of the complexes. Goddard prepared complexes from sodium alkyl sulfates and long-chain alcohols by two methods. One involved heating the two components to about 110° F for 10 min and then allowing the melt to cool overnight. In the other method the complexes were prepared in either aqueous or aqueous alcohol solutions at elevated temperatures. The complexes precipitated from the cooled solution and were isolated by filtration. Analysis of the complexes was carried out by three different methods, differential thermal analysis, infrared, and x-ray. In all cases, the ratio of alcohol to sulfate in the lauryl alcohol–sodium lauryl sulfate and the myristyl alcohol–sodium myristyl sulfate complexes was found to be 1:2. This work established beyond a doubt the existence of the long-chain alcohol–alkyl sulfate complexes.

Miles et al.,[36,37] showed that films from aqueous solutions of sodium lauryl sulfate were fast draining but the addition of lauryl alcohol resulted in films that were slow draining. This effect was considered to be due to an increase in surface viscosity due to complex formation. The slow draining films became fast draining at a specific higher temperature called the "film drainage transition temperature (FDTT)." The change occurring at the transition temperature was due to thermal disruption of the complex.

The investigation of film drainage transitions was extended to nonionic systems by Becher and Del Vecchio[38] who demonstrated by surface viscosity measurements that combinations of polyoxyethylene lauryl ethers with long-chain alcohols also exhibited film drainage transition phenomena.

Emulsion Stabilization by Molecular Complexes.   The theory that the nature and strength of the interfacial film of emulsifying agent is the major factor controlling the stability of emulsions to coalescence has been mentioned but the conditions that influence the nature of the interfacial film have not yet been discussed. Since the interfacial film is formed from the surface-active agent, one of the most important factors is the concentration of the agent at the interface. If relatively few molecules of the emulsifying agent are present at the interface so that only a portion of the total interfacial area of the droplet is covered, then the interfacial film would be expected to be weak and provide little protection against coalescence. However, a condensed, strongly coherent film, would be able to resist the displacement necessary before coalescence can occur. Molecular complexes can form strong, coherent films and this undoubtedly is one way in which they stabilize emulsions.

Many anionic (and cationic) surface-active agents by themselves produce relatively poor, unstable emulsions regardless of the fact that they impart an electric charge to the droplets. This is one of the arguments against the theory that the electric charge on the droplets is the main factor in the stabilization of emulsions. An example of an unstable emulsion of this type is the mineral oil emulsion prepared with sodium cetyl sulfate mentioned earlier. The interesting aspect about these unstable emulsions is that the electric charge of the surface-active agent creates a situation that results in a sparse interfacial film and poor emulsion stability. This apparent paradox, in which the electric charge is both a help and a hindrance in stabilizing an emulsion, can be understood by referring to Figure 16–7a which illustrates an oil droplet with an interfacial film of cetyl sulfate ions.

The oil droplets tend to repel each other since they have a similar charge but the polar heads of the cetyl sulfate molecules oriented at the interface on the same droplet will also repel each other because they too have the same charge. Therefore, the surface-active agent repels itself and the consequence of this self-repulsion is a low concentration of surface-active agent at the interface and a weak interfacial film. However, when cetyl alcohol is present, the cetyl alcohol molecules orient themselves between the sodium cetyl sulfate molecules and give a closely packed, condensed, strong interfacial film. A condensed film of this type as pictured by Schulman is illustrated in Figure 16–7b. The relatively weak complex with oleyl alcohol resulting from the steric hindrance of the double bond is shown in Figure 16–7c.

The formation of the complex between the sodium cetyl sulfate and the cetyl alcohol molecules is due to the existence of attractive forces between

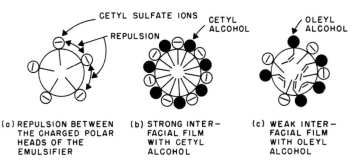

(a) REPULSION BETWEEN THE CHARGED POLAR HEADS OF THE EMULSIFIER

(b) STRONG INTER-FACIAL FILM WITH CETYL ALCOHOL

(c) WEAK INTER-FACIAL FILM WITH OLEYL ALCOHOL

Figure 16–7   Molecular complex formation in emulsions.

the molecules. The attractive forces that account for the complex formation are probably the result of ion–dipole interactions between the ionized sulfate heads and the cetyl alcohol molecules, hydrogen bonding, and possibly London–van der Waals forces of attraction between the hydrocarbon chains of the cetyl sulfate and the cetyl alcohol molecules.[33,39] The cetyl alcohol molecules may function in part as insulators by screening the ionized cetyl sulfate molecules from each other which would then reduce the repulsion between the electrically charged heads.[39,40]

There has been another theory advanced to explain the stabilization of emulsions by molecular complexes and that is that the high interfacial viscosity of the alkyl sulfate–cetyl alcohol complex prevents the interfacial film from being easily disrupted or pushed aside if the droplets do come into contact. The ineffectiveness of sodium lauryl sulfate alone as an emulsifying agent is considered to be due to the low interfacial viscosity of the sodium lauryl sulfate interfacial films.[40] The high interfacial viscosity of the complex may be due to hydration, which is discussed in the following section.

Hydration of Interfacial Films. There seems to be little doubt that the state of water in the immediate vicinity of interfaces is different from that of the bulk phase. This has a considerable bearing upon emulsion stability since it is possible that the hydration layers have sufficient strength to prevent coalescence when droplets come into contact. Some investigators such as Derjaguin[20] believe that thick layers of water are immobilized at interfaces while others consider that there are only one or two fixed layers of water with a slight disorientation extending somewhat further into the bulk phase. Davies[40] reported that several layers of water may be oriented at liquid surfaces to form a rather rigid layer of "soft ice." He suggested that such a layer would have a viscosity like that of butter with a thickness of about 10 Å. The density of the absorbed water layer was reported to be higher than normal.

Ferguson[41] stated that practically all ions in aqueous solution are hydrated. This would include ionized surface-active agents. McBain[42] reported that dipoles of the surface-active agent in the interface can orient the dipoles of water between and below the polar groups of the surface. He compared this action of the surface-active agent to a magnet that can pick up several nails. The attractive forces that account for the hydrated layers are probably ion–dipole interactions and hydrogen bonding.

Hydration is probably one of the main factors in the stabilization of emulsions by nonionic surface-active agents. As mentioned previously, the polyoxyethylene ethers can bind water molecules by hydrogen bonding with the oxygen atoms of the ether.[19]

Stabilization by Solids

It has been known for many years that emulsions can be stabilized by finely divided solids. A wide variety of solids have been reported to be effective for this purpose and included are such materials as the basic sulfates of polyvalent metals, calcium carbonate, hydroxides of the poly-valent metals, etc.

In order for a solid to act as a suitable stabilizer, the particle size of the solid must be small compared to the size of the emulsified droplet, and the particle should have the proper wettability by the aqueous and oil phases.[19] If the solid stays suspended in either the aqueous or the oil phases, then it obviously will not collect at the oil–water interface and will not act as a stabilizer. Pickering[43] has suggested that if the solid is more easily wetted by water than by oil, an oil-in-water emulsion wil be formed. There seems to be some justification for this because Schlaepfer[44] prepared concentrated water-in-oil emulsions using soot, which is wetted more by oil than water.

Becher[1] has suggested that the concentration of solid stabilizers at an interface could produce an interfacial film of considerable strength that could account for the stabilizing action of the solids. Verwey[45] has also indicated that emulsions increase in stability as the interfacial films around the droplets become more solid in character.

STABILIZATION OF WATER-IN-OIL EMULSIONS

The literature on water-in-oil emulsions is not nearly as extensive as that for oil-in-water emulsions. Schulman and Cockbain[46] investigated emul-sions of water-in-mineral oil using various emulsifying agents and came to the conclusion that in order to obtain stable emulsions the emulsifier should form uncharged interfacial films that possessed considerable rigidity. They felt that the droplets could not be charged because an electrical double layer would not be expected to exist in a nonionizing medium like mineral oil.

Albers and Overbeek,[47] however, found that there was a considerable electrokinetic potential present in water-in-oil emulsions stabilized with the oleates of polyvalent metals. They reported that there was no correlation between the electrokinetic potential and stability of the emulsion against flocculation and concluded that the stability of the emulsions was due to the formation of a thick film of the hydrolysis products of the oleates. The stabilizing activity of the hydrolysis products was explained on the basis that they could be considered as small particles, insoluble in both the oil and water phase. Because these products were mostly hydrocarbon in their structure, they were more easily wetted by the oil phase than the water phase and therefore formed water-in-oil emulsions. The particles collected at the interface and stabilized the emulsion mechanically.

Ford and Furmidge, in a recent paper[48] studied the stabilization of water-in-oil emulsions using a variety of oil-soluble emulsifiers. They concluded that in order to obtain a stable water-in-oil emulsion, the interfacial film of the emulsifier should prevent the coalescence of water droplets while allowing the coalescence of oil droplets. The emulsifier should be hydrophobic in nature.

The water droplets in many water-in-oil emulsions have an irregular shape. This was observed as early as 1915 by Briggs and Schmidt[49] and has also been reported by a number of other investigators. Schulman and Cocsbain attributed the irregularity of the water droplets in their emulsions to the stiffness of the interfacial film. However, not all dispersed water droplets are irregular in shape.[1]

DETERMINATION OF AEROSOL EMULSION STABILITY

The emulsion stability of aerosol can be determined by packaging the emulsions in clear glass bottles, shaking to emulsify the propellant, and noting the time until phase separation appears.

The method by which the propellant is emulsified is quite important. One of the most effective procedures for emulsification of aerosols in the laboratory is the Briggs intermittent method of shaking. Briggs found that intermittent shaking, with periods between the shakes, was much more effective than continuous shaking.[50] Gopal[51] reported, for example, that about 3000 shakes in a machine were necessary to emulsify 60 vol% of benzene in 1% sodium oleate. This required about 7 min. The same mixture could be completely emulsified with only 5 shakes by hand if a rest period of 20 -30 sec was allowed after each shake.

In the "Freon" Products Laboratory, samples to be checked for emulsion stability are first emulsified using an intermittent shaking procedure, allowed to stand overnight, and reshaken before the emulsion stability is determined. It is important for comparison purposes to treat all samples in exactly the same way.

VISCOSITY OF EMULSIONS

Emulsions can vary in viscosity from very thin fluids to gels. There are many factors that affect the viscosity of emulsions and References 1, 2, and 52 should be consulted for details on this subject. Sherman[52] has listed a number of variables that influence the viscosity of emulsions and included the following: (1) Viscosity of the continuous phase, (2) volume concentration of the internal phase, (3) particle size distribution of the dispersed phase, (4) type of emulsifying agent, and (5) the nature of the interfacial film.

Griffin[3] pointed out that the viscosity of emulsions is close to the viscosity of the external or continuous phase if the latter constitutes more than 50% of the emulsion by volume. However, when the concentration of the dispersed phase is increased beyond 50%, the emulsion becomes increasingly more viscous and eventually becomes a gel.

The maximum volume that can be occupied by the internal phase is 74%, assuming that all droplets are spherical and the same size.[2,3] However, emulsions with as high as 99% of the internal phase can be prepared. This is possible when the droplets of the disperse phase are nonuniform in diameter and the small droplets can be packed in between the larger droplets. In such high concentrations, the droplets become distorted and lose their spherical shape.

The type of surface-active agents present and the nature of the resulting interfacial film can have a profound effect upon the viscosity of the emulsion. An increase in viscosity has been observed in many cases with combinations of surface active agents and long-chain alcohols that form molecular complexes at interfaces. The sodium alkyl sulfates, triethanolamine salts of the fatty acids, and the polyoxyethylene fatty ethers are examples of surface-active agents that have been observed to show this effect with the fatty alcohols.[7,9,10] These complexes form strong, condensed interfacial films and hydration of these films may be a factor in the increased viscosity of the system.

The viscosity of the external phase may also be increased by the use of thickeners or gelling agents, such as sodium carboxymethyl cellulose, methyl cellulose and natural gums.

Determination of Viscosity of Aerosol Emulsions

No effective method for measuring the viscosity of aerosol emulsions quantitatively has yet been devised. Palmer and Morrow[53] reported a method for determining consistency (viscosity) of aerosol coating compositions which involved the use of a falling ball viscometer. The apparatus would not be suitable for aqueous emulsion products, however, because of the foaming that would occur during transfer of the sample to the viscometer.

Crude comparisons of the viscosity of aerosols can be made by preparing the samples in clear glass bottles, inverting the bottles, and noting the flow characteristics of the emulsion.

DROPLET SIZE AND EMULSION APPEARANCE

Emulsions vary in appearance from milky, opaque systems to products that are transparent. The appearance of the emulsions is related to the size of the dispersed droplets. Griffin[3] reported the relationship between the appearance of an emulsion and its droplet size as shown in Table 16-1.

TABLE 16-1 DROPLET SIZE AND EMULSION APPEARANCE

Droplet Size	Appearance of Emulsion
0.05 μ and smaller	Transparent
0.05–0.1 μ	Gray, semitransparent
0.1–1 μ	Blue-white
Greater than 1 μ	Milky, white

From Ref. 3. Reproduced with permission.

The appearance of the emulsion can be judged best by shaking the bottle and observing a thin film of the emulsion as it drains down the side. Most aerosol emulsions are milky white, indicating a droplet size greater than 1 μ.

REFERENCES

1. P. Becher, "Emulsions: Theory and Practice," 2nd ed. Reinhold Publishing Corp., New York, N.Y., 1965.
2. W. Clayton, "The Theory of Emulsions and Their Technical Treatment," 5th ed., The Blakiston Company, Inc., New York, N.Y., 1954.
3. W. C. Griffin, "Emulsions," Encyclopedia of Chemical Technology, Vol. 8, 2nd ed., John Wiley & Sons, Inc., New York, N.Y., 1965.
4. "Freon" Aerosol Report, FA–21, "Aerosol Emulsions with 'Freon' Propellants."
5. A. J. Schwartz, J. W. Perry, and J. Berch, "Surface Active Agents and Detergents," Vol. 2, Interscience Publishers, Inc., New York, N.Y., 1958.
6. J. J. Bikerman, "Foams and Emulsions," in S. Ross, "Chemistry and Physics of Interfaces," American Chemical Society Publications, Washington, D. C., 1965.
7. J. H. Schulman and E. G. Cockbain, Trans. Faraday Soc. 36, 651 (1940).
8. G. G. Stokes, Trans Cambridge Phil. Soc. 9 (1951).
9. P. A. Sanders, J. Soc Cosmetic Chemists 17, 801 (1966).
10. P. A. Sanders, Soap Chem. Specialties 43, 68, 70 (1967).
11. H. Schulze, Prakt. Chem. 25, 431 (1882); 27, 320 (1883).
12. S. E. Linder and H. Picton, J. Chem. Soc. (London) 61, 148 (1892).
13. R. A. Gortner, "Outlines of Biochemistry," 3rd ed. Chap. 8, John Wiley & Sons, Inc., New York, N.Y., 1949.
14. W. B. Hardy, Proc. Roy. Soc. (London) 66, 110 (1900).
15. E. F. Burton, Phil. Mag. 11, 425 (1906).

16. H. Helmholtz, *Wied Ann.* **7**, 537 (1879).
17. G. Gouy, Compt. Rend. **149**, 654 (1900).
18. O. Stern, *Z. Elektrochem.* **30**, 508 (1924).
19. J. A. Kitchener, and P. R. Mussellwhite, "The Theory of Stability of Emulsions," in P. Sherman, "Emulsion Science," Academic Press, Inc., New York, N.Y., 1968.
20. B. V. Derjaguin and L. Landau, *Acta Phys. Chim. URSS* **14**, 633 (1941).
21. E. J. W. Verwey and J. Th. G. Overbeek, "Theory of Stability of Lyophobic Colloids," Elsevier, Amsterdam, 1948.
22. A. R. Martin and R. N. Herman, *Trans. Faraday Soc.* **37**, 30 (1941).
23. M. Van den Tempel, *Rec. Trav. Chim.* **72**, 419 (1953).
24. H. Limburg, *Rec. Trav. Chim.* **45**, 772, 854 (1926).
25. A. King and G. W. Wrzeszinski, *J. Chem. Soc.* 1513 (1953).
26. L. Ya. Kremnev, L. A. Nikishechkina, and A. A. Ravdel, Dokl. Akad. Nauk SSSR **152**, 816 (1963).
27. B. V. Derjaguin, and A. S. Titievskaya, *Kolloidn. Zh.* **15**, 416 (1953).
28. J. W. McBain and M. C. Field, *J. Phys, Chem.* **37**, 675 (1933).
29. J. W. McBain and M. C. Field, *J. Chem. Soc.* **37**, 920 (1933).
30. J. W. McBain and A. J. Stewart, *J. Chem. Soc.* **37**, 924 (1933).
31. F. Atlkins, *Perfumery Essent. Oil Record* 332 (1934).
32. F. V. Ryer, *Oil Soap* **23**, 310 (1946).
33. E. D. Goddard and H. C. Kung, *Proc. 52nd Ann. Meeting,* CSMA (December 1965), p. 124.
34. J. H. Schulman and E. K. Rideal, *Proc. Roy. Soc.* (London) **B 122**, 29 (1937).
35. M. B. Epstein, A. Wilson, C. W. Jakob, L. E. Conroy, and J. Ross, *J. Phys. Chem.* **58**, 860, (1954).
36. G. D. Miles, L. Shedlovsky, and J. Ross, *J. Phys. Chem.* **49**, 93, (1945).
37. G. D. Miles, J. Ross, and L. Shedlovsky, *J. Amer. Oil Chemists Soc.* **27**, 268 (1950).
38. P. Becher, and A. J. Del Vecchio, *J. Phys. Chem.* **68**, 3511 (1964).
39. J. M. Prince, *J. Colloid Sci.* **23**, 165 (1967).
40. J. T. Davies and E. K. Rideal, "Interfacial Phenomena," 2nd ed., Chap. 5, Academic Press, New York and London, 1963.
41. L. N. Ferguson, "Electron Structures of Organic Molecules," Prentice-Hall, Inc., New York, N.Y., 1952.
42. J. W. McBain, "Colloid Science," D. C. Heath and Company, Boston, Mass., 1950.
43. S. U. Pickering, *J. Soc. Chem. Ind.* **13**, 129 (1910).
44. A. U. M. Schlaepfer, *J. Chem. Soc.* (London) **113**, 522 (1918).
45. E. J. W. Verwey, *Trans. Faraday Soc.* **36**, 192 (1940).
46. J. H. Schulman and E. G. Cockbain, *Trans Faraday Soc.* **36**, 661 (1940).
47. W. Albers and J. Th. G. Overbeek, *J. Colloid Sci.* **14**, 501, 510 (1959).
48. R. E. Ford and C. G. L. Furmidge, *J. Colloid Sci.* **22**, 331 (1966).
49. R. T. Briggs and H. F. Schmidt, *J. Phys. Chem.* **19**, 491 (1915).
50. T. R. Briggs, *J. Phys. Chem.* **24**, 120, (1920).
51. E. S. R. Gopal, "Principles of Emulsion Formation," in P. Sherman, "Emulsion Science," Chap. 1, Academic Press, Inc., New York, S.Y., 1968.
52. P. Sherman, "Rheology of Emulsions," in P. Sherman, "Emulsion Science," Chap. 3, Academic Press, New York, N.Y., 1968.
53. F. S. Palmer and R. W. Morrow, *Soap Sanit. Chemicals* **28**, 191 (1952).

17

GENERAL PROPERTIES OF FOAM

A foam consists of a coarse dispersion of a gas in a liquid in which the volume of the gas is considerably larger than that of the liquid. The gas bubbles are separated by thin liquid films called lamellae. The foam in a washer or the foam that builds up while shampooing the hair results from mixing air with an aqueous detergent solution. In aerosol foams, the dispersed gas bubbles consist of vaporized propellant instead of air. One of the most common examples of an aerosol foam is the shaving lather.

Aerosol foams are unique in that they are self generating. An aerosol foam product is normally packaged in the container as an emulsion with the liquefied gas propellant dispersed as droplets throughout the aqueous phase. When the product is discharged, the propellant vaporizes into a gas that is trapped by the aqueous solution. This forms a foam, as illustrated in Figure 17–1.

Since an aerosol foam is formed from an emulsion, the foam will inherit many of the properties of the original emulsion. The type of interfacial film around the dispersed propellant droplets in the emulsion, for example, determines the structure of the interfacial film around the vaporized propellant bubbles. The bubble size of the foam undoubtedly is determined in part by the size of the dispersed propellant droplets in the original emulsion, although there is no direct evidence for this. Generally, if an aerosol oil-in-water emulsion is fairly stable, the foam will also be stable.

There are many types of foams and they can be formulated to possess a wide variety of properties. This is a considerable advantage in the development of products for different applications where the requirements are different. Some foams are stable and show little wetting after many hours, others wet immediately and collapse after a few minutes, and some wet immediately but do not collapse for hours. All of these properties are manifestations of the stability of the foam.

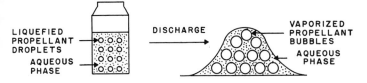

LIQUEFIED
PROPELLANT →
DROPLETS
AQUEOUS
PHASE

DISCHARGE
→

VAPORIZED
PROPELLANT
BUBBLES
AQUEOUS
PHASE

Figure 17–1   Formation of a foam from an aerosol emulsion.

FOAM STABILITY

Foams are thermodynamically unstable systems just like emulsions. The total surface area in a foam is fairly large, and, therefore, there is a considerable amount of surface energy present. When the liquid films separating the gas bubbles in a foam break and the bubbles coalesce, the liquid films form droplets and the resulting surface area of the droplets is less than that of the original system. Since this process decreases the surface area, it also decreases the surface energy. As a consequence, film rupture and bubble coalescence are spontaneous processes. Foams from pure liquids are so unstable that they seldom have a life over one second. Nevertheless, like emulsions, foams can be stabilized by the addition of surface-active agents so that they last for considerable periods of time.

There are various definitions of foam stability just as there are for emulsion stability.[1,2] In the present discussion a foam is considered to be stable for a given period only if there is no significant change in any of its properties during that time.

There are three main factors associated with foam stability. These are. (1) Drainage, (2) rupture of the liquid films followed by coalescence, and (3) change in bubble size.

Drainage causes the liquid films separating the gas bubbles to become thinner. This usually leads to film rupture. Rupture of the liquid films (lamellae) separating the bubbles leads to coalescence of the bubbles and complete collapse of the foam structure. Change in bubble size can lead to thinning of the lamellae and may cause mechanical shocks that result in film rupture.

Drainage

Drainage of a liquid in a foam is somewhat similar to the downward creaming in an emulsion except that in the foam the continuous liquid phase settles to the bottom and not the dispersed gas phase. This is illustrated in Figure 17–2.

Drainage in a foam occurs for two reasons, gravitation, and capillary action resulting from surface tension. The drainage resulting from capil-

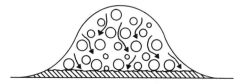

Figure 17–2   Drainage in a foam.

lary action occurs at a location in the foam known as Plateau's border. The junction between three gas bubbles in a foam and the location of Plateau's border is illustrated in Figure 17–3.

Drainage occurs at the Plateau border because liquid flows from the lamellae at location A towards Plateau's border, PB. The reason for this is a difference in pressure at locations A and PB. The pressure depends upon the curvature of the interface between the liquid and the gas. At location A the curvature is esentially zero so that the pressure in the liquid is about the same as that in the gas. However, at location PB, the pressure is lower in the liquid than in the gas because of the negative curvature towards the gas. Therefore, the pressure in the liquid at PB is lower than that at A and liquid flows from A to PB.[1,2]

Rupture of the Liquid Films and Bubble Coalescence

Drainage of liquid weakens the foam structure because the liquid films separating the bubbles becomes thinner. This allows the gas bubbles to come closer together and may proceed to the point where the liquid films rupture and the gas bubbles coalesce. Davies[2] reported that the mean thickness of the liquid films in a representative series of foams varied from about 3,300–30,000Å, as calculated from air permeability data. The foam structure may remain fairly stable during drainage until the film thickness decreases to about 200–2,000 Å. At this point, the stability of the foam depends upon whether further thinning occurs or whether the foam is subjected to mechanical shocks. If further thinning reduces the thickness of the liquid films to about 50–100 Å, then molecular attractive forces cause the films to rupture. This is followed by bubble coalescence and the

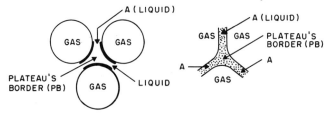

Figure 17–3   Location of a Plateau border in a foam.

entire foam structure collapses. Bubble coalescence can also result from mechanical shocks, such as rubbing the foam between the hands.

Change in Bubble Size Distribution

Another factor that decreases the stability of foams is the change in size of the bubbles as the foam ages. In any foam, the small bubbles become smaller and the large bubbles grow larger. This occurs because the pressure in a small bubble is higher than that in a large bubble and this differential in pressure causes the molecules of the gas in the small bubble to diffuse through the liquid film and into the larger bubble.

The pressure in a gas bubble in a foam is greater than atmospheric pressure P_a, by a factor, ΔP. The total pressure in the bubble, P_t, is

$$P_t = P_a + \Delta P. \tag{17-1}$$

The magnitude of the factor ΔP, is determined by the surface tension, γ, and the radius of the bubble, R,:

$$\Delta P = \frac{2\gamma}{R}. \tag{17-2}$$

The total pressure in the gas bubble is

$$P_t = P_a + \frac{2\gamma}{R} \tag{17-3}$$

Therefore, the pressure in a small bubble is higher than that in a larger bubble because the radius of the former is smaller. As the small bubble becomes smaller, its radius decreases further and the pressure increases while the opposite is true of the larger bubbles. Therefore, the difference in pressure between the two bubbles increases until the smaller bubble disappears completely. Davies' and Rideal cite the example of a foam prepared from a solution of "Teepol" (a surface-active agent) where the total number of bubbles after 15 min was only 10% of the original number, although there had been no rupture of the liquid films. It has been suggested that the continual change in bubble size and the resulting rearrangement of the bubbles in the foam could lead to an increased possibility of mechanical shock followed by film rupture and coalescence.

Some of the factors that affect the rate at which the bubble size in a foam changes are the magnitude of the pressure difference between the bubbles, the permeability of the interfacial film to the gas, and the solubility of the gas in the liquid phase.

STABILIZATION OF FOAMS

Foams are inherently unstable but they can be stabilized by the addition of surface-active agents. In order to stabilize a foam it is necessary to prevent drainage, coalescence of the bubbles, and the change in bubble size. The major role of the surface-active agents is to increase the surface and bulk viscosity of the system, which reduces drainage, and to form a strong interfacial film around the bubbles, which retards coalescence if the bubbles do come into contact. A strong interfacial film may have a low permeability to gas molecules and thus can decrease the rate of diffusion of the molecules from the smaller bubbles to the larger bubbles.

Electrical repulsion plays a comparatively minor part in the stabilization of foams. It does not stop drainage of the initial foam structure but may prevent further thinning of the liquid film once the film thickness has decreased sufficiently so that a repulsion develops between the electric double layers.

The Electric Double Layers

Anionic and cationic surface-active agents are absorbed at the gas–liquid interfaces in a foam structure in the same way that they are absorbed at an oil–water interface in an emulsion system. This results in the formation of electric double layers similar to those postulated for emulsion systems (Figure 17–4).

It can be seen from Figure 17–4 that there are two electric double layers in a single liquid film. As the liquid film thins from drainage, the two opposing electric double layers come closer together so that ultimately there is a repulsion between them. This is analogous to the situation in an emulsion when two electrically charged oil droplets approach each other. The energy barrier that develops in the film opposes further thinning. This can confer some stability on the foam structure and it can also provide protection against shocks. This type of system, where stabilization does not

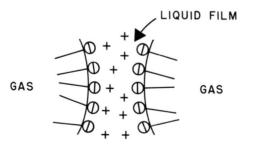

Figure 17–4   Electric double layers at gas-water interfaces.

occur until after an initial period of drainage and film thinning has taken place, is referred to as a pseudo-stable system.[2]

In the discussion on emulsions, it was pointed out that since the polar heads of the anionic surface-active agents have the same charge, they will repel each other when they are located on the same droplet. This self repulsion results in a weak interfacial film because it reduces the concentration of the surface-active agents at the interface. The same phenomenon undoubtedly occurs at the gas–liquid interfaces in a foam—one reason why many surface-active agents produce relatively poor foams by themselves.

Viscosity

Since drainage from a foam results from the flow of liquid between the gas bubbles, increasing the viscosity of the system decreases the rate of drainage. This has been demonstrated by a number of investigators. Miles, Shedlovsky, and Ross[3] studied the flow of liquid through various foams and found that the rate of flow decreased when the surface viscosities of the system increased. For example, the viscosity of a sodium lauryl sulfate solution was low and the rate of flow of liquids through the foam was high. When lauryl alcohol was added, the reverse situation occurred—the rate of flow of liquid through the foam was low and the viscosity of the system was high.

Subsequently, Miles and his co-workers[4] studied the drainage rates of aqueous films from glass frames that were suspended in a vertical position. The film from aqueous sodium lauryl sulfate drained rapidly but the addition of lauryl alcohol resulted in films that drained slowly. This difference in the drainage rates was attributed to the increase in viscosity that resulted when lauryl alcohol was added.

In addition, it was noted that when the temperature of the slow draining films from sodium lauryl sulfate–lauryl alcohol films was raised, the films became fast draining at a specific higher temperature. The temperature at which this occurred was called the film drainage transition temperature. The change from slow to fast drainage took place because the sodium lauryl sulfate–lauryl alcohol complex, which was responsible for the higher viscosity of the system, dissociated at the higher temperature.

Epstein, Ross, and Jakob[5] devised an apparatus for studying the drainage transition temperature of foams and reported that the transition temperatures for foams from an aqueous sodium lauryl sulfate–lauryl alcohol system agreed with the drainage transition temperature for films from the same system.

In a further study of the sodium lauryl sulfate–lauryl alcohol system, Epstein and his co-workers came to the conclusion that the high surface viscosity and slow film drainage was due to the presence of a crystalline molecular complex formed from two molecules of sodium lauryl sulfate and one molecule of lauryl alcohol.[6]

Brown, Thuman, and McBain devised a sensitive rotational viscometer in 1953 and reported that foams with the highest stability were produced from solutions with appreciable surface viscosity.[7]

It seems likely that the increase in surface viscosity that occurs with the formation of molecular complexes, such as that from sodium lauryl sulfate and lauryl alcohol, is due to hydration of the interfacial film of the molecular complex. The effect of hydration and the formation of layers of "soft ice" upon the stability of emulsions was discussed previously. This type of structure would increase the viscosity of the system, particularly if it extended to any degree into the bulk phase.

The Interfacial Film

The interfacial film formed by a molecular complex, at a gas–liquid interface in a foam should be as strong, if not stronger, than the interfacial film from the same complex at an oil–water interface in an emulsion. In the aqueous portion of the interfacial region, the same attractive forces, such as ion–dipole and dipole–dipole interactions, would be operative in foams and in emulsions. However, London–van der Waals forces of attraction between the hydrocarbon chains of the complex would be expected to be stronger in foams, where the chains extend into the gas phase and can be more closely associated than in oil droplets. In an emulsion there would be stronger competition between the oil molecules and the hydrocarbon chains of the surface-active agent than between the gas molecules and the hydrocarbon chains in a foam.

A strong interfacial film serves at least two functions in a foam. There is evidence that the interfacial film formed from molecular complexes is less permeable to the passage of gases. This would decrease the diffusion of gases from the smaller bubbles to the larger bubbles and thus increase the stability of the foam. The low permeability of the interfacial film to the passage of gases is considered to be due to the presence of thin layers of soft ice formed by hydration.[2]

Finally, a strong condensed interfacial film would retard coalescence of gas bubbles when they came into contact. This would be quite important in a foam where the liquid films in the initial foam were very thin because of the high concentration of the gas phase relative to the liquid phase.

EVALUATION OF AEROSOL FOAMS

The ultimate test of any product is whether or not it is suitable for the application for which it was developed. If the product has been formulated as an aerosol shaving lather, for example, then the final test is how the product performs in shaving tests.

Usually, a considerable amount of exploratory investigation is necessary before a satisfactory product is developed. In order to arrive at this objective as rapidly as possible, it is necessary to know how certain variables and additives affect the properties of foams. Therefore, it is desirable to be able to characterize a foam as precisely as possible in order to have some basis for comparison with other foams and to have a permanent record of the properties of the foam. Properties of foams, such as density, viscosity, and stability are very useful for this purpose. Not only do these properties serve to characterize a foam at a given time, but they are also useful as a means of determining to what extent a foam changes during aging (foam stability).

General Observations

A considerable amount of information can be obtained about an aerosol foam merely by discharging it onto a paper towel or the hand and observing it for a few minutes. The extent to which a foam wets a paper towel, retains its shape, or peaks when a glass rod or pencil is placed in the foam and then raised vertically, all give an indication of foam stability and viscosity. The way the foam feels in the hand, the ease with which it spreads, and the degree to which it collapses when spread or rubbed, are further indications of such properties as stability and density.

Bubble Size

Microscopic examination of foams can be very informative and photomicrographs of foams can be quite useful since they can be examined at leisure. Most professional photographers can suggest equipment which is suitable for taking photomicrographs of foams.

At the "Freon" Products Laboratory[8] observations of foam structures are made with a Bausch and Lomb Sterozoom microscope, Model BVB-73, equipped with a 10X paired widefield eyepiece and a power pod magnification of 3X. Approximate bubble sizes of foams are determined using a micrometer disk (#31-16-08) which measures intervals of 0.001 in.

Photomicrographs of foams are taken at 30 times magnification using a Spencer triocular single stage microscope manufactured by the American Optical Company. It is equipped with a 15X eyepiece and a 3.5 objective lens. The camera is focused through 10X eyepieces with 3.5 objectives. The pictures are taken with a MP-3 Polaroid Multipurpose Industrial View camera with a 4 × 5 film adapter, using surface illumination and a one second exposure. Pictures are taken 20 sec after the foam has been discharged.

Density

The density of an aerosol foam can be obtained by filling a container of known volume with the foam and determining the weight of the foam. A large enough vessel should be used so that errors due to air entrapment in the foam are minimized. A crystallizing dish with a volume of 350 cc is used at the "Freon" Products Laboratory.

In order to facilitate the discharge of the foam into the dish, a short piece of "Tygon" tubing about 3-in. long is attached to the foam actuator. The foam is layered carefully in the dish keeping the end of the tubing slightly above the surface of the foam at all times. Enough foam is added so that the surface of the foam extends above the dish. After the dish is filled, an additional 1–2 min are allowed for any residual expansion. The excess foam is then removed with a spatula and the dish is weighed.

Several equations have been developed for calculating the density of an aerosol foam. Spitzer, Reich, and Fine[9] have published the following equation:

$$D = \frac{100\,M}{24{,}000\,X + 100\,M}.$$ (17–4)

D is the density of the foam in g/cc, M is the molecular weight of the propellant, and X is the wt % of the propellant in the aerosol.

Gorman and Hall[10] have reported the folowing equation for predicting foam density:

$$d_f = \frac{100}{\%\,P\,V_r}$$ (17–5)

d_f is the density of the foam in g/cc, $\%P$ is the wt% propellant in the initial emulsions, and V_r is the volume occupied by one gram of propellant vapor. The authors report that there is fairly good agreement between the published experimental values of various foams and the calculated values.

The following equation, which gives accurate values for foam density, was developed by Becher:[11]

$$D = \frac{W_a + W_p}{(22{,}400\,W_p/M) + V_a}$$ (17–6)

W_a is the weight of the aqueous phase, W_p is the weight of the propellant, M is the molecular weight of the propellant, and V_a is the volume of the liquid phase. Becher determined foam density by discharging approximately 50 ml of foam and drawing it up into a tared 10-ml hypodermic syringe (without needle). The syringe was then weighed. A second determination on the same sample was carried out by expelling half of the foam volume and reweighing the syringe. A large discrepancy between the two densities was assumed to indicate bubble occlusion and the results were rejected.

Stability

Drainage. The comparative rates of drainage of a series of foams can be obtained by discharging the foams into glass funnels located over graduates and measuring the quantities of liquid that collect in the graduates after various times. Since the initial weight of foam in the funnel is known, the percent drainage of the foams after any specified time can be calculated and the foams compared on this basis. If the drainage is determined for several different times, a curve for each foam, which illustrates the percent drainage as a function of time, can be drawn. From this curve, the time required to produce any given percent of drainage, i.e., 50%, can be obtained. The 50% drainage times can then be used for comparing the foams.

Wetting. Wetting is a consequence of drainage and therefore correlates with drainage. Wetting can be judged by discharging the foam onto a paper towel and determining the period of time that elapses before visible wetting of the paper towel occurs. De Navarre and Lin[12] devised a simple electrical apparatus to measure wetting time. A piece of filter paper is placed over two electrical contacts in a circuit with a galvanometer. The foam is placed on the paper and the flow of current that occurs when the foam wets the paper is registered on the galvanometer. The time required to reach a predetermined galvanometer reading can be used as a basis for comparison with other foams.

Foam Collapse. Another indication of foam stability is the extent to which a foam will collapse during aging. This can be determined qualitatively for a series of foams by discharging them onto paper towels and noting their appearance after specified periods of time.

A somewhat more satisfactory procedure is to discharge the foams on paper towels and observe them against a background of horizontal lines drawn at ¼ or ½-in. intervals. If the height of the foam is noted immediately after discharge, any subsequent decrease or collapse can be determined by referring to the original height of the foam on the horizontal line background.

Miscellaneous Tests. Any measurement that can be carried out on a foam before and after aging will give an indication of stability. Johnsen[13] uses a test that consists in attaching a light metal rod to a stainless steel disk and measuring the rate of fall of the disk and rod through a 3-in.

depth of foam. The difference in rate of fall between a fresh foam and one aged for 10 min is reported to give an excellent indication of foam stability.

Stiffness and viscosity measurements on fresh and aged samples of foam can be used as an indication of foam stability.

Overrun

Overrun is a measure of the change in volume that occurs when an emulsion of a liquefied gas propellant is allowed to expand into a foam. Overrun can be calculated as follows:[14]

$$\% \text{ Overrun} = \frac{\text{Vol of Foam} - \text{Vol of Liquid}}{\text{Vol of Liquid}} \times 100. \quad (17\text{--}7)$$

One method of determining overrun is to attach a piece of glass tubing to the foam actuator with rubber tubing of sufficient length so that the assembly reaches to the bottom of a 1000-ml graduate. The container is slowly discharged with the glass tube at the bottom of the graduate. As the foam builds up in the graduate, the glass tube is raised so that the end of the tube is always slightly above the surface of the rising foam. The overrun can be calculated from the volume of foam that is obtained and the weight of the product that was discharged. The volume of the emulsion is calculated from its density.

The density of a foam changes during discharge and the percent overrun is an average value for the quantity of product discharged. Therefore, the percentage of product discharged from the container for an overrun measurement should be specified.

Viscosity

The measurement of viscosity is an excellent way of characterizing foams. Carter and Truax[15] studied the effect of a number of variables upon the properties of aerosol shave lathers. Viscosity measurements on the foams were carried out using a Brookfield Model HAT viscometer 20 rpm, T type spindle A, Helipath stand.

Richman and Shangraw[16] have investigated the effect of variables upon the rheological properties of aerosol foams in considerable detail. This publication gives an excellent review of the historical background of the rheology of foams and the types of equipment that have been used for viscosity measurements. Richman and Shangraw determined viscosity with a Haake Rotovisco Viscometer (Gebrüder Haake K.B., Berlin). They reported that the Rotovisco is a very versatile instrument that can measure apparent viscosities in the range of 5×10^{-3}—4×10^{7} poise.

The resistance of a foam to penetration or deformation is a function of viscosity and can be used to characterize foams. In the "Freon" Products Laboratory, the relative stiffness of aerosol foams is determined with a Cherry–Burrell Curd Tension meter.[17] A crystallizing dish is filled with foam and placed on a scale platform located directly underneath a specially designed curd knife. The curd knife is then driven downwards at a constant, controlled rate of speed by a synchronous motor. After the knife has contacted the foam, the resistance of the foam to penetration by the knife causes the scale platform to depress. The extent of the resistance of the foam is recorded on the scale dial in grams.

This is a relatively simple method for detecting changes in foams that occur during aging or variations that occur as the container is discharged. Stiffness measurements are also suitable for detecting changes in foam structure that occur with modifications in the aerosol formulation.

REFERENCES

1. J. J. Bikerman, 'Foams: Theory and Industrial Applications," Reinhold Publishing Corp., New York, N.Y., 1953.
2. J. T. Davies, and E. K. Rideal, "Interfacial Phenomena," 2nd ed., Academic Press, New York and London, 1963.
3. G. D. Miles, L. Shedlovsky, and J. Ross, *J. Phys. Chem.* **49**, 93 (1945).
4. G. D. Miles, L. Shedlovsky, and J Ross, *J. Amer. Oil Chem. Soc.* **27**, 268 (1950).
5. M. B. Epstein, J. Ross, and C. W. Jakob, *J. Colloid Sci.* **9**, 50 (1954).
6. M. B. Epstein, A. Wilson, C. W. Jakob, L. E. Conroy, and J. Ross, *J. Phys. Chem.* **58**, 860 (1954).
7. A. G. Brown, W. C. Thuman, and J. W. McBain, *J. Colloid Sci.* **8**, 491 (1953).
8. P. A. Sanders, *J. Soc. Cosmetic Chemists* **17**, 801 (1966).
9. J. G. Spitzer, I. Reich, and N. Fine, U. S. Patent 2,655,480, October, 1953.
10. W. G. Gorman, and G. D. Hall, *Soap Chem. Specialties* **40**, 213 (1964).
11. P. Becher, Private Communication.
12. M. G. DeNavarre and T. J. Lin, *Amer. Perfumer Cosmetics,* **78**, 36 (1963).
13. M. A. Johnsen, "Laboratory Techniques," in "Aerosols: Science and Technology," H. R. Shepherd, Interscience Publishers, Inc., New York, N.Y., 1958, 1959.
14. F. T. Reed, Proc. 39th Ann. Meeting, CSMA, 1952, p. 30.
15. P. Carter and H. M. Truax, *Proc. Sci. Sect. Toilet Goods Assoc.* **35**, 37 (1961).
16. M. D. Richman and R. F. Shangraw, *Aerosol Age* **11**, (May–November, 1966).
17. P. A. Sanders, *Aerosol Age* **8**, 33 (1963).

18

AQUEOUS AEROSOL EMULSIONS AND FOAMS

Aqueous aerosol emulsion products fall into two groups, depending upon whether they have been formulated as oil-in-water or water-in-oil systems. The oil-in-water emulsions generally produce foams and are used for products such as window cleaners and shaving lathers. The water-in-oil emulsions usually discharge as nonfoaming sprays and are used for products requiring a fairly fine droplet size, such as room deodorants and space insecticides.

Although foams can be produced from systems other than oil-in-water emulsions, the latter account for the majority of the foam products that are on the market. These products can be subdivided into two more groups, according to whether they have been designed and packaged as sprayable foams, or as conventional foams. Sprayable foam products are equipped with spray valves with mechanical breakup actuators and discharge soft sprays that foam on contact. Conventional aerosol foam products are equipped with foam actuators and usually foam during discharge.

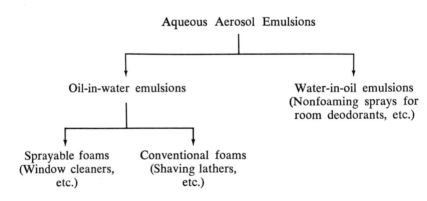

Aqueous Aerosol Emulsions

Oil-in-water emulsions

Water-in-oil emulsions (Nonfoaming sprays for room deodorants, etc.)

Sprayable foams (Window cleaners, etc.)

Conventional foams (Shaving lathers, etc.)

These products are normally referred to as aerosol foams, with the understanding that sprayable foams are designated as such.

The aqueous aerosol emulsions and the types of products for which they are used can be illustrated as shown on p. 222:

OIL-IN-WATER EMULSIONS

The use of foams for applying products has a number of advantages. Foams are usually formulated with low concentrations of propellant and this provides a high concentration of active ingredients in the container. Foams are easy to use, can be applied directly to the desired area with negligible waste, and are attractive in appearance. In addition, the properties of foams can be varied over a wide range, thus making them suitable for many different applications.

Sprayable Foams

Window cleaners, starch sprays, oven cleaners, and bathtub cleaners are typical sprayable foam products. These aerosols are designed to give a soft, light spray that foams slightly on contact. The foam serves two purposes—it holds the product on a vertical surface until it can be wiped off with a cloth or paper towel, and it shows where the surface has been sprayed. The foam should be stable enough so that it achieves the two preceding objectives but not so stable that it is difficult to break down and remove by wiping.

Surface-Active Agents. Since surface-active agents determine emulsion and foam stability and wetting properties, they are essential components of sprayable foams. Many sprayable foam products are designed for cleaning operations and, therefore, may contain appreciable quantities of additives, such as inorganic salts, ammonia, isopropyl alcohol, etc. The surface-active agent must be effective in the presence of these materials, and this requirement usually limits the agents to the nonionic class. Oven cleaners generally contain sodium hydroxide and the surface-active agent for this product must be stable to hydrolysis by caustic. This eliminates the esters.

Since most applications of sprayable foams involve wetting, the surface-active agent should possess good wetting properties. Those for glass cleaners and similar products should preferably be liquid at room temperature. Waxy surface-active agents tend to leave films on glass that are difficult to remove. Among the numerous existing nonionic agents, polyoxyethylene sorbitan esters, polyoxyethylene fatty ethers, and alkyl phenoxy polyethoxy ethanols[1] have proven to be effective.

Foams with a relatively low stability are desired for most applications, and this usually necessitates a concentration of the surface-active agent of about 1% or less. A low concentration of surface-active agent may create a problem of emulsion stability, but this can often be overcome by combinations of emulsifiers or by the use of propellant blends which have a density close to that of the aqueous phase (described in the next section). For products where a waxy surface-active agent is not objectionable, complexes from polyoxyethylene fatty ethers and the long-chain alcohols can provide systems with high emulsion stability, at very low concentrations of surface-active agent. These complexes are discussed in detail in Chapter 19. The compositions of some typical basic sprayable foam systems of this type with an emulsion stability over 30 min are given in Table 18–1.

TABLE 18-1 SPRAYABLE FOAM COMPOSITIONS WITH GOOD
EMULSION STABILITY

	Composition (wt %)		
Components	No. 1	No. 2	No. 3
Polyoxyethylene (4) lauryl ether	0.40	—	—
Polyoxyethylene (2) cetyl ether	—	0.36	—
Polyoxyethylene (2) stearyl ether	—	—	0.40
Lauryl alcohol	—	0.21	0.21
Cetyl alcohol	0.27	—	—
Water	89.34	89.43	89.39
"Freon" 12–"Freon" 114 (15/85)	10.00	10.00	10.00

Propellants. Fluorocarbons, hydrocarbons, and blends of the two are utilized as propellants for sprayable foams. Hydrocarbons, such as isobutane, are used at concentrations of about 3.5–4.0 wt% while the fluorocarbons are employed at higher concentrations of 8–10%. The fluorocarbons normally used are "Freon" 12 and "Freon" 114 which are stable in aqueous systems under most conditions. "Freon" 11 is not satisfactory for use in oil-in-water emulsions in metal containers because it decomposes in the presence of water and metal.[2]

When there is a problem of poor emulsion stability as a result of low surface-active agent concentration, it may be advantageous to use a propellant blend with about the same density as that of the aqueous phase. This will minimize creaming of the propellant. The disadvantage of these blends is that they are always flammable because of the quantity of hydrocarbon required to produce a density of 1.0 g/cc.

The compositions of two such blends are given in Table 18-2. One blend is formulated with "Freon" 12–"Freon" 114 (15/85) as one of the components and is suitable for glass bottle products. It has a calculated vapor pressure of 26.5 psig at 70°F. The other blend with "Freon" 12–"Freon" 114 (40/60) can be used for products packaged in metal containers and has a calculated vapor pressure of 35.6 psig at 70°F.

TABLE 18-2 PROPELLANT BLENDS WITH A DENSITY OF 1.0 G/CC.

	Composition (*wt %*)	
Components	*Blend #1*	*Blend #2*
"Freon" 12/"Freon" 114 (15/85)	72.2	—
"Freon" 12/"Freon" 114 (40/60)	—	72.9
Isobutane	27.8	27.1

In products where solvents, such as isopropyl myristate, odorless mineral spirits, VM & P naphtha, mineral oil, etc., are useful or can be tolerated, "Freon" 12 alone can be used as the propellant since the solvents can depress the vapor pressure to an acceptable level. For example, a "Freon" 12–mineral oil blend with a density of 1.0 g/cc would have a composition of 44.9% "Freon" 12 and 55.1% odorless mineral spirits assuming a density at 70°F of 1.32 g/cc for "Freon" 12 and 0.83 g/cc for mineral oil. Combinations of "Freon" 12 and propane have also been used as propellants in systems containing organic solvents.

Products with auxiliary solvents are prepared by first emulsifying the organic solvent in the aqueous phase. The resulting emulsion is loaded in the container and the propellant is pressure loaded. The emulsion of the solvent in the aqueous phase must have sufficient stability so that it does not separate while it is being pumped to the filling heads. This must be checked.

Conventional Aerosol Foams

These products constitute the largest group of aerosol foams that are on the market. All are equipped with foam actuators and either discharge as a foam or as a liquid that subsequently expands into a foam. The properties of the foams can be changed considerably by varying the composition. Foams can be obtained which collapse immediately after discharge or are so stable and coherent that they bounce. Some foams develop a sparkling surface as they age—others make a crackling noise when rubbed. A foam can be formulated to have almost any type of property desired.

Some typical foam products now on the market are listed in Table 18–3.

TABLE 18-3 TYPICAL COMMERCIAL AEROSOL FOAM PRODUCTS

1. After-Bath	6. Hand Cream
2. After Shave	7. Hand Disinfectant
3. Cologne	8. Shampoo
4. Deep Cleansing Foam	9. Shaving Lather
5. Face Make Up Foam	10. Skin Moisturizer

Surface-Active Agents. Aerosol emulsions and foams with adequate stability for aerosol applications can be obtained with a wide variety of anionic, nonionic, and cationic surface-active agents. Judging by the formulations that have appeared in the literature, most of the foam products are prepared with anionic surface-active agents. However, the use of nonionic agents is increasing, particularly in cosmetic and pharmaceutical products. Cationic agents, such as the quaternary ammonium compounds, are generally used for their disinfectant or textile softening properties rather than their emulsifying or foaming characteristics. However, some excellent foams can be obtained with cationic surface active agents.

Unfortunately, relatively few basic studies on the types of surface-active agents that are suitable for aerosols are available so that a considerable amount of experimental work may be required to develop a satisfactory emulsion system for a specific product. However, there are several publications that can be of assistance. Reference 3 gives the results of an evaluation of over 400 surface-active agents in a simple aerosol system consisting of 86% water, 4% agent, and 10% "Freon" 12–"Freon" 114 (40/60) propellant. The properties of the emulsions and foams obtained with these agents are tabulated and thus provide a starting point for formulating foam products. The properties of aerosol emulsions and foams prepared with the triethanolamine salts of the fatty acids are discussed in Reference 4 while those of systems formulated with the nonionic polyoxyethylene fatty ethers are listed in Reference 5.

In addition, there are a number of articles in the literature that disclose the types of surface-active agents used for various products.[1,6] The patent literature is particularly valuable in this respect and a search of the patent literature can be very rewarding. The well-known shaving lather patent,[7] discloses a wide variety of surface-active agents that are useful in producing aerosol foams. Many reference books, both aerosol and nonaerosol, disclose the types of surface-active agents that have been used for a variety of products.[8-13] McCutcheon's list of synthetic detergents and their uses, properties, structure, and source is almost a necessity for anyone involved in the formulation of emulsion products.[14]

A considerable amount of information about the surface-active agents used in various aerosol products has never been published and probably never will be. The development of a satisfactory aerosol foam product can involve considerable effort and many laboratories understandably prefer not to publicize the compositions of their products. The analysis and identification of surface-active agents is not a simple procedure and the composition of many a product remains a well-kept secret because of this factor.

ANIONIC SURFACE-ACTIVE AGENTS

Salts of the Fatty Acids. The salts of the fatty acids are the most commonly used anionic surface-active agents and are employed for a variety of products including shaving lathers, shampoos, hand creams, and hand cleaners. These agents produce excellent emulsions and foams, are low in toxicity, and are essentially noncorrosive in metal containers. Although some formulations contain sodium and potassium salts, the majority of aerosol products are formulated with the triethanolamine salts. In a few instances, the diethanolamine salts have been used.

The salts are prepared by neutralizing fatty acids with the appropriate base. Most of the commercially available fatty acids used for this purpose are obtained from natural occurring fats and oils and are composed of mixtures of the fatty acids, regardless of how they are named. *Triple-pressed* stearic acid, commonly used in aerosols, consists of a mixture of about 45% stearic acid and 55% palmitic acid. This mixture is obtained from tallow, and the triple-pressed designation refers to the series of pressing operations used to separate the solid mixture of stearic and palmitic acids from liquid *red oil,* which is primarily oleic acid.[15] Coconut fatty acids, derived from coconut oil, are also commonly employed in aerosol formulations. The composition of a commercial coconut oil is as follows:[16] 4% Capric acid; 55% Lauric acid; 19% Myristic acid; 7% Palmitic acid; 4% Stearic; 8% Oleic; and 3% Linoleic.

It can be seen that coconut oil contains a rather startling array of fatty acids. Coconut-oil fatty acids are used in shaving lather aerosols throughout the United States and give consistently good foams.

Since the compositions of commercial fatty acids depend upon the source, it is sometimes difficult to know exactly how much triethanolamine is required to neutralize the fatty acids. Usually, the composition of a commercial fatty acid or its average molecular weight is available from the supplier and this is sufficient to calculate the quantity of base required. The molecular weight can also be calculated from the acid number, which can be determined in most laboratories.

Most aerosol formulations contain mixtures of the commercial fatty acids, which means that they contain mixtures of mixtures of acids. This makes the ultimate composition a little complicated. These mixtures of the commercial fatty acids are used because experience has shown that they give products which have more desirable properties than those obtained with any one fatty acid alone. Many aerosol shaving lathers, for example, are formulated with a mixture of the triethanolamine salts of triple pressed stearic acid and coconut fatty acid in which the ratio of stearic to coconut fatty acid ranges from about 90/10–70/30.[17] Triethanolamine stearate by itself gives viscous emulsions that have a sputtery, noisy discharge and produce pasty, unattractive foams. The triethanolamine salts of the coconut fatty acids produce attractive foams but coconut oil can be irritating to a sensitive skin. Therefore, they are used in shaving lathers only to such an extent that skin irritation is not a problem.[18] The combination of triple-pressed stearic acid and coconut fatty acid gives excellent foams with a minimum of skin irritation. A number of formulations for aerosol shaving lathers are given in References 7 and 17.

Other Anionic Agents. Other anionic surface-active agents that produce aerosol foams are listed in Reference 7. These include products such as aryl-alkyl sulfonates and alkyl sulfates that have been used from time to time as components of shampoos, etc. Sometimes the results have been rather disastrous because of container corrosion.[19] Adequate storage tests should always be carried out on any product that contains agents of this type.

Self-emulsifying esters of the fatty acids, like glycerol monostearate, are often used in aerosol formulations.[8] The esters themselves are nonionic but the self-emulsifying products contain a low concentration of an anionic agent, i.e., potassium stearate, which promotes dispersion of the nonionic ester. The self-emulsifying esters are therefore, mixtures of nonionic and anionic agents.

Molecular Complexes. The property of molecular complexes in stabilizing nonaerosol emulsions and foams was discussed in Chapters 16 and 17. These complexes have a considerable effect upon the properties of aerosol emulsions and foams. Because of their importance, they are discussed separately in Chapter 19.

NONIONIC SURFACE-ACTIVE AGENTS Nonionic surface-active agents are already used to a considerable extent in aerosol cosmetic and pharmaceutical products and there is little doubt that their use will increase. Some of the advantages of nonionic surface-active agents are their generally low toxicity, noncorrosive properties and effectiveness in the presence of electrolytes. In addition, many nonionic surface-active agents, such as the polyoxyethylene ethers, are stable to either acidic or alkaline conditions. The use of nonionic surface-active agents in household products of the sprayable foam type has already been discussed.

General. The choice of any nonionic surface-active agent depends upon the type of product and application involved. Nonionic agents that are useful in nonaerosol cosmetic and pharmaceutical applications often are satisfactory for the corresponding aerosol products. Sometimes all that is necessary to convert a nonaerosol product into an aerosol foam is to add propellant. The surface-active agents already present in the nonaerosol product may be sufficient to emulsify the propellant and stabilize the foam.

Nonionic agents already used in aerosol products include the polyoxyethylene sorbitan esters, polyoxyethylene fatty ethers, alkyl phenoxy polyethoxy ethanols, fatty acid esters, and the alkanolamides.[1]

Molecular Complexes. Polyoxyethylene fatty ethers have been shown to form complexes with long-chain alcohols and acids. These complexes can have a pronounced effect upon the properties of aerosol emulsions and foams and are discussed in detail in Chapter 19.

Propellants

HYDROCARBON PROPELLANTS   Hydrocarbon propellants, such as isobutane and propane, are used extensively in foam products and are the predominant propellants for aerosol shaving lathers. They are used at concentrations of about 3.5–5.0 wt%. This low concentration is adequate because of the low molecular weight of the hydrocarbons. The propellant used for shaving lathers is mostly isobutane, but it often contains small percentages of propane.[6] Because of its high pressure, propane is not used alone.

The hydrocarbons are satisfactory propellants except for their flammability.

FLUORINATED HYDROCARBON PROPELLANTS   A number of the fluorinated hydrocarbon propellants have properties that make them desirable for aerosol foams. "Freon" 12 and "Freon" 114, for example, not only give excellent foams but are nonflammable, low in toxicity, essentially odorless, and stable in aqueous systems under most conditions. Therefore, they are particularly useful for cosmetic and pharmaceutical products.

Comments on the individual fluorinated hydrocarbon propellants are as follows:

"Freon" 12. "Freon" 12 has too high a vapor pressure (70.2 psig at 70°F). to be used alone as the propellant in aqueous systems when no other

organic solvents are present and is usually combined with a vapor pressure depressant. The vapor pressure depressant can be a high boiling organic compound, such as mineral oil, or a higher boiling propellant, such as "Freon" 114. Combinations of "Freon" 12 with a solvent like mineral oil have a number of advantages. For example, the composition of the mixture can be adjusted to have a density about the same as that of the aqueous phase, which will reduce creaming. In addition, mineral oil is more easily emulsified than "Freon" 12 and gives more stable emulsions. As a result, combinations of "Freon" 12 and mineral oil generally have superior stability. About 5–7% "Freon" 12 is sufficient to give a good foam.

"Freon" 12/"Freon" 114 Blends. Blends of "Freon" 12 and "Freon" 114 are the most commonly used propellants of the fluorinated hydrocarbon group. Mixtures with vapor pressures varying from 12.9–70.2 psig at 70°F can be obtained by varying the ratio of "Freon" 12–"Freon" 114.[20] The "Freon" 12–"Freon" 114 (15/85) blend is often used for glass bottle products and the "Freon" 12–"Freon" 114 (40/60) blend for products in metal containers.

"Freon" 114. "Freon" 114 has a relatively high boiling point and a low vapor pressure and, therefore, has properties which make it useful for products that discharge as a liquid and subsequently expand into a foam. It is also utilized in foam products containing high concentrations of propellant (50%–90%). "Freon" 114 can give very stable foams and because of its stability and low order of toxicity is useful for cosmetic and pharmaceutical foams.

"Freon" 11. "Freon" 11 is not suitable for oil-in-water emulsions in metal containers because metal catalyzes the hydrolysis of "Freon" 11 and the hydrolysis products can cause corrosion.[2] "Freon" 11 may be stable in aqueous systems packaged in glass where contact with metal is at a minimum. However, because of the known instability of "Freon" 11 in metal containers in the presence of water, it has been erroneously assumed that it was equally unstable in aqueous products in glass. As a result, there have been relatively few attempts to package aqueous products in glass bottles with "Freon" 11. Another reason why "Freon" 11 is not particularly desirable for use in foams is that it is a good solvent and in foam products prolonged contact with the skin might occur and cause defatting.

Aqueous foams prepared with "Freon" 12–"Freon" 11 propellant blends usually are much less stiff and stable, and have a lower density than corresponding foams prepared with "Freon" 12–"Freon" 114 blends with about the same vapor pressure.[4,21]

Propellant 142b and Propellant 152a. These propellants generally produce foams that have a lower density and are less stiff and stable than "Freon" 12–"Freon" 114 blends.[4,21] In some products this is an advantage and these propellants are used for this reason. As a result of their lower molecular weights, less of the propellants is required than "Freon" 12–"Freon" 114 blends. Propellants 142b and 152a are stable to hydrolysis but both have the disadvantage of being flammable.

BLENDS OF NONFLAMMABLE AND FLAMMABLE PROPELLANTS. A distinct advantage using propellant blends with a density approximating that of the aqueous phase is that creaming is minimized, and thus little shaking is required to redisperse the propellant after the product has been stored. These blends are used to an increasing extent, and there is every reason to believe that their use will increase. The compositions of some typical propellant blends with a density of approximately 1.0 g/cc were listed previously in Table 18–2.

Effect of Propellant Concentration. The fact that increasing the propellant concentration in an aqueous foam product increases foam stiffness was first reported by Reed.[22] Reed observed that high propellant loadings with a shampoo concentrate resulted in stiff, dry elastic foams while low propellant concentrations gave soft and less resilient foams.

In a subsequent experiment, the stiffness and density of a series of foams with propellant concentrations varying from 5%–50% were determined. The aerosols were prepared with a typical aqueous shaving lather concentrate and "Freon" 12–"Freon" 114 (40/60) as the propellant.[21] The results are shown in Table 18–4.

TABLE 18-4   EFFECT OF VARIATION IN PROPELLANT CONCENTRATION

Propellant Concentration (wt %)	Stiffness (g)	Density (g/cc)
5	43	0.138
10	65	0.078
15	94	0.053
25	114	0.045
50	126	0.036

The data in Table 18–4 show how increasing the propellant concentration increases foam stiffness and decreases foam density. The same type of effect can be observed with any of the other liquified gas propellants, but the magnitude of the effect depends upon the particular propellant that is used.

Effect of Discharge. Foresman,[19] in a discussion of foam properties, pointed out that there is a drop in foam strength as increasing amounts of material are dispensed. The reason for this is as follows: As an aerosol product is discharged from the container, the volume of the liquid phase decreases, and the volume of the vapor phase increases correspondingly. In order to maintain a constant pressure in the aerosol container during discharge, it is necessary for some of the propellant in the liquid phase to vaporize and migrate to the vapor phase. This decreases the concentration in the liquid phase.

In spray products containing a relatively high proportion of propellant, the loss of propellant from the liquid phase to the vapor phase has little noticeable effect upon the spray characteristics as the product is discharged, since the percentage loss of propellant is small. However, in foam products, which are usually formulated with a minor proportion of propellant, the change in concentration of propellant in the liquid phase as the product is discharged is significant and causes a noticeable change in foam properties.

From the data in Table 18–4, which illustrates the change in foam stiffness and density with a change in propellant concentration, it can be predicted that as the concentration of propellant in the liquid phase decreases during product use, the foam stiffness will decrease and the density will increase. This was verified experimentally by repeatedly discharging a 12-oz. container of an aqueous shaving lather containing 10% "Freon" 12–"Freon" 114 (40/60) propellant into a crystallizing dish for density and stiffness determinations. After each discharge the container was weighed to determine the quantity of product that had been discharged. The process was repeated until the amount of product remaining in the container was insufficient for a stiffness measurement.

The relationship between discharge and foam stiffness and density for this particular system is illustrated in Figure 18–1. The extent to which discharge affects these properties in other systems will depend upon the quantity of propellant initially present and the type of propellant that is used.

Richman and Shangraw[23] calculated the loss of foam consistency with discharge of a typical shaving lather formulation. Two propellant systems were evaluated; a single propellant consisting of Propellant 12, and a propellant blend of 57% Propellant 12 and 43% Propellant 114. They reported that the change in foam consistency during product use could

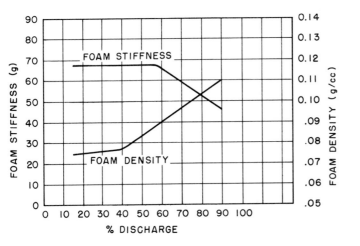

Figure 18-1   Variation of foam stiffness and density with percent of discharge.

be predicted by correlating the change in propellant concentration with known rheological and density data.

When mixtures of propellants are employed, fractionation takes place as a result of the difference in vapor pressure of the propellants. In one of the series of experiments reported by Richman and Shangraw, the composition of the Propellant 12/Propellant 114 blend changed from an initial ratio of 57/43–43/57 during discharge of 90% of the product. This caused a pressure drop from 64.5–59 psia. Therefore, there was a loss in pressure in the container as well as a decrease in propellant concentration during discharge.

If the aerosol foam product contains appreciable concentrations of an organic solvent, such as VM & P naphtha, mineral oil, isopropyl myristate or odorless mineral spirits, then the decrease in pressure with product discharge will be much more marked. The solvents have a relatively low vapor pressure and their concentration in the aqueous phase remains essentially unchanged as the product is discharged. Therefore, only the propellant decreases in concentration and the ratio of solvent to propellant in the liquid changes drastically.

WATER-IN-OIL EMULSIONS

In water-in-oil emulsions, the water is dispersed in the oil phase. The oil phase may consist only of propellant, or it may be composed of a mixture of propellants, solvents and various active ingredients.

One of the earliest references (1950) to aerosol water-in-oil emulsions is in a patent by Boe.[24] Two later publications that describe basic studies on water-in-oil emulsions with the "Freon" propellants appeared in 1956 and 1958.[25,26] Although the new system was suggested at this time for a variety of products, such as room deodorants, moth proofers, etc., there was little activity in this field until about 1962. Some of the problems in formulating room deodorants, hair sprays, snows, and insecticides were mentioned by Fowks[27] and a number of publications dealing with insecticides formulated as water-in-oil emulsions appeared shortly thereafter.[28–30] Aerosol insecticides have since become one of the major water-based products. Other water-in-oil emulsion products that appeared on the market included room deodorants and furniture polishes.

There are still many areas in which the use of aerosol water-in-oil emulsions has been limited. This is particularly true in the cosmetic and pharmaceutical fields. Undoubtedly, the advantages of water-in-oil emulsion systems for products in these fields will be recognized in the future and the systems will be utilized to a much greater extent.

Advantages

Some of the advantages of water-in-oil emulsions for aerosol products are as follows:

1. Nonfoaming sprays which range from soft, space sprays to wet, residual sprays can be obtained by proper formulating techniques. Water-in-oil systems therefore can be used for a variety of products.

2. Water-in-oil emulsions utilize substantial proportions of water, which is inexpensive, low in toxicity and an excellent solvent. There are many materials which are soluble in water but not in organic solvents, and the water-in-oil system provides a way of spraying these materials.

3. The sprays are warm and do not have an undesirable cooling effect when applied to the skin.

4. Corrosive characteristics are low. Even "Freon" 11 can be used as one of the propellants in the water-in-oil system.

Surface-Active Agents

The surface-active agent for this system must be oil soluble. Probably the most widely used agent is polyglycerol oleate.[25,26,30,31] Other agents found to be effective in this system were sorbitan monolaurate and polyethylene glycol 400 ditriricinoleate. The suppliers of surface-active agents may be able to suggest additional agents.

Polyglycerol oleate is used at concentrations from about 0.5%–4.0% based upon the entire aerosol. Generally as little as possible is used and the most effective concentration for any product can only be determined by experimentation. The normal concentration is 1.0–2.0%.

Propellants

Hydrocarbons and Hydrocarbon–Fluorocarbon Blends. Hydrocarbons, or blends of hydrocarbons with Propellant 12 are used extensively as the propellants in commercial water-in-oil emulsion products. Moore and Baker[28] used an isobutane–propane (87/13) mixture for insecticides at concentrations from 30–35%. Glessner[29] mentions that either hydrocarbons or Propellant 12–hydrocarbon blends are satisfactory for use with water-based insecticides. Jones and Weaning[30] used "Freon" 12–butane (70/30) or (65/35) mixtures for water-based insecticides with a propellant concentration from about 39%–44% while Scott and Clayton[31] suggest Propellant 12–isobutane (30/70) or (25/75) mixtures at concentrations of 45–48% as a starting point in formulating aqueous products.

The main disadvantage of hydrocarbons as propellants for water-in-oil emulsion products is their flammability. The major problem of flammability occurs mostly in the handling and storage of the propellants in the aerosol plants. However, the proportion of propellant in water-in-oil emulsions is fairly high and therefore these products present a greater hazard to the consumer than foam products which contain a much smaller concentration of propellant.

Fluorinated Hydrocarbon Propellants. Although the hydrocarbon propellants are used to a considerable extent in water-based products, nevertheless, little basic information on water-in-oil systems containing these propellants has been published. Since much more fundamental information on the use of "Freon" propellants in water-in-oil systems is available, the discussion of these systems will of necessity be centered around these propellants.

EMULSION STABILITY WITH THE INDIVIDUAL PROPELLANTS   The stability of water-in-oil emulsions prepared with "Freon" 12, "Freon" 114, "Freon" 11, and "Freon" 113 is given in Table 18–5.[26] ("Freon" 11 and "Freon" 113 are considered as propellants to simplify the discussion in the present case). The emulsions were prepared at four different propellant–water ratios using about 4% polyglycerol oleate ("Emcol" 14, Witco Chemical Corp., or "Soleonic" PGO, Hodag Chemical Corp.) as the emulsifying agent. Emulsion stability was determined by noting the time after the bottles had been shaken before phase separation was observed.

The relative stability of the emulsions prepared with the various propellants corresponds roughly to the solvent properties of the propellants, as

TABLE 18-5   EMULSION STABILITY OF WATER-IN-OIL EMULSIONS
WITH "FREON" PROPELLANTS

| | Kauri-Butanol | Emulsion Stability | | | |
| | Value of | Propellant–Water Ratio (wt %) | | | |
Propellant	Propellants	90/10	80/20	60/40	40/60
"Freon" 12	18	< 1 min	< 1 min	< 1 min	1–5 min
"Freon" 114	12	< 1 min	< 1 min	< 1 min	15–30 min
"Freon" 11	60	15–30 min	15–30 min	> 1 hr	> 1 hr
"Freon" 113	32	1–5 min	1–5 min	> 1 hr	> 1 hr

indicated by their Kauri–Butanol values (see Chapter 10). "Freon" 12 and
"Freon" 114 are comparatively poor solvents, and evidently the solubility
of the polyglycerol oleate in these propellants is too low to form a good
emulsion. Water-in-oil emulsions prepared with "Freon" 12, "Freon" 114
or blends of the two have such poor stability when no other solvents are
present that they are impractical. However, by using another compound
with better solvent properties, such as "Freon" 11, "Freon" 113, or an
organic solvent, in conjunction with "Freon" 12 or "Freon" 114, emulsions
with satisfactory stability can be obtained.

"FREON" 12–"FREON" 11–WATER EMULSIONS   Very good emulsions can
be obtained by combining "Freon" 11, with its relatively good solvent
powers, with "Freon" 12. The spray characteristics of the emulsions may
be varied from fine to very wet by changing either the ratio of "Freon" 12
to "Freon" 11 or the ratio of the propellant to the water.
Variation in the "Freon" 12–"Freon" 11 Ratio.   The effect of varying the

TABLE 18-6   VARIATION IN "FREON" 12–"FREON" 11 RATIO

(60/40 Propellant–Water Ratio)		
"Freon" 12–"Freon" 11 Ratio (wt %)	Emulsion Stability	Spray Characteristics
100/0	< 1 min	--------
70/30	30–60 min	Fine
50/50	30–60 min	Medium fine
30/70	30–60 min	Soft medium
0/100	> 1 hr	No spray

"Freon" 12–"Freon" 11 ratio upon emulsion stability and spray characteristics (standard valves and actuators) at a propellant–water ratio of 60/40
is shown in Table 18–6.[26] The concentration of polyglycerol oleate was
about 4%. The data show that emulsion stability increased as the proportion of "Freon" 11 in the propellant increased.

The spray characteristics and the emulsion stability may be affected by
the addition of other components to the aerosol.

Variation in the Propellant–Water Ratio.   The effect of varying the propellant–water ratio is shown in Table 18–7.[26] The propellant in this case
was "Freon" 12–"Freon" 11 (30/70).

TABLE 18-7   VARIATION IN PROPELLANT–WATER RATIO

("Freon" 12–"Freon" 11 (30/70) Propellant)

Propellant–Water Ratio (wt %)	Emulsion Stability	Spray Characteristics
90/10	1–5 min	Very fine
80/20	5–15 min	Fine
60/40	30–60 min	Medium
40/60	> 1 hr	Very coarse— almost foams
20/80	< 1 min	Stream-foams

The increase in emulsion stability that occurs with an increase in the
concentration of water probably is the result of an increase in viscosity.
This is discussed later. For a basic aerosol room deodorant system, a combination of 68% "Freon" 12–"Freon" 11 (30/70) 2% polyglycerol oleate,
and 30% water was found to be satisfactory.

Water-in-oil emulsions with "Freon" 12–"Freon" 11 propellant alone
may exhibit slight bubbling when sprayed on a surface. This probably
results from vaporization of the "Freon" 11 that is retained in the emulsion. For some uses this may be undesirable. It can usually be eliminated
by incorporating a small amount of an auxiliary solvent into the formulation.

Water-in-Oil Emulsions with Auxiliary Solvents

Hydrocarbon and Hydrocarbon–Fluorocarbon Blends.   Many of the publications on water-based products are concerned primarily with aerosol insecticides. In order to obtain the requisite particle size in the spray, the

presence of an organic solvent is necessary.[28] The solvents that were used include "Chlorothene" NU,[28,29] methylene chloride,[28] petroleum distillate,[28,29] and kerosene and isoparaffinic oils.[30] Scott and Clayton[31] evaluated a series of water-in-oil emulsions prepared with a variety of organic solvents, including dipropylene glycol, hexylene glycol, isopropyl alcohol, mineral oil and triethylene glycol. Mineral oil was the preferred solvent on the basis of the type of spray that was obtained.

Fluorinated Hydrocarbon Propellants. Some of the auxiliary solvents, which have been used in water-in-oil emulsions with fluorocarbon propellants, are mineral oil, isopropyl myristate, cottonseed oil, odorless mineral spirits, and VM & P naphtha. It must be emphasized that in products formulated to have a fine spray, the use of high boiling organic solvents should be avoided if there is any chance of inhalation of the small droplets in the spray. High boiling, water insoluble compounds such as mineral oil, can cause lipoid pneumonia if a sufficient quantity in the form of small droplets is inhaled.

Some of the advantages of using auxiliary solvents are as follows:

a. Added solvents generally increase emulsion stability. This is particularly noticeable in emulsions with "Freon" 12 or "Freon" 12/"Freon" 114 blends as the propellant. In fact, in these cases, the presence of an auxiliary solvent is almost a necessity in order to have sufficient emulsion stability for the product to be practical.

In addition, the composition of the propellant–solvent combinations can be adjusted to have a density about the same as that of the aqueous phase. This may increase the emulsion stability even more. Often, "Freon" 12 is used alone as the propellant because of the vapor pressure depressant effect of the solvents.

The density adjustment is one of the reasons why solvents with a density less than that of water are preferred to solvents such as methyl chloroform or methylene chloride that have a density greater than water.

b. Some "Freon" 12–"Freon" 11 water-in-oil emulsions tend to bubble when sprayed on a surface. The addition of a small quantity of an auxiliary solvent may minimize or eliminate the bubbling.

c. The efficiency of emulsification may be increased by the use of solvents. For example, the solvent may be used to dissolve the water insoluble surface-active agent. A water-in-oil emulsion is then prepared by adding water to the solution of surface-active agent and this emulsion is loaded into the container. The addition of the propellant merely extends the continuous oil phase.

The following properties of "Freon" 12–odorless mineral spirits–water emulsions are typical of those containing auxiliary solvents. These give nonfoaming sprays. The spray characteristics may be varied from fine to very coarse by changing the ratio of "Freon" 12 to odorless mineral spirits (OMS) as shown in Table 18–8.[26]

TABLE 18-8 "FREON" 12-ODORLESS MINERAL SPIRITS-WATER EMULSIONS

"Freon" 12–OMS–Water Ratio *(wt %)*	*Emulsion* *Stability*	*Spray* *Characteristics*
60/20/20	1–5 min	Fine
40/20/40	> 1 hr	Medium
20/20/60	> 1 hr	Wet

The emulsions were prepared with about 4% polyglycerol oleate as the emulsifying agent. The concentration of odorless mineral spirits was kept constant while the "Freon" 12 concentration was varied.

Similar data for water-in-oil emulsions containing combinations of odorless mineral spirits with "Freon" 12–"Freon" 114 or "Freon" 12–"Freon" 11 blends are given in Reference 26.

Effect of Ethyl Alcohol

Some water-in-oil emulsions can tolerate a fairly high concentration of alcohol without an appreciable effect upon stability. For example, concentrations of ethyl alcohol in the aqueous phase up to 30% had no effect upon the emulsion stability of "Freon" 12–"Freon" 11–odorless mineral spirits–water systems emulsified with polyglycerol oleate. Higher concentrations caused a definite decrease in stability. The effect of ethyl alcohol upon "Freon" 12–"Freon" 11–water systems was not determined.[26]

Effect of Sodium Chloride

Concentrations of sodium chloride of 10% and 25% in the aqueous phase decreased emulsion stability in systems formulated with odorless mineral spirits. In most cases, however, the emulsion stability was sufficient for practical purposes.[26]

Viscosity of Water-in-Oil Emulsions

One of the most noticeable factors that affects the viscosity of water-in-oil emulsions is the propellant–water ratio. As the concentration of the aqueous phase is increased, there is an accompanying increase in the viscosity of the emulsion.[26] This increases the stability of the emulsions because of the increased difficulty of movement of the dispersed water droplets.

When the concentration of the aqueous phase is increased to 60–80%, inversion of the emulsion from water-in-oil to oil-in-water generally occurs. Inversion is accompanied by a sharp decrease in emulsion viscosity. Electrical conductivity measurements now indicate the emulsion to be oil-in-water rather than water-in-oil, and creaming occurs from the top rather than from the bottom. Other indications of the phase change are the foaming that occurs in the bottles when they are shaken in contrast to the lack of foaming when water-in-oil emulsions are shaken. In addition, the emulsion now produces a stream that foams on contact where the previous water-in-oil emulsion gave a nonfoaming spray. Thus, the change from water-in-oil emulsion to an oil-in-water emulsion is unmistakable.

These results appear to be similar to those obtained by Sherman[32] who noted that the viscosities of water-in-oil emulsions prepared with distilled water in mineral oil reached a maximum value between 77% and 82% water. At higher concentrations of water, inversion of the emulsion occurred. The specific concentration at which inversion occurred was a function of the amount of emulsifying agent present.

Occasionally it was observed that of two, supposedly duplicate samples prepared with 20% "Freon" propellant and 80% water, one was a very viscous water-in-oil emulsion while the other was the less viscous oil-in-water type. The emulsion stability of the former was greater than an hour with creaming starting from the bottom while the emulsion stability of the latter was less than a minute with creaming starting from the top. This was probably a case of the formation of dual emulsions (emulsions with the same compositions but having opposite phase types where the system is such that the type of emulsion obtained depends upon how the bottle is shaken. This phenomenon was first observed by Ostwald[33] and later confirmed by Cheesman and King[34]).

Storage Stability

Emulsions with Hydrocarbons or Hydrocarbon-Fluorocarbon Blends. Aerosol insecticides had adequate storage stability when properly formulated, according to the data reported by Glessner[29] and by Jones and Weaning.[30] The insecticide itself had to be chosen with care since some insecticides are affected by water and will produce chlorides or acids which in turn cause corrosion.

The fact that, since this initial work, millions of cans of room deodorants and space insecticides have been packaged as water-in-oil emulsions and successfully marketed shows that basically the water-in-oil emulsions formulated with hydrocarbons or hydrocarbon–fluorocarbon blends have sufficient storage stability.

Emulsions with the Fluorocarbon Propellants. The storage stability of basic fluorocarbon water-in-oil systems prepared with polyglycerol oleate appears to be excellent. The following emulsions were evaluated:[26]
1. "Freon" 12–kerosene–water—60/20/20 ratio
2. "Freon" 12–"Freon" 114 (15/85)–mineral oil–water—60/20/20 ratio
3. "Freon" 12–"Freon" 11 (30/70)–mineral oil–water—60/20/20 ratio
4. "Freon" 12–"Freon" 11 (30/70)–water—80/20 and 60/40 ratios

The emulsions were stored under the following conditions:
1. Crown lacquer-lined tinplate containers equipped with Precision nylon–brass valves.
       Aging Periods: 2 mo at 130°F
                              1 yr at room temperature
2. Continental tinplate containers with Precision nylon–brass valves.
       Aging Periods: 6 mo at 130°F
                              1 yr at room temperature

The first three emulsion systems produced no observable corrosion in any of the containers under any of the storage conditions. The fourth system, the "Freon" 12–"Freon" 11 (30/70)/water emulsions, did not cause any corrosion in the lacquer-lined containers under any conditions or in the tinplate containers at room temperature. After 6 mo at 100°F, slight detinning on the can shoulders was observed.

The general lack of corrosion observed with the water-in-oil emulsions is very encouraging. It should be recognized, however, that the presence of active ingredients may introduce corrosive properties into the system. Therefore, any practical formulation should be shelf tested.

The lack of corrosion of water-in-oil emulsions with "Freon" 11 was quite a contrast to the excessive corrosion that occurs in metal containers with oil-in-water emulsions with "Freon" 11. Boe[24] has suggested that water-in-oil systems might be expected to show less corrosion than oil-in-water emulsions because in the former the dispersed water has less tendency to come into contact with the container and valve.

It certainly is true that in a water-in-oil emulsion the dispersed water droplets are surrounded by an interfacial film of surface-active agent which would minimize contact of the water droplets with metal; and it has been shown that it is metal that catalyzes the decomposition of "Freon" 11 by water.[2] However, since it is the interfacial film which prevents contact with metal, the surface active agent itself is the most important factor. The interfacial film surrounding the droplets must have sufficient coherence and strength so that coalescence of the droplets does not occur during storage, even if creaming occurs. Otherwise, a separate water layer will form after coalescence and contact with the metal will then take place. This is borne out by the fact that some water-in-oil emulsions prepared with agents other than polyglycerol oleate were found to cause corrosion.

Valves

The choice of a valve and actuator for a water-based product is based upon such factors as the particular application involved and the type of spray desired. Some of the variables in valves and actuators that affect the spray characteristics of aqueous products have been discussed by Pizzurro.[35]

Vapor tap valves with capillary dip tubes have been recommended for aerosol insecticides.[28] The valves should be equipped with mechanical breakup or large chamber actuators. Similar combinations have been used for room deodorants.

PREPARATION OF AEROSOL EMULSIONS

An emulsion can be considered to consist simply of water, oil, and an emulsifying agent, and it would seem a relatively simple matter to combine these ingredients. However, the way in which they are combined can have a considerable effect upon the properties of the resulting emulsion. No basic studies on the preparation of aerosol emulsions have appeared in the literature. Therefore, some of the information already available on the techniques of preparing conventional, nonaerosol emulsions is reviewed at this point.

General Discussion

Becher, in his discusion of the techniques of emulsification, lists the following four standard procedures for combining the oil phase, aqueous phase, and emulsifying agent to give an emulsion.[13]

Agent-in-Water Method. The emulsifying agent is dissolved in the water and the oil is added to the aqueous phase with vigorous agitation. This forms an oil-in-water emulsion. At very high concentrations of the oil phase, inversion of the oil-in-water emulsion to a water-in-oil emulsion may occur.

This is one of the simplest and easiest procedures, but is the least preferred of the four methods because it tends to give coarse emulsions with a wide range of droplet sizes. Subsequent passage of the coarse emulsion through a colloid mill or homogenizer decreases the droplet size and produces a more stable emulsion.

Agent-in-Oil Method.   The emulsifying agent is dissolved or dispersed in the oil phase which is then added to the water with agitation thus forming an oil-in-water emulsion. If the order of mixing is reversed, with the water being added to the oil phase, a water-in-oil emulsion may form initially. Continued addition of water may cause inversion to an oil-in-water emulsion.

The agent-in-oil method, also called the Continental Method, is one of the preferred procedures for preparing emulsions. According to de Navarre,[36] it usually gives good emulsions with a uniform distribution of droplet sizes. This is particularly true if the inversion procedure is used.

Nascent Soap (in situ) Method.   This method is used in the preparation of emulsions stabilized with salts of the fatty acids and is particularly applicable to aerosol systems, many of which are formulated with this type of emulsifying agent. In this method, the fatty acid is dissolved in the oil phase and the base in the aqueous phase. When the two phases are mixed, the salt of the fatty acid is formed immediately at the oil–water interface. This is the preferred method when salts of the fatty acids are used as the emulsifying agent.

Alternate Addition Method.   This is also called the English Method and involves adding portions of the aqueous phase and oil phase alternately to the emulsifying agent. This procedure is reported to be particularly advantageous for the preparation of food emulsions containing vegetable oil.

The droplet size of an emulsion usually can be reduced by passage through a colloid mill or homogenizer. However, a discussion of this type of equipment is beyond the scope of this chapter. Reference 13 is suggested for those who are interested in a detailed treatment of this subject.

Aerosol Emulsions

The properties of the liquefied gas propellants cause some rather unique problems in the preparation of aerosol emulsions and limit the methods that can be used. The liquefied gas propellants must be loaded into the containers either under pressure or at temperatures below the boiling points of the propellants. The boiling points of most propellants are considerably below the freezing point of water; therefore, the cold fill procedure generally is not satisfactory because most of the propellants would flash off and propellant loss would be excessive. Practically all aerosol emulsion products are packaged at room temperature and the propellant is pressure loaded.

Most commercial loaders are not equipped to disperse or dissolve the emulsifying agent in the propellant before it is loaded into the containers and consequently such procedures as the agent-in-oil method are not practical when the propellants are the only constituents of the oil phase.

Oil-in-Water Emulsions. The efficiency of emulsification depends to a considerable extent upon whether oil soluble materials other than the propellants are present. Both cases are considered in the following discussion.

OTHER OIL-SOLUBLE MATERIALS PRESENT   When the aerosol contains oil soluble materials other than the liquefied gas propellants, such as mineral oils, mineral spirits, esters, etc., then the concentrate can be treated as a conventional emulsion and prepared using any one of the four methods previously outlined. The resulting emulsion concentrate can be homogenized or passed through a colloid mill if necessary. The emulsion concentrate is loaded into the aerosol containers, and the propellant is pressure filled. The concentrate must have sufficient emulsion stability so that creaming or coalescence does not occur before it is added to the aerosol containers.

When triethanolamine salts of the fatty acids are used as the emulsifying agents, the nascent soap (in situ) method is utilized. The oil soluble materials and the fatty acids are mixed together and heated to about 130–175°F in order to melt the fatty acids and dissolve them in the oil phase. The aqueous phase containing the triethanolamine and other water soluble ingredients is heated separately to about the same temperature as that of the oil phase and added slowly with stirring to the oil phase. After addition is complete, the mixture is allowed to cool to room temperature with stirring.

The triethanolamine salt of the fatty acid is formed at the oil–water interface regardless of whether the oil phase is added to the aqueous phase or the latter is added to the oil phase. However, there is a reason for adding the aqueous phase to the oil phase instead of the reverse. When the aqueous phase is added to the oil phase, initially, at least, there is an excess of fatty acid with respect to the triethanolamine salt that is formed. This provides the conditions for the formation of a molecular complex between the free fatty acid and the triethanolamine salt. Molecular complexes from triethanolamine salts and free fatty acids have been shown to enhance emulsion and foam stability of aerosol products (see Chapter 19). If the oil phase is added to the aqueous phase, the triethanolamine base initially is in excess of the salt that is formed and this discourages complex formation.

When nonionic agents or combinations of the agents with other compounds that form complexes, such as the fatty alcohols or acids, are used, a variety of methods can be utilized to prepare the concentrate. For example, the nonionic surface active agent, fatty alcohol or acid, and other oil soluble ingredients can be heated to about 130–175°F to melt the solid materials. The aqueous phase, heated to about the same temperature in a separate container, is added slowly with stirring to the oil phase. An alternate procedure is to heat the mixture of oil soluble materials along with any fatty alcohols or acids in one container and the mixture of nonionic surface-active agent and water in a separate container and mix the two solutions slowly with stirring. The simplest procedure of any is to weigh all of the ingredients into one container and heat the mixture to 150–175°F while stirring.

Unfortunately there are no experimental data available to indicate which, if any, of the above procedures gives the best emulsion. However, by analogy with the nascent soap procedure, it might be expected that adding the hot aqueous phase containing the nonionic agent to the oil phase containing the fatty alcohol or acid might be the preferred method.

OTHER OIL SOLUBLE MATERIALS ABSENT   These systems consist of a mixture of propellant, emulsifying agent and an aqueous phase. Since the propellant is added last and there is no other oil phase, the procedure for preparing the aerosol emulsion is, of necessity, the agent-in-water method. The only step that can be varied is the method of mixing the emulsifier and the aqueous phase. As yet, there is no information available to indicate whether the method of mixing the emulsifier and aqueous phase has any effect upon the aerosol emulsion obtained after the propellant is added.

When the triethanolamine salts of the fatty acids are used, the hot aqueous solution of triethanolamine is usually added with stirring to the fatty acid, which has been heated to a temperature of about 130–175°F. Another procedure is to weigh all the ingredients into the same container and heat the mixture with stirring until the fatty acids melt and form the triethanolamine salts. Whether either of these procedures gives better aerosol emulsions is not known.

When nonionic surface-active agents are used, there are a number of methods of mixing the emulsifier and the aqueous phase. The components can be heated in separate containers and mixed while hot, or they can be combined initially in the same container and heated together. Any new aerosol product should be prepared by several different methods to determine if any specific procedure gives a superior product.

Water-in-Oil Emulsions.   Water-in-oil emulsions generally contain either an auxiliary oil soluble solvent, such as mineral oil, isopropyl myristate, odorless mineral spirits, etc., or a comparatively high boiling propellant like "Freon" 11. In this system the oil-soluble emulsifying agent is dissolved in the solvent, the "Freon" 11, or the mixture of solvent and "Freon" 11, and the aqueous phase is added to the organic phase with stirring. This gives a concentrate which is a water-in-oil emulsion. The concentrate is loaded into containers and the propellant pressure filled. The added propellant merely extends the continuous oil phase. The concentrate, which is itself an emulsion, must have sufficient stability so that neither creaming nor coalescence occurs before it is loaded into the aerosol containers.

Some very crude water-in-oil emulsion products have been marketed which consist primarily of a hydrocarbon propellant, an emulsifying agent, and an aqueous phase. These products are prepared by dispersing the emulsifying agent in the aqueous phase, loading the resulting concentrate in the containers, and pressure loading the propellant.

REFERENCES

1. M. J. Root, *Amer. Perfumer Aromatics* **71**, 63 (1958).
2. P. A. Sanders, *Soap Chem. Specialties* **41**, 117 (1965).
3. P. A. Sanders, *Soap Chem. Specialties* **39**, 63 (1963).
4. P. A. Sanders, *J. Soc. Cosmetic Chemists* **17**, 801 (1966).
5. P. A. Sanders, *Soap Chem. Specialties* **43**, 68, 70, (1967).
6. J. W. Hart and H. C. Cook, *Amer. Perfumer Aromatics,* **77**, 49, (1962).
7. J. G. Spitzer, I. Reich, and N. Fine, U. S. Patent 2,655,480 (1953).
8. H. R. Shepherd, "Aerosol Cosmetics" in E. Sagarin, "Cosmetics: Science and Technology," Interscience Publishers, Inc., New York, N.Y., 1957.
9. A. Herzka, and J. Pickthall, "Pressurized Packaging," Academic Press, Inc., 1958.
10. H. R. Shepherd, "Aerosols: Science and Technology," Interscience Publishers, Inc., New York, N.Y., 1961.
11. A. Herzka, International Encyclopaedia of Pressurized Packaging (Aerosols), Permagon Press, New York, N.Y. 1966.
12. A. M. Schwartz, J. W. Perry, and J. Berch, "Surface Active Agents and Detergents," Vol. 2, Interscience Publishers, Inc., New York, N.Y., 1958.
13. P. Becher, "Emulsions: Theory and Practice," Reinhold Publishing Corp., New York, N.Y., 1965.
14. McCutcheon's Detergents and Emulsifiers, D & E., 1968 Annual, John W. McCutcheon, Inc., Morristown, N. J.
15. S. J. Strianse, "Hand Creams and Lotions," in E. Sagarin, "Cosmetics: Science and Techology," Interscience Publishers, Inc., New York, N.Y., 1957.
16. Drew Chemical Corp., New York.
17. P. Carter, and H. M. Truax, *Proc. Sci. Sect. Toilet Goods Assoc.* **35**, 37 (1961).

18. H. H. Guest, "Shaving Soaps and Creams," in E. Sagarin, "Cosmetics: Science and Technology," Interscience Publishers, Inc., New York, N.Y., 1957.
19. R. A. Foresman, Proc. 39th Ann. Meeting, CSMA, 32, (December 1952).
20. "Vapor Pressure and Liquid Density of 'Freon' Propellants," "Freon" Aerosol Report, FA–22.
21. P. A. Sanders, Aerosol Age 8, 33 (1963).
22. F. T. Reed, Proc. 39th Ann. Meeting, CSMA, 30 (December 1952).
23. M. D. Richman and R. F. Shangraw, Aerosol Age 11, July–November, 1966).
24. C. F. Boe, U. S. Patent 2,524,590.
25. "Aerosol Emulsions with 'Freon' Propellants," "Freon" Aerosol Report FA–21.
26. P. A. Sanders, J. Soc. Cosmetic Chemists 9, 274 (1958).
27. M. E. Fowks, Proc. 49th Ann. Meeting, CSMA, 42 (December 1962). Aerosol Age, 8, 25 (1963).
28. C. J. Baker and J. B. Moore, Proc. 50th Mid-Year Meeting, CSMA, 63 (May 1964). Aerosol Age 9, 28 (1964).
29. A. S. Glessner, Proc. 50th Mid-Year Meeting, CSMA, 58 (May 1964). Aerosol Age 9, 98 (1964).
30. C. D. G. Jones and A. J. S. Weaning, Aerosol Age 11, 21 (1966).
31. M. E. Clayton and R. J. Scott, Proc. 50th Mid-Year Meeting, CSMA, 68 (May 1964).
32. P. Sherman, J. Soc. Chem. Ind. 69, 570, 574 (1950).
33. W. O. Ostwald, Kolloid-Z, 6, 103 (1910).
34. D. F. Cheesman and A. King, Trans. Faraday Soc. 34, 594 (1938).
35. J. C. Pizzurro, Soap Chem. Specialties 40, 163 (1963).
36. M. G. de Navarre, "Chemistry and Manufacture of Cosmetics," p. 191, D. van Nostrand Co., New York, N.Y., 1941.

19

MOLECULAR COMPLEXES IN AEROSOL EMULSIONS AND FOAMS

The concept of molecular complexes formed between water-soluble surface-active agents and long-chain polar compounds, and the way in which they can alter the properties of conventional, nonaerosol colloidal systems, was discussed in Chapters 16 and 17. One of the most common examples of a molecular complex is that formed from sodium lauryl sulfate and lauryl alcohol.

The evidence for the existence of stoichiometric molecular complexes (also called association complexes) is both indirect and direct. The presence of the complexes had been suspected and postulated for many years on the basis of the properties of monolayers of long-chain alcohols spread on the surface of solutions of surface-active agents and upon the properties of emulsions and foams containing mixed emulsifiers. Subsequently, direct experimental proof for the existence of the complexes was obtained by the isolation and analysis of many complexes which showed that these complexes were chemical entities with definite stoichiometric ratios of the surfactants and long-chain polar compounds. This led to the conclusion that in many stable emulsion systems, the interfacial films, whether monomolecular or polymolecular, contained complexes resulting from the interaction between the water-soluble surface-active agents and the oil-soluble components.

In recent years, the investigation of molecular complexes has been extended to aerosol systems, and it was found that certain combinations of water-soluble ionic surfactants and long-chain alcohols, which were known from previous investigations to form molecular complexes in nonaerosol systems, also had a marked influence on the properties of aerosol emulsions and foams. Emulsion and foam stability, and viscosity, were increased and the rate of foam drainage was decreased. These effects were analogous to those that had been observed previously in nonaerosol systems, and on this basis it was concluded that complexes between the water-soluble and oil-soluble components were present in the interfacial films of aerosol emulsions and foams.

There is no direct experimental evidence to indicate that stoichiometric molecular complexes are formed in nonaerosol systems between the nonionic polyoxyethylene fatty ethers and fatty alcohols or acids. However, in aerosol systems, the effect of combinations of the polyoxyethylene fatty ethers and fatty alcohols or acids upon emulsions and foam properties was similar to that produced by molecular complexes in the ionic surfactant systems. Judging by this, it seemed reasonable to assume that complexes between the polyoxyethylene fatty ethers and fatty alcohols and acids were present in the interfacial films in aerosol emulsions and foams. Whether or not these complexes have a stoichiometric or indefinite composition in the interfacial films is not known.

The use of molecular complexes provides a powerful method for controlling and altering the properties of aerosol systems so that they can be tailored to meet the requirements of a variety of applications. Thus, by the proper choice of the surface-active agent and long-chain alcohol or acid, aerosol foams can be obtained which have many different properties, such as: (1) Immediate wetting after discharge followed by collapse; (2) immediately wetting after discharge but retention of foam structure; and (3) no wetting or collapse for extended periods.

In addition, because complexes often are so much more effective than the surface-active agent by itself, the use of complexes may permit a marked reduction in the concentration of the surface-active agent without any decrease in the quality of the product. Many aerosol formulations contain combinations of compounds that form molecular complexes but the existence of the complexes and their effect upon the properties of the product have not been generally recognized. For all these reasons, molecular complexes are considered to be sufficiently important to be discussed as a separate subject in the present chapter.

The results of molecular complex studies in aerosol systems have been reported in a series of three papers.[1-3] The material presented in this chapter is essentially a condensation of the data in these papers.

MOLECULAR COMPLEXES WITH ANIONIC AGENTS

The triethanolamine salts of the fatty acids are among the most commonly used surface-active agents in aerosol emulsions and foams and for that reason were investigated first.

Triethanolamine Fatty-Acid Salt—Fatty-Alcohol Complexes

The investigation of complex formation between the triethanolamine salts of the fatty acids and long-chain alcohols is reported in Reference 1. The triethanolamine salts were prepared from the following fatty acids:

Lauric acid	Palmitic acid
Myristic acid	Stearic acid

Each of the triethanolamine salts was evaluated by itself and in combination with the following alcohols in an aerosol system consisting of 90% of an aqueous concentrate and 10% "Freon" 12–"Freon" 114 (40/60) propellant:

Lauryl alcohol	Stearyl alcohol
Myristyl alcohol	Oleyl alcohol
Cetyl alcohol	Cholesterol

Cholesterol was included in the series because it had been shown by Schulman to form complexes with the sodium alkyl sulfates.[4] The concentration of triethanolamine laurate and triethanolamine myristate in the aqueous phase was 0.10 M (about 3.5% on a weight basis). The addition of sufficient alcohol to produce a 1:1-M ratio with the surface-active agent increased the weight percent to about 4.5%. Triethanolamine palmitate and stearate were normally used at concentrations of 0.025 M in the aqueous phase (about 1% on a weight basis). The addition of the alcohol increased the total concentration of the complex to about 2%, which is fairly low for aerosol emulsion systems.

General Discussion. Selected examples that illustrate the effect of the triethanolamine salt–fatty alcohol complexes upon various properties of aerosol emulsions and foams are given in the following sections. Some general comments are listed first.

a. The molecular complexes usually had an effect on the properties of both the aerosol emulsions and the corresponding foams. This indicated that complexes were formed initially at the propellant–water interface in the aerosol emulsions and subsequently influenced the properties of the foam after the emulsion was discharged.

b. Microscopic examination of the foams showed that the presence of complexes usually decreased the bubble size of the foams. It is tempting to assume that this indicates that the droplet size of the dispersed propellant in the emulsion was also smaller in the systems with complexes but there is no direct experimental evidence to support this assumption.

c. The effect of the complexes in decreasing foam drainage was quite marked and was consistent with the results reported previously in nonaerosol systems. The decrease in foam drainage is considered to result from an increase in surface viscosity. There were many instances in which an obvious decrease in foam drainage was accompanied by an increase in foam stiffness, which is a function of foam viscosity. In other cases, however, a decrease in foam drainage occurred without any apparent increase in foam stiffness. It seems likely that the foam stiffness measurements are less sensitive than the foam drainage tests and do not indicate an increase in foam viscosity unless the over-all effect upon viscosity is fairly large.

d. The extent to which complex formation affected the properties of the aerosol systems depended upon such factors as the type and concentration of the long-chain alcohol, the triethanolamine salt, and the propellant. Specific examples of the effect of these variables are given in the following sections.

Variation in Alcohols. The effectiveness of any specific alcohol was a function of the particular triethanolamine salt with which it was complexed. It seemed to be a rather general rule throughout the triethanolamine salt–fatty alcohol series that the straight-chain alcohols with about the same chain length as that of the fatty acid salt usually had the most noticeable effect upon emulsion and foam properties. Thus, lauryl and myristyl alcohols were more effective in triethanolamine laurate systems than stearyl alcohol while the reverse was true in the triethanolamine stearate systems. Oleyl alcohol generally had much less effect than the straight-chain alcohols, indicating weak complex formation. This is due to the steric hindrance caused by the double bond in oleyl alcohol. This steric factor was discussed in Chapter 16.

TABLE 19-1   VARIATION IN ALCOHOLS—TRIETHANOLAMINE
LAURATE[a] SYSTEM

| | | Foam Properties | |
| | | % Drainage | Stiffness |
Alcohol[b]	Emulsion Stability[c]	(30 min)	(g)
None	< 1 min	82	8
Lauryl	> 5 hr	2	20
Myristyl	> 5 hr	0	33
Cetyl	> 5 hr	0	18
Stearyl	> 5 hr	67	10
Oleyl	15–30 min	2	12

[a] Triethanolamine laurate concentration = 0.10 M.
[b] Soap–alcohol ratio (molar) = 1:1.
[c] Time to observable phase separation after the bottles had been shaken.

The effect of various alcohols in the triethanolamine laurate system is illustrated in Table 19–1. Similar data for triethanolamine myristate, palmitate and stearate systems are given in Reference 1.

The change in emulsion stability and foam drainage upon addition of the alcohols was very pronounced. Foam drainage results correlated well with wetting properties, and this was generally true in the other systems. The foams from triethanolamine laurate alone or those with stearyl alcohol present had the highest drainage rates and wetted paper almost immediately after discharge. Foams with lauryl, myristyl, or cetyl alcohols did not wet for over an hour.

Foam stability, as judged by foam collapse, was increased by the presence of the alcohols. The foams from triethanolamine laurate alone or those containing stearyl alcohol started to collapse within 30 min after discharge. Foams with lauryl, myristyl, or cetyl alcohols retained their structure for over an hour.

The presence of complexes usually resulted in a smaller bubble size in the foam. This is illustrated in Figure 19–1 by photomicrographs of foams from triethanolamine stearate and the triethanolamine stearate–stearyl alcohol complex. The smaller bubble size of the foam containing the complex is apparent.

Cholesterol is considered separately because it is unique in some respects. There seems to be sufficient evidence that it forms complexes and yet it has an opposite effect from that of the straight-chain alcohols upon some properties. For example, cholesterol increased the bubble size of foams[1] and decreased foam stiffness. The effect of cholesterol in various triethanolamine fatty-acid salt systems, in comparison with that of the salt alone and a salt-straight-chain alcohol complex, is shown in Table 19–2.

The increase in emulsion stability and the decrease in the foam drainage that occurred upon addition of cholesterol to the triethanolamine laurate and myristate systems indicates that cholesterol complexed with the triethanolamine salts. In contrast to this is the decrease in foam stiffness and the increase in foam bubble size that accompanied the change in emulsion stability and drainage. However, these apparently anomalous effects can be explained by the nature of the interfacial films that are formed with the various triethanolamine salt–alcohol complexes. Schulman[4] showed that the interfacial film formed by sodium cetyl sulfate–cholesterol complexes was very fluid in contrast to the solid, condensed films from the sodium cetyl sulfate–cetyl alcohol complexes. If the triethanolamine salt–cholesterol interfacial film is also very fluid and plastic, it could expand considerably during discharge of the aerosol emulsion without destruction of the film structure, and this could account for both the increased bubble size and the lower foam stiffness. It seems likely that the interfacial films

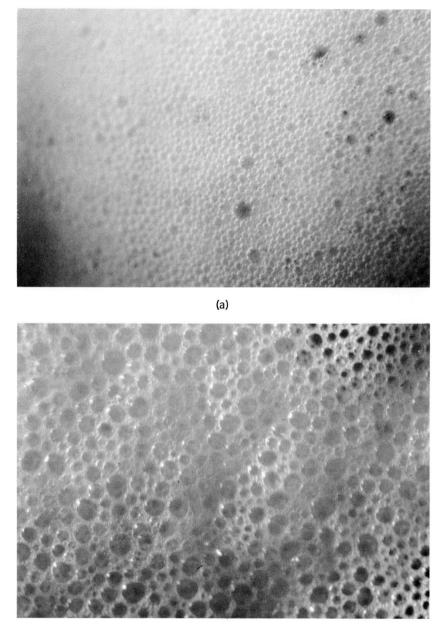

(a)

(b)

Figure 19-1   (a) Photomicrograph of a triethanolamine stearate–stearyl alcohol foam. (b) Photomicrograph of a triethanolamine stearate foam.

formed by the triethanolamine salt–straight-chain alcohols are the solid, condensed type.

TABLE 19-2   EFFECT OF CHOLESTEROL IN TRIETHANOLAMINE
FATTY-ACID SALT SYSTEMS

Fatty Acid[a]	Alcohol[b]	Emulsion Stability	Foam Properties	
			% Drainage (60 min)	Stiffness (g)
Lauric	None	1–5 min	84	6
Lauric	Lauryl	> 24 hr	11	20
Lauric	Cholesterol	> 24 hr	36	3
Myristic	None	1–5 min	34	34
Myristic	Myristyl	> 24 hr	0	50
Myristic	Cholesterol	> 24 hr	5	17
Palmitic	None	5–15 min	0	39
Palmitic	Cetyl	> 24 hr	0	42
Palmitic	Cholesterol	> 24 hr	2	12
Stearic	None	5–15 min	0	30
Stearic	Stearyl	> 24 hr	0	88
Stearic	Cholesterol	> 24 hr	2	16

[a] Triethanolamine-fatty acid concentration = 0.10 M.
[b] Soap–alcohol ratio (molar) = 1:1.

Variation in Alcohol Concentration.   The relationship between the concentration of lauryl alcohol in the triethanolamine laurate system and the change in emulsion and foam properties is shown in Table 19–3. Although

TABLE 19-3   VARIATION IN LAURYL ALCOHOL CONCENTRATION—
TRIETHANOLAMINE LAURATE SYSTEMS

Laurate[a]–Lauryl Alcohol Ratio (Molar)	Emulsion Stability	Foam Properties	
		% Drainage (60 min)	Stiffness (g)
1:0	1–5 min	86.0	8
1:¼	30–60 min	83.0	10
1:½	> 60 min	65.0	16
1:¾	> 16 hr	40.0	18
1:1	> 16 hr	11.0	22

[a] Triethanolamine laurate concentration = 0.10 M.

low concentrations of lauryl alcohol affected the properties, maximum efficiency occurred at a 1:1-mole ratio of triethanalomine salt to lauryl alcohol. In other triethanolamine salt systems, the maximum appeared at a 1:½-mole ratio.[1]

Variation in Complex Concentration.   This is a relatively unknown factor. The properties of triethanolamine laurate–lauryl alcohol and triethanolamine myristate–myristyl alcohol systems at two different concentrations of the complexes are given in Reference 1. In both cases, increasing the concentration of the complex increased emulsion stability and viscosity and decreased foam drainage. There probably is an optimum concentration at which any particular triethanolamine salt–fatty alcohol complex is most efficient, but this has not been determined experimentally for any of the complexes.

Variation in Propellants.   Although there is no doubt that various propellants affect the properties of aerosol emulsions and foams differently, the difference between the propellants seems dependent to a considerable extent upon the particular aerosol system used for the evaluation.[1,2,5] It is therefore, difficult to draw any firm conclusions about the relative effect of the different propellants. However, in general the foams prepared with "Freon" 12, "Freon" 114 or "Freon" 12–"Freon" 114 blends were more stable than those with "Freon" 12–"Freon" 11 (50/50), Propellant 142b or Propellant 152a.

Triethanolamine Salt–Fatty Acid Complexes.   An aqueous system of a triethanolamine salt of a fatty acid, which has been prepared by neutralizing a fatty acid with the equivalent amount of base, contains not only the salt but also free fatty acid and free base. The latter two components result from hydrolysis of the salt. For example, in an aqueous system containing sodium stearate, free stearic acid and sodium hydroxide are also present. Likewise, in an aqueous system containing triethanolamine myristate, free myristic acid and triethanolamine are present. This effect of hydrolysis is important with regard to molecular complexes because free fatty acids can complex with their salts. One of the best known of this type of complex is the sodium stearate–stearic acid complex that was discussed in Chapter 16.

    The experimental results obtained in the study of the aerosol systems also indicate that complex formation occurs between the triethanolamine salts and their corresponding free acids. This was determined by preparing two samples from each triethanolamine salt. One had an excess of triethanolamine, which would be expected to minimize hydrolysis and thus decrease the possibility of complex formation because of the decreased concentration of fatty acid. The other had an excess of fatty acid with respect to the triethanolamine, which would promote complex formation if it occurred. The difference in properties between the emulsions and foams from the two samples should be an indication of the extent of complex formation in the sample with the excess of fatty acid.

The data obtained with three systems, triethanolamine laurate, triethanolamine myristate, and triethanolamine palmitate, are given in Table 19–4. These data indicate that complex formation between the triethanolamine salt and the free fatty acid occurred in the triethanolamine myristate and palmitate systems but not in the triethanolamine laurate system. The reasons for the lack of complex formation in the latter system is not clear. It is possible that the complex did form but was too water soluble to be effective.

TABLE 19-4 VARIATION IN FATTY ACID–TRIETHANOLAMINE RATIO

| | | | Foam Properties | |
Acid	Acid–Base Ratio (Molar)	Emulsion Stability	% Drainage (60 min)	Stiffness (g)
Lauric	1:1½ [a]	< 1 min	85	9
	1½:1 [b]	< 1 min	86	10
Myristic	1:1½ [a]	< 1 min	71	21
	1½:1 [b]	> 1 hr	5	37
Palmitic	1:1½ [a]	15–30 min	0	23
	1½:1 [b]	> 1 hr	0	42

[a] 1:1½ ratio = 0.10-M acid/0.15-M base.
[b] 1½:1 ratio = 0.15-M acid/0.10-M base.

The change in foam properties that results with a variation in the acid–base ratio (i.e., complex formation) is important in the formulation of foam products such as shaving lathers and shampoos. A system with the strongest complex might not necessarily be the most desirable since the foam might not have sufficient wetting properties.

It is evident from the preceding discussion that aqueous aerosol systems containing mixtures of the triethanolamine salts and long-chain alcohols are not quite as simple as they may have appeared since they contain mixtures of the triethanolamine salts, free fatty acids, and long-chain alcohols. Complex formation between the salts and fatty acids may occur as well as between the salts and the long-chain alcohols. Mixed complexes involving the salts, acids and alcohols may also be present. However, the fact that the addition of the long-chain alcohols to a triethanolamine salt prepared from equivalent quantities of acid and base produces such a pronounced effect upon emulsion and foam properties indicates that the triethanolamine salt–alcohol complex is the most important factor in these systems.

Sodium Lauryl Sulfate–Fatty Alcohol Complexes

Sodium lauryl sulfate was included in the investigation of complexes in aerosol systems because the sodium alkyl sulfate–fatty alcohol complexes had been widely studied in nonaerosol systems.

The change in properties that occurred upon addition of various alcohols to aerosol systems containing sodium lauryl sulfate is shown in Table 19–5. The effect upon emulsion stability, foam drainage and foam stiffness is particularly noticeable with lauryl and myristyl alcohols, thus indicating fairly strong complex formation.

TABLE 19-5   VARIATION IN ALCOHOLS–SODIUM LAURYL SULFATE SYSTEMS[a]

		Foam Properties	
Alcohol[b]	*Emulsion Stability*	*% Drainage* (*60* min)	*Stiffness* (g)
None	< 1 min	82	11
Lauryl	> 5 hr	0	40
Myristyl	> 5 hr	0	38
Cetyl	> 5 hr	2	14
Stearyl	< 5 min	71	12
Oleyl	30–60 min	84	10
Cholesterol	1–5 min	86	13

[a] Sodium lauryl sulfate concentration = 0.10 M.
[b] Sodium lauryl sulfate–alcohol ratio (molar) = 1:1.

Microscopic examination of the sodium lauryl sulfate foams showed bubble sizes ranging from about 0.001–0.01 in. with laminae thickness of about 0.001–0.003 in. The foams containing lauryl alcohol had a smaller bubble size and thinner laminae. Photomicrographs of the foams confirmed the difference in bubble size between the two foams (Figure 19–2).

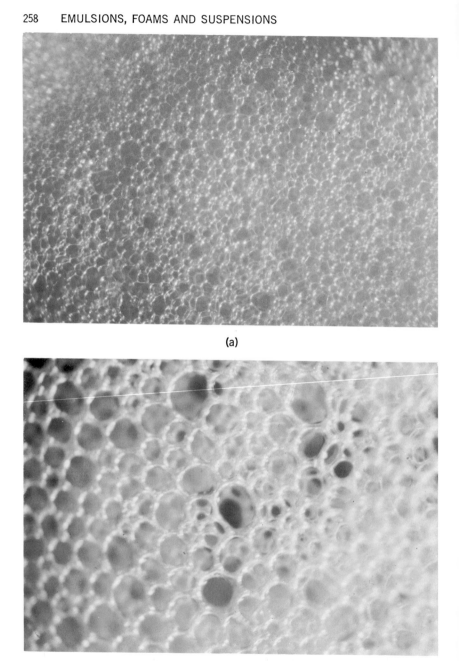

(a)

(b)

Figure 19-2   (a) Photomicrograph of a sodium lauryl sulfate-lauryl alcohol foam. (b) Photomicrograph of a sodium lauryl sulfate foam.

MOLECULAR COMPLEXES WITH NONIONIC SURFACE-ACTIVE AGENTS[2]

Polyoxyethylene Fatty Ethers

The nonionic polyoxyethylene (POE) fatty ethers are one of the few classes of surface-active agents that have been investigated in some detail in aerosol systems. The advantage of studying this class of compounds is that a series of the ethers is commercially available in which the members range from those that are predominantly lipophilic, with low HLB numbers, to those that are hydrophilic, with high HLB numbers.[5] These products are prepared by condensing varying proportions of ethylene oxide with the fatty alcohols, and the number of moles of ethylene oxide that have been reacted with 1 mole of the fatty alcohol is indicated in parenthesis in the name of the agent. For example, polyoxyethylene (4) lauryl ether indicates that the product was prepared by reacting 4 moles of ethylene oxide with 1 mole of lauryl alcohol. The compounds become more hydrophilic as the number of ethylene oxide units in the molecule increases.

The properties of aerosol emulsions and foams prepared with the polyoxyethylene fatty ethers by themselves were first determined in a simple aerosol system consisting of 90% aqueous phase and 10% "Freon" 12–"Freon" 114 (40/60) propellant. The concentration of the POE ethers in the aqueous phase was 0.05 M, assuming an average molecular weight corresponding to the empirical formula. On a weight percent basis, the concentrations ranged from 1.6% for the low molecular weight POE ethers to 6.0% for those with the highest molecular weight. The results of the evaluation are summarized in Table 19–6.

Only two of the eight polyoxyethylene fatty ethers were effective as emulsifying and foaming agents. These were POE (2) cetyl ether and POE (2) stearyl ether, the products with the lowest HLB numbers and the only POE ethers that were insoluble in water. This suggests that these materials act as solids in stabilizing aerosol emulsions and foams. The mechanism by which this occurs was discussed in Chapter 16. Present evidence indicates that two other classes of aerosol foams, aqueous alcohol and nonaqueous foams, are stabilized in the same way (see Chapters 20 and 21). It seems likely that the stabilization of aerosol emulsions and foams by materials that act as solids may be one of the major mechanisms by which these systems are stabilized.

Polyoxyethylene Fatty Ether–Fatty Alcohol Complexes

The investigation of polyoxyethylene fatty ether–fatty alcohol combinations was first reported by Becher and Del Vecchio.[6] They observed that

TABLE 19-6 COMPARISON OF POLYOXYETHYLENE ETHERS

Polyoxyethylene ether	HLB Value	Emulsion Stability	Foam Properties		
			Stability[a] (min)	Stiffness (g)	% Drainage (30 min)
Polyoxyethylene (4) lauryl	9.7	< 1 min	< 5	5	95
Polyoxyethylene (23) lauryl	16.9	< 1 min	< 5	5	90
Polyoxyethylene (2) cetyl	5.3	> 30 min	> 120	22	0
Polyoxyethylene (10) cetyl	12.9	< 1 min	< 15	8	87
Polyoxyethylene (20) cetyl	15.7	< 1 min	< 5	10	96
Polyoxyethylene (2) stearyl	4.9	> 30 min	> 120	25	10
Polyoxyethylene (10) stearyl	12.4	< 1 min	< 15	5	92
Polyoxyethylene (20) stearyl	15.3	< 5 min	< 5	8	99

[a] Foam stability is indicated by the time before observable collapse occurred after discharge.

the addition of fatty alcohols to polyoxyethylene lauryl ethers resulted in systems which showed film drainage transition phenomena typical of that previously noted in anionic systems. Subsequent investigations in the aerosol field showed that the addition of fatty alcohols and fatty acids to systems containing polyoxyethylene fatty ethers often resulted in a considerable effect upon foam properties. Although there was no direct experimental evidence to show that stoichiometric association complexes formed between the relatively complex polyoxyethylene fatty ethers and the fatty alcohols, the effect upon foam properties was interpreted as an indication of complex formation in the interfacial film between the polyoxyethylene fatty ethers and the long-chain polar compounds. In order to have an effect upon emulsions and foam properties, there is no necessity for the complexes in the interfacial film to have a definite and fixed composition, but there must be an association between the polyoxyethylene fatty ethers and the long-chain polar compounds.

The investigation in aerosol systems was carried out with combinations of the eight polyoxyethylene fatty ethers (listed in Table 19–6) with lauryl, myristyl, cetyl, stearyl, and oleyl alcohols.

General Discussion.   Some general comments about the results of the study are given below.

a. The addition of fatty alcohols to the POE fatty ether systems caused a marked change in one or more foam properties. Myristyl and cetyl alcohols generally had the greatest effect.

b. The effect of the alcohols was most pronounced with the three polyoxyethylene fatty ethers having intermediate HLB values, i.e., POE (4) lauryl ether, POE (10) cetyl ether, and POE (10) stearyl ether. In systems containing POE (2) cetyl ether and POE (2) stearyl ether the effect of the alcohols was less noticeable because these two surface-active agents gave good emulsions and foams by themselves.

c. Emulsion stability was increased in only two out of six systems.

d. Complex formation decreased the bubble size of the foams. The effect was similar to that observed with foams stabilized with anionic complexes.

Variation in Alcohols.   The specific properties of the aerosols affected by the alcohols and the extent to which they were changed were determined both by the particular alcohol and POE fatty ether present. The effect of the various alcohols upon the properties of POE (4) lauryl ether systems is shown in Table 19–7. Data for the other seven POE fatty ether systems are given in Reference 2.

Both emulsion and foam stability were increased by the addition of the alcohols. Myristyl, cetyl, and stearyl alcohols caused the greatest change in properties. Oleyl alcohol had little effect and this was generally true with the other POE fatty ethers.

Foam Bubble Size.   The correlation between bubble size of the foams and other foam properties was very marked in the POE fatty ether series. Whenever strong complex formation occurred and affected other foam properties, there was a corresponding decrease in the foam bubble size. This is illustrated in Figure 19–3 by the photomicrographs of the POE (4) lauryl ether foams listed in Table 19–7. The bubble size of the foams containing myristyl, cetyl, and stearyl alcohols was considerably smaller than that of the other foams. This correlates with the effect of these alcohols upon the foam properties listed in Table 19–7.

Variation in Fatty Alcohol Concentration.   The relationship between the concentration of cetyl alcohol in POE (4) fatty ether systems and emulsion and foam properties is shown in Table 19–8. The first addition of cetyl alcohol, which gave an ether–alcohol mole ratio of 1:¼, caused a significant change in properties. The effect upon foam properties became

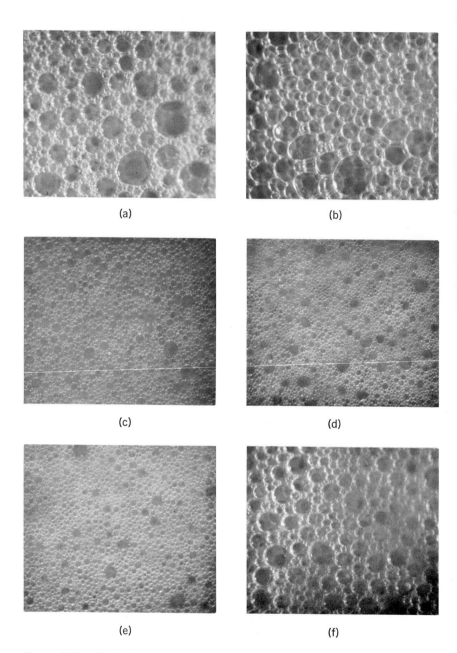

(a)

(b)

(c)

(d)

(e)

(f)

Figure 19–3 Photomicrographs of polyoxyethylene (4) lauryl ether foams. Effect of alcohols. (a) No alcohol. (b) Lauryl alcohol. (c) Myristyl alcohol (d) Cetyl alcohol. (e) Stearyl alcohol. (f) Oleyl alcohol.

most pronounced when the ether–alcohol ratio reached 1:½. Continued addition of cetyl alcohol had relatively less effect, although foam stiffness continued to increase. Therefore, a 1:1 ratio of alcohol to POE ether may not be necessary in many cases.

TABLE 19-7 POLYOXYETHYLENE (4) LAURYL ETHER SYSTEMS—
EFFECT OF ALCOHOLS

Alcohol	Emulsion Stability	Foam Properties			
		Stability (min)	Stiffness (g)	% Drainage (30 min)	Wetting (min)
None	< 1 min	< 5	5	95	< 1
Lauryl	> 30 min	60–120	12	6	< 1
Myristyl	> 30 min	> 120	23	0	> 120
Cetyl	> 30 min	> 120	20	0	> 120
Stearyl	> 30 min	> 120	8	0	> 120
Oleyl	< 5 min	< 15	5	93	< 1

TABLE 19-8 POLYOXYETHYLENE (4) LAURYL ETHER SYSTEMS—
VARIATION IN CETYL ALCOHOL CONCENTRATION

Ether–alcohol ratio (molar)	Emulsion Stability	Foam Properties			
		Stability (min)	Stiffness (g)	% Drainage (30 min)	Wetting (min)
1:0	< 1 min	< 5	5	89	< 1
1:¼	> 30 min	15–30	7	11	15–30
1:½	> 30 min	> 120	13	0	> 120
1:¾	> 30 min	> 120	21	0	> 120
1:1	> 30 min	> 120	26	0	> 120
1:2	> 30 min	> 120	31	0	> 120

Polyoxyethylene Fatty Ether–Fatty Acid Complexes

Combinations of the eight polyoxyethylene fatty ethers with four fatty acids, lauric, myristic, palmitic, and stearic acids, were also evaluated. Like the fatty alcohols, the fatty acids had a considerable effect upon the properties of the emulsions and foams in many cases, indicating complex formation between the polyoxyethylene ether and the fatty acid.

A survey of the data in Reference 2 shows that the systems most affected by the acids were those containing the POE ethers with intermediate HLB values. These results are similar to those observed with the fatty alcohols. The effect upon systems with the two ethers with low HLB numbers was less pronounced, but the fatty acids increased foam stiffness. Systems containing the POE ethers with high HLB numbers were least affected by the fatty acids.

Variation in Fatty Acids. As a general rule, myristic, palmitic, and stearic acids appeared to be somewhat more effective than lauric acid. As with the alcohols, however, the extent to which any particular fatty acid affected the properties of an aerosol system depended upon the polyoxyethylene fatty ether that was present. The results obtained upon the addition of various fatty acids to the POE (10) cetyl ether system are shown in Table 19–9. The effect produced by most of the fatty acids is quite apparent. The data for the other POE fatty ethers are given in Reference 2.

TABLE 19-9   POLYOXYETHYLENE (10) CETYL ETHER SYSTEMS—
EFFECTS OF ACIDS

		Foam Properties			
Acid	*Emulsion Stability*	*Stability* (min)	*Stiffness* (g)	*% Drainage* (30 min)	*Wetting* (min)
None	< 1 min	5–15	8	87	< 1
Lauric	1–5 min	15–30	9	91	< 1
Myristic	15–30 min	> 120	17	0	15–30
Palmitic	> 30 min	> 120	25	0	> 120
Stearic	> 30 min	> 120	19	0	> 120

Variation in Fatty Acid Concentration. The effect of different concentrations of myristic acid in a polyoxyethylene (4) lauryl ether system is illustrated in Table 19–10. The first addition of myristic acid, which gave an ether–acid mole ratio of 1:¼, had a considerable effect upon the properties of the system. Subsequent additions had very little effect. Thus, in many of the systems a 1:1 mole ratio of POE ether to acid is not needed and lower concentrations of acid are sufficient.

Fatty Alcohols versus Fatty Acids. When compared in the same system, the long-chain alcohols generally had a greater effect upon such properties as foam stiffness, foam drainage, and wetting than the corresponding fatty acids. However, there were some instances where the fatty acids were more effective. For example, the fatty acids were superior to the alcohols in improving the emulsion stability of the POE (10) cetyl ether and the POE (10) stearyl ether systems. This suggests that for some products, a combination of all three components, the polyoxyethylene fatty ether, fatty alcohol, and fatty acid, might provide properties superior to those from either the POE ether–fatty alcohol or POE ether–fatty acid complexes alone.

TABLE 19-10   POLYOXYETHYLENE (4) LAURYL ETHER SYSTEMS—
VARIATION IN MYRISTIC ACID CONCENTRATION

Ether–acid ratio (molar)	Emulsion Stability	Foam Properties			
		Stability (min)	Stiffness (g)	% Drainage (30 min)	Wetting (min)
1:0	< 1 min	< 5	5	89	< 1
1:¼	> 30 min	> 120	19	0	> 120
1:½	> 30 min	> 120	18	0	> 120
1:¾	> 30 min	> 120	15	0	> 120
1:1	> 30 min	> 120	18	0	> 120
1:2	> 30 min	> 120	24	0	> 120

The POE ether–fatty acid complexes had far less effect upon the bubble size of the foams than those containing the fatty alcohols—additional evidence that the complexes containing the fatty acids are weaker than those with the fatty alcohols.

PEARLESCENT AND IRIDESCENT COMPLEXES[3]

During the investigation of the polyoxyethylene fatty-ether complexes, certain combinations of the components were observed to give highly pearlescent aqueous concentrates and aerosols. These systems subsequently were studied in some detail because they had considerable aesthetic appeal owing to their optical properties and should be of interest for the formulation of cosmetic and pharmaceutical products. The results of the study of the pearlescent and iridescent complexes are reported in the following sections.

General Discussion

Some general conclusions that were drawn from the data are listed below:

1. The degree of pearlescence in the aqueous concentrates was deter-

mined both by the particular polyoxyethylene fatty ether and the specific fatty alcohol or acid that were present. Only polyoxyethylene fatty ethers that were water dispersible or water soluble gave pearlescent concentrates. The most effective alcohols for producing pearlescence were myristyl and cetyl alcohols. The fatty acids were relatively ineffective in producing pearlescent structures, except for one specific combination involving lauric acid.

2. The pearlescence was affected by the concentrations of the POE fatty ether and the fatty alcohol in the aqueous phase. In general, a mole ratio of the POE fatty ether to alcohol of 1:1 was preferred.

3. The pearlescence of the aqueous concentrates was dependent upon the temperature. In all cases, increasing the temperature ultimately caused the pearlescence to disappear. The loss of pearlescence occurred over a temperature range rather than at a specific temperature. This property, in conjunction with the optical characteristics, suggests that these pearlescent complexes belong to the class of substances known as *liquid crystals*.

4. The type of propellant added to the aqueous pearlescent concentrates affected the optical properties. Several propellants had an adverse effect upon pearlescence.

Optical Properties of the Aqueous Concentrates

Polyoxyethylene Fatty Ether–Alcohol Combinations. Eleven of the concentrates prepared in this group developed high pearlescence. Many of the concentrates were iridescent with a predominantly bluish cast. The POE fatty ether–fatty alcohol complexes that produced pearlescent concentrates are listed in Table 19–11.

TABLE 19-11 POE FATTY ETHER–FATTY ALCOHOL COMBINATIONS PRODUCING HIGH PEARLESCENCE

1. POE (10) Cetyl Ether–Lauryl Alcohol
2. POE (10) Cetyl Ether–Myristyl Alcohol
3. POE (10) Cetyl Ether–Cetyl Alcohol
4. POE (10) Cetyl Ether–Stearyl Alcohol
5. POE (20) Cetyl Ether–Cetyl Alcohol
6. POE (10) Stearyl Ether–Myristyl Alcohol
7. POE (20) Stearyl Ether–Myristyl Alcohol
8. POE (20) Stearyl Ether–Cetyl Alcohol
9. POE (20) Oleyl Ether–Myristyl Alcohol
10. POE (20) Oleyl Ether–Cetyl Alcohol
11. POE (20) Oleyl Ether–Stearyl Alcohol

All of the POE fatty ethers listed in Table 19–11 are either water soluble or water dispersible, and all have HLB values above 12.

Polyoxyethylene Fatty Ether–Fatty Acid Combinations.   The POE (4) lauryl ether–lauric acid combination was the only complex in this group that gave a concentrate of any interest but it was the most iridescent concentrate obtained. Unfortunately, the product was very sensitive and difficult to reproduce.

Variation in Concentration.   The concentration of the POE fatty ethers and other components in the aqueous phase usually was 0.05 m. Increasing the concentrations of any of the components above 0.05 m generally decreased pearlescence. Often the decrease in pearlescence was accompanied by an increase in viscosity.

The intensity of pearlescence was usually highest when the mole ratio of the polyoxyethylene fatty ether and fatty alcohol was 1:1. This dependence of pearlescence upon component concentration is illustrated by the data in Table 19–12 for the POE (20) oleyl ether–cetyl alcohol complex.

Effect of Temperature.   All of the aqueous concentrates lost their pearlescence at higher temperatures. The pearlescence disappeared gradually over a temperature range (the transition temperature range) rather than at a specific temperature. This range was determined more by the fatty alcohol present than by the type of polyoxyethylene fatty ether.

The transition temperature ranges for complexes from two polyoxyethylene fatty ethers with four different alcohols are given in Table 19–13 along with the melting points of the alcohols. The data indicate that there may be a relationship between the melting point of the alcohol and the transition temperatures. The higher the melting points of the alcohols the higher are the transition temperatures.

Examples of the temperature stability of complexes with the same alcohol but different polyoxyethylene fatty ethers are shown in Table 19–14. Although there appears to be a slight trend towards more stable complexes as the molecular weight of the ether increases, the effect of the different POE fatty ethers upon the transition temperature range is not nearly as evident as with the different alcohols.

Optical Properties of Aerosol Formulations

Many of the aerosol formulations exhibited high pearlescence in glass bottles but none of the foams obtained by discharging the aerosols ex-

TABLE 19-12   VARIATION IN CONCENTRATION OF COMPONENTS—
POLYOXYETHYLENE (20) OLEYL ETHER-CETYL ALCOHOL MIXTURE

Concentration (molar)		Mole Ratio POE Ether– Cetyl Alcohol	Pearlescence Rating*	Comments
POE (20) Oleyl Ether	Cetyl Alcohol			
0.025	0.025	1/1	3	Somewhat iridescent—bluish silver predominates
0.05	0.025	2/1	2	White—develops pearly effect upon swirling
0.05	0.05	1/1	3	Pearlescent—effect heightened by swirling—concentrate shows phase separation
0.10	0.05	2/1	—	Opaque white
0.05	0.10	1/2	—	Opaque white
0.10	0.10	1/1	—	Opaque white

* Pearlescence Rating
— No pearlescence.
1. Possible pearlescence but slight.
2. Definite pearlescence—concentrates were white or silvery.
3. Very high pearlescence—many concentrates were iridescent with a predominantly bluish cast.

TABLE 19-13 TEMPERATURE DEPENDENCE OF PEARLESCENT STRUCTURES—
VARIATION IN FATTY ALCOHOLS

Polyoxyethylene Fatty Ether	Fatty Alcohol		
	Name	Melting Pt. (°F)	Transition Temp. Range (°F)
POE (10) Cetyl Ether	Lauryl	Liquid	77–82
"	Myristyl	91–95	93–101
"	Cetyl	121	95–115
"	Stearyl	136	100–125
POE (20) Oleyl Ether	Myristyl	91–95	89–97
"	Cetyl	121	105–117
"	Stearyl	136	123–132

TABLE 19-14 TEMPERATURE DEPENDENCE OF PEARLESCENT STRUCTURES—
VARIATION IN POLYOXYETHYLENE FATTY ETHERS

Polyoxyethylene Fatty Ether	Fatty Alcohol	Transition Temp. Range (°F)
POE (10) Cetyl Ether	Lauryl	77–82
POE (10) Stearyl Ether	Lauryl	79–85
POE (10) Cetyl Ether	Myristyl	93–101
POE (10) Stearyl Ether	Myristyl	98–103
POE (20) Cetyl Ether	Cetyl	99–125
POE (20) Stearyl Ether	Cetyl	101–126
POE (20) Oleyl Ether	Cetyl	123–132

hibited any pearlescence. Therefore, in order to have any aesthetic utility, the aerosol pearlescent products would have to be packaged in clear glass bottles.

Preferred Pearlescent Aerosol Systems.   Not all of the aqueous pearlescent concentrates described in the preceding sections could be formulated as pearlescent aerosols. In some cases, addition of the propellant to the concentrate destroyed the pearlescence. A possible reason for this is discussed in the next section.

Some of the preferred pearlescent aerosol formulations and their characteristics are listed in Table 19–15. All of the aerosols have poor emulsion stability and must be shaken immediately before use. As a result of the

TABLE 19-15 CHARACTERISTICS OF PEARLESCENT AEROSOL* FORMULATIONS

Components	Pearlescence	Product Characteristics
POE (10) Cetyl Ether, Myristyl Alcohol	3	Quiet liquid discharge which expands into a nice foam
POE (10) Cetyl Ether, Cetyl Alcohol	3	Noisy; semiliquid, foamy discharge—expands to medium dense foam
POE (20) Stearyl Ether, Myristyl Alcohol	3	Slightly noisy liquid discharge—expands to a medium dense foam
POE (20) Stearyl Ether, Cetyl Alcohol, Palmitic Acid	3	Noisy, bubbly discharge that expands into a good, creamy foam
POE (20) Oleyl Ether, Myristyl Alcohol	2	Quiet discharge, expands to a fairly dense foam
POE (20) Oleyl Ether, Cetyl Alcohol	3	Slightly noisy liquid discharge—expands to a fairly nice foam
POE (20) Oleyl Ether, Cetyl Alcohol, Palmitic Acid	2	Slightly noisy semiliquid discharge—expands to a medium dense foam

* 90% aqueous concentrate, 10% "Freon" 12–"Freon" 114 (15/85) propellant.

poor dispersion of propellant, the systems generally have a noisy, sputtery discharge and the product is emitted as a foamy liquid. The liquid usually builds up into a foam.

Effect of Propellants. "Freon" 12–"Freon" 114 (15/85) is the preferred propellant for pearlescent aerosols because it had the least effect upon the pearlescent structures. Initially, as a result of the poor emulsion stability that resulted with this propellant, several other propellants were evaluated. One was a "Freon" 12–"Freon" 114–isobutane blend with a density about the same as that of the aqueous phase. The other was isobutane. Both of these alternate propellants had an adverse effect upon the pearlescence of the aqueous concentrates.

It seems that for a propellant to be satisfactory for use with pearlescent systems, it must resist emulsification or solubilization by the aqueous pearlescent complex. Certainly, for either emulsification or solubilization to occur, the liquid crystalline structure of the complex would have to be disrupted which undoubtedly would affect the pearlescence. Apparently the "Freon" 12–"Freon" 114 (15/85) propellant is less easy to emulsify or solubilize than either the blend or isobutane, possibly because it has less affinity for the pearlescent complex than the other two propellants.

Structure of the Pearlescent Complexes

The properties of the pearlescent polyoxyethylene fatty ether–fatty alcohol complexes in aqueous systems indicate that these complexes belong to the class of materials called *liquid crystals*. Liquid crystals have structures that fall between those of liquid and solids. These structures are also referred to as mesomorphic structures.[7] A true solid or crystalline material has a single, sharp melting point and when the compound is at the melting point, the temperature remains constant until the compound is completely melted. In contrast, a liquid crystalline material exhibits two transition or melting points and thus the transition from a solid to a liquid takes place over a temperature range.

The fact that all three types of surface-active agents, anionic, cationic, and nonionic, including the polyoxyethylene fatty ethers, form mesomorphic phases has been established. Rosevear[7] has reported that one of the mesomorphic phases of surface-active agents, the so-called *neat phase,* consists of flexible layers where the molecules are arranged essentially parallel to each other in the layers. The layers are packed with crystalline regularity according to x-ray studies, but the lateral arrangement is random, thus producing the liquid crystal structure. In the presence of water, the layers are separated by water molecules with the polar heads of the surface-active agent oriented towards the aqueous phase. X-ray studies of the fatty alcohols have also shown that these compounds exist in a number of different solid phases.[8] Therefore, it is not surprising that the molecular complexes from the polyoxyethylene fatty ethers and fatty alcohols form mesomorphic phases.

The pearlescence and iridescence of the polyoxyethylene fatty ether complexes in aqueous systems undoubtedly are a consequence of the liquid crystal structure of the complexes. The colors are due to interference phenomena resulting from the reflection of light from different planes or layers in the liquid crystal structure. This effect is similar to that observed with soap bubbles, where the light is reflected from two film surfaces, or from wet pavements, where colors result from the reflection of light from thin films of oil.

The disappearance of pearlescence over a specific temperature range evidently results from the gradual disruption or *melting* of the liquid crystalline structures responsible for the pearlescence. Thus, the behavior of these polyoxyethylene fatty ether systems with their two transition temperatures is similar to that of typical liquid crystalline materials.

Emulsion Stability and Pearlescence

During the previous investigation of the effect of the polyoxyethylene fatty ether complexes on the properties of aerosol systems, a somewhat puzzling aspect of emulsion stability developed for which there appeared to be no reasonable explanation. The subsequent investigation of the pearlescent complexes provided a possible explanation for this apparently anomalous behavior.

The previous study of the polyoxyethylene fatty ethers showed that only two out of the eight compounds by themselves gave stable aerosol emulsions and foams (see Table 19–6). The addition of fatty alcohols to the other six polyoxyethylene fatty ethers improved foam stability markedly in every case but emulsion stability was improved in only two out of the six systems. The increase in foam stability indicated that complex formation between the polyoxyethylene fatty ethers and the fatty alcohols had occurred in all six cases but it was not clear as to why complex formation was not reflected by an increase in emulsion stability.

A study of the data in References 2 and 3 showed that there was a relationship between the lack of emulsion stability and pearlescence. The polyoxyethylene fatty ether complexes which did not produce stable emulsions were the same complexes that gave the most pearlescent products. Conversely, the polyoxyethylene fatty ether complexes that gave stable emulsions did not give pearlescent products. The two classes of polyoxyethylene fatty ethers are listed in Table 19–16.

The correlation between the emulsion stability and the lack of pearlescence is quite striking. Evidently, the intermolecular forces responsible for complex formation are so strong in some cases that the resulting pearlescent structures are very stable. When propellant is added, the structures retain their identity instead of dissociating in order to emulsify or solubilize the propellant. These systems therefore show poor emulsion stability.

TABLE 19-16 EMULSION STABILITY AND PEARLESCENCE WITH
POLYOXYETHYLENE FATTY ETHER–FATTY
ALCOHOL COMBINATIONS

POE Ethers Producing Stable Emulsions (> 30 min) but Nonpearlescent Systems	POE Ethers Producing Unstable Emulsions (< 1 min) but Pearlescent and Iridescent Systems
POE (4) lauryl ether	POE (35) lauryl ether
POE (2) cetyl ether	POE (10) cetyl ether
POE (2) stearyl ether	POE (20) cetyl ether
	POE (20) stearyl ether

However, when the systems with poor emulsion stability were discharged, there was obvious stabilization of the resulting foams by the complexes. It can be postulated that the vaporization of the liquefied gas propellants, which occurred during discharge, broke up the complexed pearlescent structures sufficiently so that the complexes became available for foam stabilization. Other possibilities are that the pearlescent structures are highly hydrated and increase foam stability by their effect on the viscosity of the foam system, or else provide a mechanical barrier to the coalescence of the gas bubbles.

Discharge Characteristics of Aerosol Emulsions

Although complex formation was consistently helpful in improving foam stability, increasing foam stiffness and decreasing the rate of drainage, its effect upon discharge characteristics varied considerably. In some instances the addition of alcohols improved discharge, changing it from a noisy to a quiet discharge, and in others alcohols had an adverse effect. This is illustrated by the data in Table 19–17, which shows the effect of various alcohols upon the properties of the POE (23) lauryl ether system.

In this particular system, lauryl alcohol improved the discharge and appearance of the foam while cetyl, stearyl, and oleyl alcohols had an adverse effect upon these properties. Cetyl alcohol, however, was quite effective in increasing foam stability and stiffness and decreasing the drainage. These data show that the selection of an alcohol as a complexing agent has to be based upon a number of factors, including the effect upon discharge as well as upon emulsions and foam properties. In the POE (23) lauryl ether systems, myristyl alcohol or a mixture of myristyl and lauryl alcohols would probably be the most desirable for complexing.

TABLE 19-17   POLYOXYETHYLENE (23) LAURYL ETHER SYSTEMS—
EFFECT OF ALCOHOLS

			Foam Properties		
Alcohol	Discharge	Appearance	Stability (min)	Stiffness (g)	% Drainage (30 min)
None	Slightly noisy	Slightly bubbly	< 5	5	90
Lauryl	Quiet	Smooth	15–30	29	28
Myristyl	Slightly noisy	Slightly cratered	> 120	44	0
Cetyl	Very noisy	Poor—very bubbly	> 120	38	14
Stearyl	Very noisy	Poor—very bubbly	15–30	6	82
Oleyl	Very noisy	Icing	< 15	4	94

Notes: 1. All emulsions had a low viscosity except that with cetyl alcohol, which was medium.
2. All emulsions had a stability less than 1 min.
3. All foams wet paper immediately after discharge.

TABLE 19-18   VARIATION IN MYRISTYL ALCOHOL CONCENTRATION

Triethanolamine Myristate Systems

Myristate[a]– Alcohol Ratio (Molar)	Emulsion Properties		Foam Properties		
	Viscosity	Stability	% Drainage (3 hr)	Stiffness (g)	Type of Discharge
1:0	Low	15–30 min	56	42	Quiet
1:¼	Low	> 5 hr	2	45	Quiet
1:½	High	> 5 hr	1	42	Slightly noisy
1:¾	High	> 5 hr	0	52	Noisy
1:1	High	> 5 hr	0	56	Noisy

[a] Triethanolamine myristate concentration = 0.10 M.

It may be possible in many cases to retain the advantages of the complexes with respect to such properties as emulsion and foam stability, stiffness, drainage, etc., and eliminate the adverse effect upon discharge and foam appearance merely by decreasing the concentration of the fatty alcohol or acid. Most of the complexes that caused a noisy discharge were prepared with a 1:1 mole ratio of the surface-active agent to fatty alcohol or acid. The data in Table 19–18, which illustrate the effect of various concentrations of myristyl alcohol in triethanolamine myristate systems, show that a mole ratio of the salt to myristyl alcohol of 1:¼ to 1:½ increases emulsion stability and decreases drainage to about the same extent as the 1:1 ratio but does not have the adverse effect upon discharge that the 1:1 ratio does. Other instances of the same effect may be found in References 1 and 2.

There undoubtedly are a number of factors that contribute to a poor discharge. However, until suitable microscopic equipment is available for examination of the emulsions, most explanations of poor discharge have to be based more upon speculation than scientific data. It is generally assumed that a noisy, sputtery discharge is the result of a poor emulsion of the liquefied gas propellant where many of the droplets have a comparatively large diameter. The circumstantial evidence for this is that many aerosol emulsions with poor stability also have a sputtery discharge. However, poor emulsion stability per se does not necessarily mean a noisy discharge. There are many emulsions listed in References 1 and 2 that have extremely poor stability and yet have a quiet discharge. Actually, there is no contradiction in this. A system can have low enough interfacial tension so that the propellant is dispersed with ease by shaking and still have poor emulsion stability. Also, good emulsion stability does not necessarily insure a quiet discharge. Aerosols containing cholesterol complexes have excellent emulsion stability but a noisy discharge (see Table 8 in Reference 1). This was attributed to a large droplet size in the initial emulsion that resulted from the fluid or plastic nature of the interfacial film of the cholesterol complex.

There is a certain amount of evidence which indicates that too strong a molecular complex may cause a noisy discharge. Strong complex formation may result in a poor emulsion with large and nonuniform droplets, because the complex maintains its structure rather than emulsify the propellant. In contrast, strong complex formation can also produce very stable emulsions that have a noisy discharge. (See Tables 6 and 7 in Reference 1 and Table 4 and 7 in Reference 2.) It seems likely in this case that complex formation results in the formation of a solid, condensed interfacial film around the dispersed propellant droplets which resists the rupture that has to occur when the propellant droplet vaporizes. The rupture of the interfacial films could conceivably cause a noisy discharge.

REFERENCES

1. P. A. Sanders, *J. Soc. Cosmetic Chemists* **17**, 801 (1966).
2. P. A. Sanders, *Soap Chem. Specialties* **43**, 68, 70, (1967).
3. P. A. Sanders, *J. Soc. Cosmetic Chemists,* **20**, 577 (1969).
4. J. H. Schulman and E. G. Cockbain, *Trans. Faraday Soc.,* **36**, 651 (1940).
5. Cosmetic Bulletin: "Brij" Surfactants (polyoxyethylene fatty ethers), Atlas Chemical Industries, Inc., November, 1961.
6. P. Becher and A. J. Del Vecchio, *J. Phys. Chem.* **68**, 3511 (1964).
7. F. B. Rosevear, *J. Soc. Cosmetic Chemists* **19**, 58 (1968).
8. D. G. Kolp and E. S. Lutton, *J. Amer. Chem Soc.* **73**, 5593 (1951).
9. F. Atkins, *Perfumery Essent. Oil Record* 332 (November, 1934).
10. A. M. Schwartz, J. W. Perry, and J. Berch. "Surface Active Agents and Detergents," Volume 2, p. 450, Interscience Publishers Inc., New York, N.Y., 1958.

20

AQUEOUS ALCOHOL FOAMS

Aqueous alcohol foams consist basically of a mixture of water, alcohol, surface-active agent, and propellant. Although this may not seem sufficiently unusual to warrant the separate classification given these products, the aqueous alcohol foam system is unique in several respects. Generally, the combinations of water, ethyl alcohol, and propellant are mutually soluble so that the aerosol is not an emulsion system. The commonly used surface-active agent, "Polawax,"* is soluble in the mixture of water, alcohol, and propellant so that the basic aerosol system is a clear solution and does not have to be shaken before use as does an emulsion or suspension. In addition, the system can produce foams with over 50% ethyl alcohol, and this has made possible the development of products which previously could not be formulated as foams using conventional aqueous systems.

Products based upon the aqueous ethyl alcohol foam system have appeared on the market only recently, but the number is constantly increasing and eventually these foams will constitute a considerable part of the aerosol foam market. Some types of commercial products based upon the aqueous ethyl alcohol foam system are listed in Table 20–1.

TABLE 20-1 SOME COMMERCIAL, AQUEOUS ETHYL ALCOHOL FOAM PRODUCTS

1. After-Shave	5. Hair Setting Foam
2. Athlete's Foot Relief	6. Insect Repellent
3. Cologne	7. Skin Moisturizer
4. Hair Dressing	8. Sunscreen

Aqueous ethyl alcohol foams are discussed in detail in References 1 and 2. Unless otherwise indicated, the data in the subsequent sections were obtained from these two sources.

* An ethoxylated fatty alcohol mixture (Croda, Inc.)

COMPOSITIONS OF AQUEOUS ALCOHOL FOAMS

Water–Alcohol–Propellant Ratios

Most combinations of water, ethyl alcohol, and propellant are not miscible and form two liquid layers when mixed. Aqueous alcohol foams can be prepared from these immiscible combinations, but the products are emulsions and do not have the advantages of a homogeneous system. However, there is a much smaller region in the water–alcohol–propellant system where the three components are mutually soluble and form a clear solution. Although foams from both the immiscible and miscible regions are considered, the miscible region is of particular interest for the formulation of aqueous alcohol foams since these products do not require shaking prior to use. The solubility characteristics of water–alcohol–propellant combinations are, therefore, of considerable importance because they determine whether a given composition will be in the miscible or immiscible region.

Solubility Characteristics with Different Propellants. A wide variety of propellants has been proposed for use with aqueous ethyl alcohol foams[1,3,4] including fluorocarbons, hydrocarbons, and mixtures of the two. The solu-

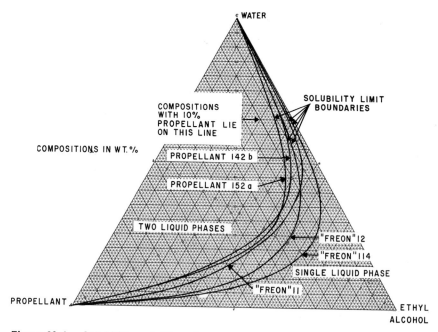

Figure 20–1   Solubilities of "Freon" compounds in water-ethyl alcohol solutions at ca. 70°F.

bility characteristics of water–ethyl alcohol–propellant systems vary considerably depending upon the particular propellant that is present; therefore, the location of the solubility limit boundary that divides the imimscible and the miscible regions depends upon the propellant.

The solubility characteristics of a number of water–ethyl alcohol–fluorocarbon propellant mixtures are illustrated on the triangular coordinate chart in Figure 20–1. Compositions to the left of the solubility limit boundaries are not miscible and form two liquid phases. Products with compositions in these regions have to be formulated as emulsion systems. Compositions to the right of the curves are mutually soluble and form a single liquid phase. This is the region of the most interest. The solubility diagrams indicate only whether a given mixture of water–ethyl alcohol, and propellant will be immiscible and form two liquid layers or will be miscible and form a single liquid phase. They do not indicate the areas in either region where foams can be obtained.

Solubility Characteristics with Different Alcohols and Acetone.   Although ethyl alcohol is the only alcohol used to any extent in aqueous alcohol foam products, foams can be obtained with other alcohols and acetone. The particular alcohol present in a water–alcohol–propellant system has a considerable effect upon the solubility properties of the system and thus affects the location of the miscible and immiscible regions.

The solubility characteristics of three-component mixtures of water and "Freon" 12–"Freon" 114 (40/60) propellant with methyl alcohol, ethyl alcohol, isopropyl alcohol, and acetone are illustrated on the triangular coordinate chart in Figure 20–2. The largest miscible region is obtained with isopropyl alcohol but it gives the poorest foams. Methyl alcohol gives the smallest miscible area while ethyl alcohol and acetone are in between with fairly similar solubility properties.

Foam Compositions in the Miscible Regions

Foams with adequate stability can be obtained only in a very limited part of the miscible region of each water–alcohol–propellant system. The boundaries of this limited area depend upon the surface-active agent, alcohol, and propellant that are present but they also depend upon the criteria used for a foam with acceptable stability. Lanzet[3] believes that foams should be of the self-collapsing type with durations of from 10–60 sec and reports that consumers prefer foams of this type to the more stable foams. He has obtained foams that fall within his classification from homogeneous aqueous alcohol systems containing as little as 30.4 vol% ethyl alcohol.

Klausner[4] refers to aqueous alcohol foams as foams of limited stability

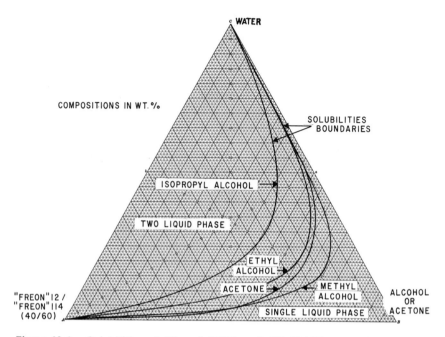

Figure 20–2   Solubilities of "Freon" 12/"Freon" 114 (40/60) propellant in water-alcohol and water-acetone solutions at ca. 70°F.

and defines them as foams that do not become completely liquefied when exposed to the atmosphere for 15 min or longer but which revert to a liquid within about 2 sec when rubbed. Klausner gives the following critical ratios of components for aqueous foams that meet his specifications (Table 20–2).

TABLE 20-2   CRITICAL RATIOS FOR AQUEOUS ALCOHOL FOAMS

Component	Composition (wt %)
Alcohol	46–66
Water	28–42
Surfactant	0.5–5
Propellant	2–15

From Klausner.[4]

When an aqueous alcohol foam system is limited to a specific alcohol, surface-active agent, and propellant, then the proportions of the components that give both a homogeneous system and a satisfactory foam be-

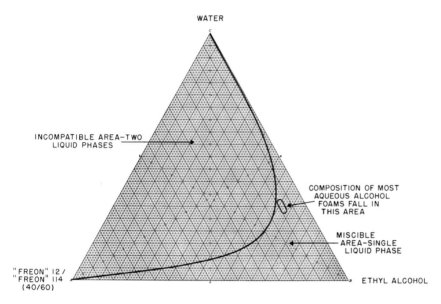

Figure 20-3 Solubility characteristics of the water-ethyl alcohol-"Polawax"-"Freon" 12-"Freon" 114 (40/60) propellant system at ca. 70°F.

come even more limited. This is illustrated by the small area of acceptable foam compositions on the triangular coordinate chart in Figure 20–3. It can be seen from Figure 20–3 that products formulated as aqueous alcohol foams have to be prepared with considerable accuracy. Only water–alcohol ratios in the range from about 38/62 wt % to 30/70 wt % give acceptable foams. If, with this particular system, the water–alcohol–propellant ratios deviate from the limited area of satisfactory compositions, the product will end up either as an emulsion system or a nonfoaming aerosol.

The composition of a typical basic aqueous ethyl alcohol foam that falls within the specified limits in Figure 20–3 is given in Table 20–3.

TABLE 20-3 COMPOSITION OF A TYPICAL, BASIC AQUEOUS ALCOHOL FOAM

Component	Composition (wt %)
Water	30.5
Anhydrous ethyl alcohol	56.5
"Polawax"	3.0
"Freon" 12–"Freon" 114 (40/60) propellant	10.0

The water–alcohol ratio in the above formulation is 35/65.

Excellent foams can be obtained with compositions falling in the miscible regions of aqueous systems based upon methyl alcohol, ethyl alcohol, or acetone, but not from isopropyl alcohol. The stability of the aqueous alcohol or acetone foams appears to be related to the relative areas of the miscible regions of the water–alcohol–propellant systems. The solubility data in Figure 20–2 show that isopropyl alcohol produces the largest miscible region and methyl alcohol the smallest.

Foam Compositions in the Immiscible Regions

Stable foams can be obtained over a wide range of water–alcohol ratios throughout the immiscible regions. Products formulated in the immiscible regions are emulsions and have to be shaken before use. Quite often these products have an initial wet, sloppy discharge. In these cases, shaking emulsifies the propellant in the bulk phase but not in the dip tube. This problem can be minimized by using a capillary dip tube or a valve without a dip tube.

Surface-Active Agents

Surface-active agents for producing foams from aqueous alcohol systems can be divided into two groups—those that are effective for the homogeneous compositions in the miscible region and those that are satisfactory for the emulsion systems in the immiscible region. The group of surfactants for the miscible region is much smaller than that for the emulsion systems.

Surfactants for the Miscible Region. Although a number of surface-active agents have been reported to be satisfactory for homogeneous aqueous ethyl alcohol foams,[1,3,4] the most effective agents are the fatty alcohols, ethoxylated fatty alcohols, and mixtures of the alcohols with the ethoxylated derivatives.

"POLAWAX" "Polawax" is considered separately because it was one of the first surface-active agents reported for the aqueous ethyl alcohol foams[5,6] and is still one of the most commonly used agents. As a result, much of the basic information on aqueous alcohol foams has been obtained with "Polawax" as the surface-active agent. "Polawax" is described as an ethoxylated stearyl alcohol with auxiliary emulsifying agents, but its composition has not been disclosed by Croda, Inc. It is probably the best single product for aqueous ethyl alcohol foams. It not only produces good foams but also is soluble in the water–ethyl alcohol–propellant mixtures and gives clear aerosol systems. Two different grades of "Polawax" are available, Grade A-31 and Grade A-22. Grade A-31 gives much stiffer foams and was used exclusively in the foams discussed in the following sections.

FATTY ALCOHOL COMBINATIONS   Cetyl alcohol–stearyl alcohol combinations gives excellent, pressure sensitive foams but the combinations have the disadvantage that generally they are not completely soluble in the water–alcohol–propellant mixtures. This produces a hazy product or one with suspended solids. If the aerosol product is packaged in an opaque bottle, as most aqueous ethyl alcohol foam products are, the haziness is of little consequence. However, the formulation must be checked to be certain that the suspended particles do not cause valve clogging.

One of the factors that controls the stiffness of foams with these combinations is the ratio of cetyl alcohol to stearyl alcohol. This is shown in Table 20–4. The data were obtained with an aerosol formulation of 88% water–ethyl alcohol (35/65) solution, 2% total surface-active agent, and 10% "Freon" 12–"Freon" 114 (40/60) propellant.

TABLE 20-4   CETYL ALCOHOL–STEARYL ALCOHOL COMBINATIONS
AND FOAM STIFFNESS

Fatty Alcohol Concentration in Aerosol (wt %)		Foam Stiffness (g)
Cetyl Alcohol	Stearyl Alcohol	
2.0	0.0	0
1.5	0.5	14
1.0	1.0	52
0.5	1.5	36
0.0	2.0	0

One of the most interesting features of the cetyl alcohol–stearyl alcohol surfactant combination is that neither fatty alcohol by itself is effective. The purity and composition of commercial cetyl and stearyl alcohols varies, depending upon the source, and therefore fatty alcohols from different suppliers may give different results.

ETHOXYLATED FATTY ALCOHOLS   The only ethoxylated fatty alcohol which by itself gave a satisfactory foam was "Siponic" E-O. This surfactant is the condensation product of one mole of ethylene oxide with one mole of stearyl alcohol and is manufactured by the American Alcolac Company. Other ethoxylated fatty alcohols containing higher proportions of ethylene oxide were less effective, possible because of their increased water solubility.

FATTY ALCOHOL–ETHOXYLATED FATTY ALCOHOL COMBINATIONS   Very stiff, stable foams can be obtained with cetyl alcohol–stearyl alcohol–

ethoxylated fatty alcohol combinations. The main disadvantage of these combinations is their general lack of complete solubility in the water–alcohol–propellant system so that suspended solid material is present.

The stiffness of the foams depends upon the ratios of the three components in the surfactant blend. The variation in foam stiffness that results with a change in the composition of surfactant mixtures prepared with cetyl alcohol, stearyl alcohol and different ethoxylated fatty alcohols is illustrated on triangular coordinate charts in Reference 1.

Surfactants for the Immiscible Region (Emulsion Systems).   In the immiscible region of emulsions, the water–alcohol ratio varies from 100/0 to about 38/62 with a "Freon" 12–"Freon" 114 (40/60) propellant concentration of 8–10%. The ethyl alcohol concentration (wt%) in the aerosols varies from 0% to slightly over 53%. At higher alcohol concentrations the water–ethyl alcohol–propellant system becomes miscible in all proportions.

At low alcohol concentrations, the emulsion system is essentially an aqueous system and most of the surface-active agents that are effective in aqueous systems can be used. As the concentration of ethyl alcohol in the emulsion system increases, the number of surface-active agents that give adequate foams decreases. Surface-active agents that gave acceptable foams from an aerosol system consisting of 86% water–ethyl alcohol (50/50) solution, 4% surfactant, and 10% "Freon" 12-"Freon" 114 (40/60) are listed in Table 20–5.

TABLE 20-5 SURFACE-ACTIVE AGENTS FOR AQUEOUS ALCOHOL (50/50) EMULSION SYSTEMS

1. Hydroxylated Lecithin	6. Polyoxyethylene (2) stearyl ether
2. Emulsifying Wax #215	7. Propylene glycol monostearate (SE)
3. Glycerol Monostearate (SE)	8. Sorbitan monopalmitate
4. Kesscowax A-33	9. Sorbitan monostearate
5. "Polawax"	

Propellants

The most commonly used propellants for the aqueous ethyl alcohol foam products are blends of Propellant 12 and Propellant 114 with ratios varying from about 15/85 to 40/60. Pressures in the bottles generally do not exceed 30 psig at 70°F with these propellants as long as the water–ethyl alcohol–propellant mixtures are miscible. Propellant 142b is also used to a limited extent to provide quick breaking foams.

FOAM FORMATION AND STABILIZATION

Foams from the homogeneous, aqueous ethyl alcohol systems are very sensitive to changes in such variables as the water–alcohol ratio and temperature. In order to understand why these variables have such a pronounced effect upon foam properties it is necessary to have a basic comprehension of how foams are formed from these systems and stabilized.

The formation of a foam from a clear aqueous ethyl alcohol aerosol can be explained by two concepts—the stabilization of foams by solid materials and solubility parameters. The data that have been accumulated on the aqueous alcohol foam system indicate that these products belong to the class of foams known as *three-phase foams*. In this type of foam, stability is achieved by the presence of solid materials that are attracted or deposited at the gas–liquid interface. According to Bikerman,[7] the solid materials are partially immersed in the liquid phase and thus prevent the gas bubbles from coming into contact with each other and coalescing. The solid material that performs this function in the aqueous ethyl alcohol foams is "Polawax," the solubility of which is responsible for controlling the properties of the foams.

The application of solubility parameters to the aqueous ethyl alcohol systems is necessary in order to understand how solid "Polawax" is deposited at the gas–liquid interface during discharge of the aerosols. The theory of solubility parameters is covered in some detail in Chapter 10 and in References 2, 8, and 9; however, it might be helpful to mention a few points about solubility parameters at this time. A solubility parameter is a fundamental physical property of a compound, just like density, boiling point or melting point. Solubility parameters for a large number of liquids and solids have been calculated or determined experimentally and are available in tables.[8,9] Solubility parameters are useful because they can be used to predict if an amorphous solid will dissolve in a liquid. If the solubility parameters of the solid and liquid are close to each other, the solid will usually dissolve in the liquid.

One important feature of the solubility parameter concept that applies to the aqueous ethyl alcohol foam system containing "Polawax" is that it explains why two liquids, neither of which by itself is a solvent for a given compound, may dissolve the compound when mixed together. For example, although neither ether nor ethyl alcohol alone is a solvent for nitrocellulose, the mixture of the two liquids will dissolve nitrocellulose. The explanation for this is that ethyl alcohol has too high a solubility parameter to dissolve the nitrocellulose while ether has too low a solubility parameter. However, the average solubility parameter of the mixture of ether and alcohol is sufficiently close to that of nitrocellulose so that the mixture becomes a solvent for the nitrocellulose.

Solubility Properties of "Polawax"

The solubility properties of "Polawax" in a basic aqueous ethyl alcohol foam system explain why it dissolves in the water–ethyl alcohol–propellant mixture but precipitates from solution during discharge of the aerosol. The solubility of "Polawax" in the components of a typical aqueous ethyl alcohol foam aerosol with a composition of 87% water–ethyl alcohol (35/65) solution, 3% "Polawax," and 10% "Freon" 12–"Freon" 114 (40/60) is shown in Table 20–6.

TABLE 20-6 SOLUBILITY PROPERTIES OF "POLAWAX"

Insoluble	Soluble
Water–ethyl alcohol (35/65) solution "Freon" 12–"Freon" 114 (40/60) propellant	Mixture of water–ethyl alcohol solution (35/65) plus "Freon" Propellant

"Polawax" is insoluble in either the water–ethyl alcohol (35/65) solution or the "Freon" 12–"Freon" 114 (40/60) propellant. However, "Polawax" is soluble in the mixture of water–ethyl alcohol (35/65) solution and the "Freon" propellant. This is analogous to the ether–alcohol–nitrocellulose system described previously. In the present case, the solubility parameter of the water–ethyl alcohol (35/65) solution is too high to dissolve the "Polawax" and that of the "Freon" propellant is too low. Apparently the average solubility parameter of the mixture of the water–alcohol solution and the "Freon" propellant is close enough to that of "Polawax" so that the combination of the three liquids dissolves the "Polawax."

In an attempt to verify this theory, the solubility of "Polawax" was determined in a variety of solvents with known solubility parameters. Burrell[8] classified solvents into three groups according to their hydrogen bonding capacity and the solubility of "Polawax" in these three groups of solvents is shown in Table 20–7.

Thus, "Polawax" was soluble in solvents having solubility parameters in the range of 7.5–14.5, depending upon the hydrogen bonding capacity of the solvent. It is insoluble in "Freon" 12–"Freon" 114 (40/60) propellant because the propellant has a solubility parameter of about 6.2 (Chapter 10) which is too low. It is also insoluble in the water–ethyl alcohol (35/65) solution, which indicates that the solution has a solubility parameter greater than 14.5 (the solubility parameter of ethyl alcohol is

12.7 and that of water is 23). When propellant is added to the water–
ethyl alcohol (35/65) solution, the solubility parameter of the resulting
mixture is shifted downward to where the mixture becomes a solvent for
the "Polawax." The solubility parameter of the water–ethyl alcohol (35/
65) solution apparently is very close to the upper limit for that of "Polawax" so that only a comparatively minor change in solubility parameter is
enough to cause the mixture to become a solvent for "Polawax."

TABLE 20-7 SOLUBILITY OF "POLAWAX" IN VARIOUS SOLVENTS

Hydrogen Bonding Capacity of Solvents	Insoluble in Solvents with these Solubility Parameters	Soluble in Solvents with these Solubility Parameters
Low hydrogen bonding	7.4, 10.7	7.5–10.0
Medium hydrogen bonding	14.8	8.4–10.8
High Hydrogen bonding	9.1, 16.5	9.5–14.5

The Mechanism of Foam Formation

The mechanism of foam formation can now be understood on the basis
of the solubility properties of "Polawax" in the water–ethyl alcohol–propellant system assuming that solid "Polawax" is necessary to stabilize the
foam structure. As shown in the previous section, the presence of both
the aqueous alcohol solution and the propellant are necessary in order to
dissolve "Polawax." However, when the aerosol is discharged, the propellant changes from a liquid to a gas and is no longer available as a solvent.
Since "Polawax" is not soluble in the remaining water–ethyl alcohol
(35/65) solution, the "Polawax" precipitates, thus providing the solid
form of the surfactant required for stabilization of the foam (Figure 20–4).
The mechanism of foam formation and stabilization described above

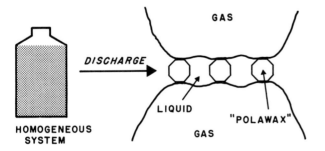

Figure 20–4 Foam formation from aqueous alcohol systems.

applies to a clear homogeneous aerosol system. It should be emphasized, however, that it is not necessary for the "Polawax" to be in solution initially in order to obtain a stable foam. All that is required is for "Polawax" to be available in solid form after the propellant has been vaporized. If the "Polawax" is insoluble in the aerosol initially, foam stabilization will still occur because insoluble "Polawax" is present for foam stabilization.

FACTORS AFFECTING FOAM STABILITY

The evidence in the previous section indicated that one of the requirements for a stable foam was the presence of a solid surfactant after the product had been discharged. If this is true, then if the surfactant is soluble in the aqueous alcohol concentrate and remains in solution after the product has been discharged, no foam will be obtained. This explains why changes in the water–alcohol ratio and the temperature affect foam stability and it also explains why the addition of certain solvents to the aqueous ethyl alcohol foam system destroys the foam.

Variation in Alcohols

Ethyl Alcohol. Aqueous ethyl alcohol systems prepared with 4% "Polawax" and 10% "Freon" 12–"Freon" 114 (40/60) propellant gave stable foams throughout the entire immiscible region of emulsions regardless of the water–ethyl alcohol ratio, but in the miscible region where homogeneous products are obtained, the water–ethyl alcohol ratio is one factor that determines foam stability (Table 20–8).

Stable foams were obtained with water–ethyl alcohol ratios varying from 100/0 to about 40/60. At lower water–alcohol ratios, the alcohol concentration becomes very important and no foam is obtained at a lower ratio than 30/70.

The solubility of "Polawax" in both the aerosols and the aqueous ethyl alcohol concentrates is also listed in Table 20–8. The data indicate that the solubility of "Polawax" in the aqueous ethyl alcohol concentrate is the factor that controls the foam stability. Throughout the immiscible region, "Polawax" is insoluble in both the aqueous ethyl alcohol concentrates and the aerosols, and, therefore, solid surfactant is always available to stabilize the foam after the product is discharged.

In the miscible region, "Polawax" is soluble in the aerosol with the water–alcohol 35/65 ratio but insoluble in the concentrate and precipitates out when the product is discharged and stabilizes the foam. However, it is almost completely soluble in the concentrate with the 30/70

TABLE 20-8 FOAM STABILITY AND STIFFNESS AS A FUNCTION OF THE WATER–ALCOHOL RATIO

Water–Ethyl Alcohol Ratio (wt. %)	Foam Stiffness (g)	Foam Stability	Type of System	Solubility of "Polawax"	
				Aerosol	Aqueous Alcohol Concentrate
100/0	50	Stable	Emulsion	Insoluble	Insoluble
75/25	47	"	"	"	"
50/50	61	"	"	"	"
40/60	72	"	"	"	"
35/65	61	Medium stability	Clear, homogeneous	Soluble	"
30/70	25	Low stability	"	"	Almost soluble
25/75	0	No foam	"	"	Soluble

* The dotted line separates the immiscible and miscible regions.

ratio and therefore most of the "Polawax" remains in solution when the product is discharged. The resulting foam is thin and unstable. No foam is obtained with the 25/75 water–alcohol ratio because "Polawax" is soluble in this concentrate.

Other Alcohols and Acetone.   Stable foams may be obtained from aqueous solutions of methyl alcohol and acetone at the same water–alcohol or water–acetone ratios as those with ethyl alcohol or even lower. Isopropyl alcohol gives only thin foams at very high water–alcohol ratios.

The stability and stiffness of foams from the three alcohols and acetone at four water–solvent ratios are given in Table 20–9. Although foam stiffness values do not correlate directly with foam stability, they are an indication of foam stability to the extent that a foam has to have a certain degree of stability in order to have a stiffness value.

TABLE 20-9   FOAM STIFFNESS WITH VARIOUS ALCOHOLS AND ACETONE*

Water–Alcohol or Water–Acetone Ratio (*wt %*)	Foam Stiffness (g)			
	Methyl Alcohol	*Acetone*	*Ethyl Alcohol*	*Isopropyl Alcohol*
35/65	46	96	61	No foam
30/70	52	51	25	No foam
20/80	44	No foam	No foam	No foam
15/85	No foam	No foam	No foam	No foam

* Aerosol Formulation = 86% Water-alcohol or acetone mixture, 4% "Polawax," 10% "Freon" 12–"Freon" 114 (40/60) propellant.

Although no data are available regarding the solubility of "Polawax" in the aqueous methyl alcohol, acetone or isopropyl alcohol concentrates, it is probably this factor that determines the stability of foams from systems containing these solvents.

Temperature

In the miscible region, increasing the temperature at which the aerosol is discharged decreases foam stability and conversely, lowering the temperature increases foam stability. This is illustrated by the data in Table 20–10, obtained with an aerosol formulation of 87% water–ethyl alcohol solution, 3% "Polawax" and 10% "Freon" 12–"Freon" 114 (40/60) propellant.

TABLE 20-10 EFFECT OF TEMPERATURE UPON FOAM STABILITY

Water–Ethyl Alcohol Ratio (wt %)	Foam Stability	
	Room Temperature	Other Temperatures
35/65	Good	No foam at 110°F
30/70	Poor	Fair at 60°F
25/75	No foam	Unstable, large bubble size foam at 32°F

The effect of temperature upon foam stability is related to its effect upon the solubility of "Polawax" in the water–ethyl alcohol concentrate. The formulation with the water–alcohol ratio of 35/65 does not foam at 110°F because "Polawax" is soluble in the concentrate at this temperature and consequently does not precipitate when the product is discharged. This is also why these aqueous alcohol foams break so rapidly when discharged on the skin. The warmth of the skin causes the solid "Polawax" to dissolve in the aqueous ethyl alcohol phase in the foam and this leads to collapse of the foam.

When the temperature of the products with the 30/70 or 25/75 water–alcohol ratios is decreased, foam stability is increased because the solubility of "Polawax" is less at the lower temperatures. Therefore, more "Polawax" precipitates from the aerosol when it is discharged at the lower temperatures and increased foam stabilization results.

TABLE 20-11 POLAWAX CONCENTRATION AND FOAM STIFFNESS

Components	Composition (wt %)			
	#1	#2	#3	#4
Water–ethyl alcohol (35/65) solution	86.0	88.0	88.5	89.0
"Polawax"	4.0	2.0	1.5	1.0
"Freon" 12–"Freon" 114 (40/60) propellant	10.0	10.0	10.0	10.0
Foam stiffness (g)	73	50	42	Foam too thin and unstable to measure

Concentration of Surface-Active Agent

Since the presence of a surface-active agent is necessary in order to obtain a stable foam, the concentration of the agent would be expected to be one of the factors controlling foam stability. The variation of foam stiffness with a change in concentration of "Polawax" in aqueous ethyl alcohol systems is shown in Table 20–11.

Most formulators try to use as low a concentration of surface-active agent as possible.

Variation in Propellants

The type of propellant can have a considerable effect upon foam properties. A number of propellants were evaluated in an aerosol system consisting of 86% water–ethyl alcohol (35/65) solution, 4% "Polawax" and 10% propellant with the following results:

"Freon" 12. This gave a sputtery discharge and a very dense, cake icing type of foam. The foam wet paper rapidly but retained its structure.

"Freon" 114. This propellant gave a product with an excellent discharge and a stable foam with a stiffness of 78 g.

"Freon" 12–"Freon" 114 (40/60). The discharge characteristics were excellent and the stable foam had a stiffness of 68 g.

"Freon" 12–"Freon" 11 (50/50). This propellant produced a thin, unstable foam that collapsed in a few minutes on a paper.

Propellant 142b. The aerosol with this propellant produced a thin foam with a stiffness of approximately 19 g. The foam collapsed in about 20 min.

Propellant 152a. The effect of propellant 152a was similar to that of "Freon" 12. It produced a very dense, thick foam that caused rapid wetting of paper but exhibited good stability. "Polawax" was not soluble in the aerosol formulation and crystallized out, resulting in an unattractive appearing product.

Effect of Other Solvents

The presence of other solvents may be necessary in order to change foam properties or add active ingredients. The solvent that is added can either enhance foam properties or destroy them. The effect of various solvents upon the foam properties of an aqueous alcohol system with a composition of 76% water–ethyl alcohol (35/65) solution, 10% solvent, 4%

"Polawax," and 10% "Freon" 12–"Freon" 114 (40/60) propellant is shown in Table 20–12.

All solvents that gave good foams had two characteristics in common— they were soluble in the aqueous ethyl alcohol concentrate, and their presence in the concentrates did not give water–alcohol–solvent mixtures in which "Polawax" was soluble. Solubility in the "Freon" propellant was not a determining factor since some of the solvents were soluble in the propellants while others were not. Some of the solvents produced exceptionally stiff foams.

TABLE 20-12 EFFECT OF SOLVENTS UPON FOAM PROPERTIES

Solvents Giving Good Foams	Solvents Having an Adverse Effect upon Foam Formation and Stability
Acetone	Chloroform
Diethylene glycol	Hexane
Ethylene glycol	Isopropyl myristate
Ethyl acetate	Methyl isobutyl ketone
Glycerine	Odorless mineral spirits
Methyl ethyl ketone	Silicone fluid DC 200
	(500 cps)
Propylene glycol	Toluene
	Trichloroethylene

The adverse effect of the other solvents was due to several reasons. In some cases the "Polawax" became soluble in the concentrate after the addition of the solvent so that the "Polawax" remained in solution after the product was discharged. In other cases the solvents were not miscible with the water–alcohol concentrate, but the "Polawax" dissolved in the added solvent and, therefore, did not precipitate when the product was discharged. The addition of some solvents resulted in the conversion of the homogeneous system to an emulsion system.

PREPARATION OF AQUEOUS ALCOHOL FOAMS

The insolubility of surface-active agents, such as "Polawax," in the aqueous ethyl alcohol concentrate at room temperature necessitates special precedures for the commercial production of these aerosols. There are essentially three methods by which the concentrate can be loaded into the containers.

1. "Polawax" can be dissolved in the aqueous ethyl alcohol solution at about 110°F, and the warm solution of "Polawax" in the aqueous alcohol mixture can be added to the containers in one step. This would require equipment to keep the solution hot until it was added to the containers.

2. "Polawax" can be dissolved in ethyl alcohol at room temperature, and the concentrate loaded into the containers in two stages, where the aqueous phase is loaded in one step and the alcoholic solution of "Polawax" is loaded in the second step. The "Polawax" will precipitate in the container when the aqueous and ethyl alcohol solutions are mixed but the surfactant will redissolve after the propellant has been added.

3. "Polawax" can be dispersed in the aqueous alcohol concentrate at room temperature and the dispersion added to the containers. The dispersion would have to be sufficiently stable so that it remained homogeneous during the addition. The use of other surfactants as an aid in obtaining a stable dispersion of "Polawax" might be helpful.

STORAGE STABILITY OF AQUEOUS ETHYL ALCOHOL FOAMS

High spot storage stability tests with a typical aqueous ethyl alcohol after-shave foam indicated that tinplate containers should not be used with this product as a result of corrosion and discoloration. In view of this, extensive storage stability tests should be carried out with any aqueous alcohol products packaged in tinplate containers. Most of the aqueous alcohol foam products on the market are packaged in glass and a few in aluminum containers.

SOME PROBLEMS WITH AQUEOUS ALCOHOL FOAMS

Normally, "Polawax" is soluble in the aerosol formulations at room temperature and gives a clear, homogeneous product. However, if the aqueous ethyl alcohol foam products are cooled slightly below room temperature, the "Polawax" may precipitate from solution. This causes an unsightly appearance and is the reason why practically all aqueous alcohol foam products are packaged in opaque glass bottles. The precipitated "Polawax" does not redissolve easily unless the samples are warmed up to room temperature or above and shaken. If the aerosol is used before the surfactant is redissolved, valve clogging may occur.

Attempts to solve this problem by adding additional solvents which would prevent the "Polawax" from precipitating at the lower temperatures generally have been unsuccessful. Any solvent that achieved this objective also increased the solubility of "Polawax" in the aqueous alcohol concentrate at room temperature and this, of course, destroyed the foam stability. One partial solution to this problem may be the use of additional dispersing agents to keep the precipitated "Polawax" in a finely divided form so that valve clogging does not occur.

Another problem that arises occasionally is precipitation of "Polawax" after most of the product has been used. The reason for this is as follows: as the vapor phase in the container increases (a result of product use), propellant vaporizes from the residual liquid phase and enters the vapor phase. This reduces the concentration of propellant in the liquid phase, and ultimately it becomes too low to keep the "Polawax" in solution. This problem can be minimized by keeping the surfactant concentration as low as possible, or by increasing the propellant concentration in the aerosol.

AEROSOL FORMULATIONS

A number of formulations for various aqueous ethyl alcohol foam products have been published and are available in References 1, 3, 4, 5, and 6.

REFERENCES

1. P. A. Sanders, *Drug Cosmetic Ind.* **99,** (August and September 1966).
2. P. A. Sanders, Aerosol Report/a e r, **8,** 202 (May, 1969) ("Freon" Aerosol Report A-75).
3. M. Lanzet, 1,121,563.
4. K. Klausner, U. S. Patent 3,131,152.
5. "Quick-Breaking Foam Aerosols," *Aerosol Age* **5,** (May 1960).
6. T. A. Wallace, *Amer. Perfumer Cosmetics* **77,** 85 (1962).
7. J. J. Bikerman, "Foams, Theory and Industrial Applications," p. 184, Reinhold Publishing Corp., New York, N.Y., 1953.
8. H. Burrell, *Offic. Dig. Federation Paint Varnish Prod. Clubs* **27,** 726 (1955).
9. J. D. Crowley, G. S. Teague, and J. W. Lowe, *Paint Technol.* **38,** 269, (1966).

21

NONAQUEOUS FOAMS

In contrast to the large number of investigations of conventional aqueous emulsions and foams, relatively little information has been published on nonaqueous systems. Most of the work in the nonaqueous field has involved studies of the formation and behavior of micelles and evaluations of mixtures of surfactants in organic solvents for applications such as dry cleaning and lubrication. The surfactants used for these systems are discussed by Schwartz et al.[1]

King[2] studied the foaming properties of a considerable number of organic liquids containing commercial surface-active agents. Methyl alcohol, ethylene glycol, glycerol, benzene, nitrobenzene and chlorinated naphthalene were typical of the liquids investigated. Several combinations of surface-active agents and organic liquids gave foams that compared favorably in volume and stability with those from aqueous systems.

Recently, Petersen and Hamill[3] investigated nonaqueous emulsions of olive oil with glycerine, propylene glycol, and polyethylene glycol 400. Typical anionic, nonionic, and cationic surface-active agents were used as emulsifying agents. Both olive oil-in-polyol and polyol-in-olive oil emulsions were obtained and some emulsions were quite stable at very low surfactant concentrations.

In the aerosol field, studies of nonaqueous systems have been equally limited. Klausner[4] obtained nonaqueous foams from various combinations of alcohols, dialkyl ketones, glycerol, alkylene glycols, surfactants and propellants. Nonaqueous foams have also been produced from combinations of surface-active agents and fluorocarbon propellants with glycols, glycol derivatives, or mineral oils. These investigations have been reported in References 5 and 6. Unless otherwise noted, the data and results on nonaqueous foams presented in the following sections of this chapter were obtained from the latter two publications.

Thus far, there are essentially no aerosol products on the market based on the nonaqueous foam system, possibly because the nonaqueous foams are still relatively new. However, they have a number of advantages and should be useful for many applications in the cosmetic and pharmaceutical fields. Foam properties can be varied over a wide range, and extremely stable, as well as quick breaking foams, can be obtained. The nonaqueous foam systems are particularly useful for active ingredients that are sensitive to moisture or are more soluble in glycols or mineral oils than water. Mineral oil foams provide hydrophobic films on the skin that resist removal by water and, therefore, are desirable for products such as suntans, etc. Compositions for a variety of products formulated as nonaqueous foams are given in References 4 and 5.

GLYCOL FOAMS

Glycol foams are usually formulated as emulsion systems with the propellant dispersed in the glycol or glycol derivative. The surface-active agent is normally present as a suspended solid. In a few instances, the propellant may be soluble in the glycol concentrate, and in some formulations the surfactant may be soluble in the glycol–propellant mixture, but in all cases the surfactant is insoluble in the glycol concentrate.

The glycol foams belong to the class of three phase, or solid-stabilized foams, thus being similar in this respect to the aqueous ethyl alcohol foams. Some of the glycol foams are much stiffer than either aqueous or aqueous ethyl alcohol foams. Stiffness, as well as other properties of the glycol foams, depends upon the type of glycol, surface-active agent, and propellant that are present. The influence of these variables upon foam properties is discussed in the following sections.

Type of Glycol (or Glycol Derivative)

The type of glycol is one of the major factors that determines whether a foam is produced. For example, the stability of foams from 25 different glycols and glycol derivatives varied from over 48 hr to less than 1 min. Seven glycols gave no foam. The glycols were evaluated in a basic system with a composition of 86% glycol, 4% propylene glycol monostearate, and 10% "Freon" 12–'Freon" 114 (40/60) propellant.

The specific properties of a glycol that determine whether it will foam with a given surfactant and propellant have not been completely clarified. A considerable amount of evidence indicates that the surfactant must be insoluble in the glycol concentrate for foam formation and, therefore, the solubility of the surfactant in the glycol is one of the controlling factors.

However, this is not the only factor because insolubility of the surfactant in the glycol is not sufficient itself to ensure foam formation. Some surfactants do not produce foams although they are insoluble in the glycol. The extent to which the surfactant is wetted by the glycol and propellant is probably one of the major factors that determines the effectiveness of the surfactant.

The solubility of the propellant in the glycol may play a part in foam formation, since glycols in which the propellants are least soluble give the most stable foams. This can be seen by comparing the stability of foams from different glycols with the solubility of the propellant in the glycols.[1,7] However, this apparent correlation may be deceptive. It is possible that glycols with the least solubility for the propellants also have the least solubility for the surfactants and that the latter is the only important factor. The fact that propellant and surfactant solubility in the glycols sometimes appear to go hand-in-hand is shown by the data in Table 21–1, which lists the foaming characteristics of nine glycols in an aerosol formulation consisting of 86% glycol, 4% "Polawax" (an ethoxylated stearyl alcohol) and 10% "Freon" 12–"Freon" 114 (40/60). These data do not indicate whether the important factor in foam stability is the solubility of the surfactant in the glycol or the solubility of the propellant. However, other evidence indicates that it is the solubility of the surfactant.

TABLE 21-1   VARIATION IN GLYCOLS AND FOAM PROPERTIES*

Glycol	Solubility of "Polawax" in Glycol	Solubility of Propellant in Glycol	Foam** Stability
Propylene Glycol	Insoluble	Insoluble	> 16 hr
Diethylene Glycol	Insoluble	Insoluble	> 16 hr
Triethylene Glycol	Insoluble	Insoluble	> 16 hr
Tetraethylene Glycol	Insoluble	Insoluble	> 16 hr
Polyethylene Glycol 400	Insoluble	Insoluble	> 16 hr
Dipropylene Glycol	Soluble	Soluble	No foam
Tripropylene Glycol	Soluble	Soluble	No foam
2-Ethyl Hexanediol-1, 3	Soluble	Soluble	No foam
Hexylene Glycol	Soluble	Soluble	No foam

* Aerosol Formulation: 86% Glycol, 4% "Polawax," 10% "Freon" 12–"Freon" 114 (40/60) Propellant.
** Time before observable collapse occurred.

The choice of a particular glycol for a product depends not only upon the foam stability and stiffness desired but also upon other factors such as the toxicity of the glycol and its solubility for the active ingredients. Toxicity considerations are extremely important, particularly for topical applications. Propylene glycol, 1,3 butylene glycol and the polyethylene glycols are relatively low in toxicity, judging by the data which have been reported for these compounds.[8-12] This is fortunate since these glycols also give stable foams.

Glycerine, although low in toxicity, is not as satisfactory by itself for nonaqueous foams as some of the glycols. Glycerine forms extremely viscous systems or gels with many surfactants and this makes it difficult to load the concentrate into the containers unless the concentrate is hot. Also, the propellant should be added to the hot concentrate; otherwise, the mixture of propellant and concentrate is difficult to emulsify because of the viscosity of the concentrate. Glycerine has many excellent properties, however, and is used widely in cosmetics and pharmaceuticals. It certainly could be used in combination with the glycols in the nonaqueous foam systems.

Emulsions prepared with propylene glycol, 1,3 butylene glycol or the low molecular weight polyethylene glycols have relatively low viscosities. This permits easy redispersion of the surfactant and propellant if phase separation occurs during storage. The discharge characteristics of the emulsions are satisfactory.

Surface-Active Agents

Many of the surface-active agents that are effective in the aqueous ethyl alcohol foam system are also good foaming agents for the nonaqueous glycol foams. The ethoxylated alcohols and the fatty acid esters are particularly useful. "Polawax," an ethoxylated stearyl alcohol, is one of the preferred surfactants.

Ethoxylated Fatty Alcohols. The efficiency of the ethoxylated fatty alcohols as foaming agents in the nonaqueous systems is related to the number of ethylene oxide groups in the surfactant molecule. Foam stability decreases as the ethylene oxide content increases; therefore, the most hydrophobic surfactants are the most effective. This is illustrated by the data in Table 21-2, which gives the properties of foams obtained with a series of ethyoxylated fatty alcohols in an aerosol system consisting of 86% propylene glycol, 4% ethoxylated fatty alcohol, and 10% "Freon" 12–"Freon" 114 (40/60) propellant. The number in parenthesis in the designation of the ethoxylated fatty alcohol surfactant indicates the average number of moles of ethylene oxide per mole of the fatty alcohol in the surfactant.

As a general rule, the least stable foams had the lowest stiffness values. Fairly stiff, stable foams can be obtained at surfactant concentrations as low as 1%, as shown by the data in Table 21–3. A typical aqueous shaving lather foam will have a stiffness value in the range of about 60–100 g.

TABLE 21-2 ETHOXYLATED FATTY ALCOHOLS IN PROPYLENE GLYCOL FOAMS

Ethoxylated Fatty Alcohol	Trade Name**	Foam Stiffness (g)	Foam Stability
Ethoxylated stearyl alcohol	"Polawax"	222	> 24 hr
POE* (1) stearyl–cetyl alcohol	"Siponic" E-0	133	> 24 hr
POE (2) stearyl–cetyl alcohol	"Siponic" E-1	134	> 24 hr
POE (4) stearyl–cetyl alcohol	"Siponic" E-2	112	> 24 hr
POE (6) stearyl–cetyl alcohol	"Siponic" E-3	27	< 10 min
POE (8) stearyl–cetyl alcohol	"Siponic" E-4	17	< 10 min
POE (2) cetyl ether	"Brij" 52	58	> 24 hr
POE (2) stearyl ether	"Brij" 72	126	> 24 hr
POE (10) cetyl ether	"Brij" 76	9	< 10 min

* POE = Polyoxyethylene.
** The "Siponic" and "Brig" series of surfactants are manufactured by the Alcolac Chemical Corp., and Atlas Chemical Industries, Inc., respectively. "Polawax" is supplied by Croda, Inc.

TABLE 21-3 CONCENTRATION OF "POLAWAX" AND FOAM PROPERTIES

	Compositions (wt %)				
"Polawax"	Propylene Glycol	"Freon" 12– "Freon" 114 (40/60)	Foam Stiffness (g)	Foam Stability
1	89	10	88	< 8 hr
2	88	10	154	> 16 hr
3	87	10	240	> 16 hr
4	86	10	222	> 16 hr

The solubility of the ethoxylated fatty alcohol in the glycol appears to be one of the major factors that controls the stiffness and stability of nonaqueous foams. This is indicated by the data in Table 21–4, which gives the properties of foams obtained from a nonaqueous system of 86% polyethylene glycol 400, 4% ethoxylated fatty alcohol and 10% "Freon" 12–"Freon" 114 (40/60) propellant. Only the surfactants that were insoluble in the polyethylene glycol 400 gave stable foams.

TABLE 21-4 FOAM STABILITY AND SURFACTANT SOLUBILITY IN POLYETHYLENE GLYCOL 400 FOAMS

Surfactant	Trade Name*	Solubility of Surfactant in Glycol	Foam Stiffness (g)	Foam Stability
Ethoxylated stearyl alcohol	"Polawax"	Insoluble	192	4-16 hr
POE (4) lauryl ether	"Brij" 30	Soluble	—	< 3 min
POE (23) lauryl ether	"Brij" 35	Insoluble	26	4-16 hr
POE (2) cetyl ether	"Brij" 52	Insoluble	162	> 16 hr
POE (10) cetyl ether	"Brij" 56	Soluble	—	< 3 min
POE (20) cetyl ether	"Brij" 58	Insoluble	84	4-16 hr
POE (2) stearyl ether	"Brij" 72	Insoluble	118	> 16 hr
POE (10) stearyl ether	"Brij" 76	Insoluble	38	> 16 hr
POE (2) oleyl ether	"Brij" 92	Soluble	—	< 3 min
POE (10) oleyl ether	"Brij" 96	Soluble	—	< 3 min

Stiffness measurements could not be carried out on foams with a stability less than 3 min. The densities of the foams ranged from 0.12–0.20 g/cc.

* "Polawax," Croda, Inc., "BRIJ" compounds, Atlas Chemical Ind.

Other Surfactants. Propylene glycol monostearate (self-emulsifying) is an excellent surfactant for glycol foams but a little more difficult to disperse in the glycols than compounds such as "Polawax." Other surfactants that give stable foams are listed in Reference 5 and include products such as sorbitan monostearate and cetyl alcohol.

Propellants

Type of Propellant. Almost any "Freon" 12–"Freon" 114 blend will give excellent foams from the glycols and the choice of a particular blend is usually determined by pressure and economic considerations. Generally, it is advisable to use a blend with at least 15% "Freon" 12 so that the vapor pressure will be high enough to obtain a satisfactory discharge. There appears to be a tendency for foam stability to increase with increasing concentrations of "Freon" 12 in the blend.

"Freon" 114 propellant gave a slow discharge as a result of its low vapor pressure and produced a shiny, pasty foam that continued to expand after discharge. "Freon" 12–"Freon" 11 combinations gave less attractive and less stable foams than the "Freon" 12–"Freon" 114 blends and were generally of little interest.

The vapor pressures of the glycol emulsion systems were slightly less than that of the pure propellants.

Concentration of Propellant. Propellant concentrations of 8–10% are generally used for nonaqueous foams. These concentrations give a satisfactory discharge and excellent foams. At lower propellant concentrations, such as 5%, the product may discharge as a liquid but the liquid will subsequently expand into a good foam. At propellant concentrations higher than 10%, the coherence of the foam increases markedly and at concentrations of 60–90%, the nonaqueous systems can be sprayed to give a product similar to cold cream. This product subsequently expands into a stable foam.

Effect of Ethyl Alcohol

Ethyl alcohol increases the pressure sensitivity of the foams. For example, a nonaqueous system consisting of 76% polyethylene glycol 400, 10% ethyl alcohol, 4% propylene glycol monostearate or "Polawax," and 10% "Freon" 12–"Freon" 114 (40/60) gave an excellent foam and one in which the pressure of the fingers caused an immediate breakdown of the foam structure with liquefaction. The extent to which the foam is pressure sensitive can be regulated by the concentration of ethyl alcohol and this depends upon the glycol and surfactant present. In the foam system described above, increasing the concentration of ethyl alcohol beyond 15% destroyed the foam.

Type of Discharge

Some glycol systems foam as soon as the product leaves the container while others discharge a liquid stream that subsequently expands into a foam. The way in which the products are discharged is determined by the propellant and surfactant present.

Surfactants that are insoluble in both the glycol concentrate and the glycol–propellant mixture will produce foams as soon as the product is discharged. When the surfactant is insoluble in the glycol concentrate but soluble in the glycol–propellant mixture, the product discharges as a liquid which expands into a foam later. For example, the surfactant "Siponic" E-O is insoluble in propylene glycol but essentially soluble in the glycol–propellant mixture. The glycol–propellant emulsion is hazy at room temperature but there is no other evidence of solid surfactant. This product discharges as a liquid which slowly expands into an extremely dense and stable foam. Expansion is essentially complete in a few minutes.

The type of discharge with combinations of "Siponic" E-O with various glycols is shown in Table 21–5. In both cases where the "Siponic" E-O was soluble in the glycol–propellant mixture, the product discharged as a liquid which foamed later.

TABLE 21-5 EFFECT OF "SIPONIC" E-0 IN GLYCOL FOAMS*

| | Solubility of "Siponic" E-0 | | |
Glycol	In Glycols	In Glycol–Propellant Mixture	Type of Discharge
Propylene Glycol	Insoluble	Essentially sol.	Liquid
Diethylene Glycol	Insoluble	Insoluble	Foam
Triethylene Glycol	Insoluble	Insoluble	Foam
Tetraethylene Glycol	Insoluble	Essentially sol.	Liquid
Polyethylene Glycol 400	Insoluble	Some solid pres.	Liquid-foam

* Aerosol Formulation: 86% Glycol, 4% "Siponic" E-0, 10% "Freon" 12–"Freon" 114 (40/60) Propellant.

The "Siponic" E-O–propylene glycol–"Freon" propellant system will exhibit either type of discharge, depending upon the temperature. If this system, which discharges a liquid at room temperature, is cooled a few degrees below room temperature, some of the surfactant separates from solution. The cooled product with solid surfactant now foams immediately during discharge. If the product is warmed so that most of the surfactant dissolves, the product again discharges a liquid stream that slowly expands into a foam. This is further evidence that solid surfactant must be present to stabilize the foam.

The substitution of "Freon" 114 for a "Freon" 12–"Freon" 114 blend in a system that normally produces a foam discharge, usually results in a product which gives a slow foamy liquid type of discharge. The slow discharge is due to the low vapor pressure of the "Freon" 114.

Formation and Stabilization of Foams

All evidence indicates that the aerosol glycol foams are three-phase, solid-stabilized foams, which puts them in the same category as the aqueous ethyl alcohol foams. The results listed in Table 21–4, which show the effect of various surfactants upon the stability of foams from polyethylene glycol 400, indicate that foams of significant stability are obtained only when the surfactant is insoluble in the glycol. The effect of the solubility of the surfactant in the glycol–propellant mixture upon the type of discharge is a further indication that solid surfactant must be present for foam formation.

If the surfactant is insoluble in both the glycol and the glycol–propellant mixture, foam formation occurs immediately during discharge because solid surfactant is present to stabilize the foam. If the surfactant is insoluble in the glycol but soluble in the glycol–propellant mixture, the vaporization of the propellant during discharge causes the surfactant to separate from solution. The solid surfactant resulting stabilizes the foam. This is similar to the way in which foams are formed and stabilized in the aqueous ethyl alcohol system. The fact that glycol systems of this type discharge as liquids that expand later into foams may be due to the higher viscosity of the glycols and the slower separation of the solid surfactant from solution.

Preparation of Nonaqueous Foams

The surfactants are insoluble in the glycols and as a result the glycol concentrate consists of a dispersion of the surfactant in the glycol. The surfactant should be as finely divided as possible in order to minimize the possibility of valve clogging when the aerosol is discharged.

The standard laboratory procedure for preparing concentrates with ethoxylated alcohols, such as "Polawax," is to heat the mixture of surfactant and glycol until the surfactant dissolves. The solution is allowed to cool to room temperature with gentle stirring, during which time the surfactant separates from solution. In many cases, the dispersion of the surfactant is sufficiently finely divided so that the resulting concentrate can be loaded directly into the aerosol containers. Some surfactants, however, gave a coarse dispersion in the glycol. These concentrates were passed through a hand homogenizer that reduced the particle size of the surfactant sufficiently to give a fine dispersion.

In some systems, the concentrates gelled at room temperature or were too viscous to pour into glass bottles. Concentrates of this type had to be loaded into bottles while still warm enough to be fluid. The propellant was added to the warm concentrate and the samples were shaken while they were cooling to room temperature in order to emulsify the propellant. If the concentrates gelled before the propellant was added, emulsification by shaking was difficult to achieve. When this happened, it was necessary to repeat the process of heating and shaking.

Propylene glycol monostearate (self-emulsifying) was an excellent surfactant for glycol foams; but sometimes, the standard procedure of preparing the concentrates did not give a finely divided form of the surfactant. A modified procedure, which usually gave a satisfactory dispersion, involved heating the surfactant with a portion of the glycol. After the surfactant had dissolved, the remainder of the glycol at room temperature was added to the hot solution with stirring. For example, in the preparation of an aerosol with a composition of 86% propylene glycol, 4% propylene glycol monostearate, and 10% propellant, 10 g of the glycol and 4 g of the surfactant were heated until the surfactant dissolved. The remaining 76 g of the propylene glycol at room temperature was added slowly with stirring to the hot surfactant solution and stirring continued until the concentrate reached room temperature. The resultant dispersion could be homogenized if necessary.

MINERAL OIL FOAMS

Aerosol foams may also be obtained from mineral oils. Since the mineral oils have low toxicity, the foams are suitable for topical applications. Potential cosmetic and pharmaceutical products that might be formulated with the mineral oil foam systems include suntans, hair products, baby oils, hormone creams, vitamin creams, and various other ointments.

In most cases, the propellants are soluble in the mineral oils but the surfactants must be insoluble in the mineral oil in order to obtain a foam. The surfactants can either be soluble or insoluble in the "Freon" propellant–mineral oil solution. The properties of the mineral oil foams vary widely, depending upon such factors as the viscosity of the mineral oil, the type of surfactant, and the type and concentration of the propellant. As a general rule, the foams from the mineral oils have a fairly large bubble size and low stiffness values.

Type of Mineral Oil

High viscosity mineral oil gives stiffer foams than low viscosity oil. This is shown by the data in Table 21–6 which lists the stiffness of foams from both high and low viscosity mineral oils containing three different surfactants.

TABLE 21-6 MINERAL OIL VISCOSITY AND FOAM STIFFNESS*

Mineral Oil			Foam
Sabolt Viscosity (100°F)	Trade Name	Surfactant	Stiffness (g)
55–65	"Klearol"	Polawax"	12
345–355	"Kaydol"	"	34
55–65	"Klearol"	POE (2) stearyl ether	14
345–355	"Kaydol"	"	28
55–65	"Klearol"	Cetyl alcohol–stearyl alcohol (50/50)	16
345–355	"Kaydol"	"	24

* Aerosol Formulation: 79% mineral oil, 6% surfactant, 15% "Freon" 12.

Most of the mineral oil foams are fairly thin and have low stiffness values. The mineral oils were obtained from the Sonneborn Division of Witco Chemical Co.

Surfactants

The ethoxylated fatty alcohols or mixtures of cetyl and stearyl alcohol are the most effective surfactants for mineral oil foams. In order to obtain a foam from mineral oil, the surfactant must be insoluble in the mineral oil (illustrated in Table 21–7). Insolubility in the mineral oil is not the only factor that determines foam stability since in several cases surfactants that were insoluble in the mineral oil did not give foams.

Aerosols prepared with either cetyl alcohol or stearyl alcohol alone gave unstable foams while those prepared with a 50/50 mixture gave fairly good foams. In fact, the cetyl alcohol–stearyl alcohol (50/50) mixture is one of the preferred surfactants for the mineral oil foams. The cetyl alcohol–stearyl alcohol mixture also gave better foams than either of the components alone in the aqueous ethyl alcohol system.

Although the data in Tables 21–6 and 21–7 were obtained with 6%

TABLE 21-7 SURFACTANT SOLUBILITY AND FOAM FORMATION*

| | | Solubility of Surfactant | | |
Surfactant	Trade Name**	In Mineral Oil	In Mineral Oil–Propellant Solution	Foam
Ethoxylated stearyl alcohol	"Polawax"	Insoluble	Almost soluble, hazy	Yes
POE (4) lauryl ether	"Brij" 30	Soluble	Soluble	No
POE (2) cetyl ether	"Brij" 52	Soluble	Almost soluble, hazy	No
POE (10) cetyl ether	"Brij" 56	Insoluble	Insoluble	No
POE (2) stearyl ether	"Brij" 72	Insoluble	Almost soluble, hazy	Yes
POE (10) stearyl ether	"Brij" 76	Insoluble	Insoluble	No
POE (2) oleyl ether	"Brij" 92	Soluble	Almost soluble, hazy	No

* Aerosol Formulation: 79% Mineral oil ("Klearol"), 6% Surfactant, 15% "Freon" 12.
** "Polaway", Croda, Inc., "BRIJ" compounds, Atlas Chemical Ind.

surfactant, subsequent experiments have indicated that good foams can be obtained with a surfactant concentration of 4% or less.

Propellants

Mineral oils are miscible with the propellants and therefore depress the vapor pressure of the propellants. Generally, about 10–15% "Freon" 12 is required to give an adequate foam with good discharge. "Freon" 114 gave essentially liquid discharges as a result of its low vapor pressure.

TABLE 21-8   MINERAL OIL FOAMS

Baby Oil Foam

	wt. %
Mineral Oil	78.3
Anhydrous Lanolin	0.5
Hexachlorophene	0.1
Perfume	0.1
Cetyl Alcohol	3.0
Stearyl Alcohol	3.0
"Freon" 12	15.0

Cleansing Cream Foam

	wt. %
Petrolatum	15.0
Mineral Oil	53.9
Beeswax	4.0
Paraffin	8.0
Perfume	0.1
Cetyl Alcohol	2.0
Stearyl Alcohol	2.0
"Freon" 12	15.0

Brilliantine

	wt. %
Mineral Oil	80.9
Perfume	0.1
Cetyl Alcohol	2.0
Stearyl Alcohol	2.0
"Freon" 12	15.0

Hair Pomade

	wt. %
Mineral Oil	63.9
Stearic Acid	8.0
Spermaceti	3.0
Paraffin	10.0
Perfume	0.1
"Freon" 12	15.0

"Freon" 12–"Freon" 114 (40/60) propellant gave good foams and the discharge was somewhat quieter than that with "Freon" 12 alone in the one system that was tested. Propellant 142b gave a noisy discharge while Propellant 152a gelled the mineral oil.

Foam Formation and Stabilization

Although the data are somewhat limited, the evidence indicates that the mineral oil foams are also three-phase, solid-stabilized foams as are the aqueous ethyl alcohol and glycol foams. The mineral oil systems are quite similar to the aqueous ethyl alcohol systems since in both cases the propellant is miscible with the concentrate, resulting in a homogeneous system.

Aerosol Formulations

No formulations for products based on mineral oil foams are available in the literature. The composition of four potential products are listed in Table 21–8. Several of these are very viscous and must be loaded hot.

COMBINATIONS OF ALCOHOLS OR KETONES WITH GLYCERINE OR GLYCOLS

Klausner[4] has reported that foams of limited stability can be obtained from homogeneous liquid compositions consisting of an alcohol or ketone, glycerine or glycol, surfactant, and propellant. A foam of limited stability is defined as a foam that does not become completely liquefied when exposed to the atmosphere for about 15 min or longer but which will revert to a liquid in about 2 sec when heated or rubbed.

The following proportions of the compounds are considered to be critical for achieving the homogenous systems that produce the foams of limited stability:

Alcohol or dialkyl ketone	26–64 wt %
Glycerol or alkylene glycol	28–64
Surface active agents	0.5–5
Propellant	2–30

The surface-active agent may be anionic, nonionic, or cationic. It can be soluble in either the alcohol or dialkyl ketone, or in the glycerol or alkylene glycol, but not in both. Nonionic agents, such as the polyoxyethylated fatty alcohols and the fatty acid esters, are the preferred surfactants. Formulations are given for a suntan lotion, pre-electric shave lotion, body cologne, and paint remover.

REFERENCES

1. A. M. Schwartz, J. W. Perry and J. Berch, "Surface Active Agents and Detergents," Vol. 2, Interscience Publishers, Inc., New York, N.Y., 1958.
2. E. G. King, *J. Phys. Chem.* **48**, 141 (1944).
3. R. V. Petersen, and R. D. Hamill, *J. Soc. Cosmetic Chemists* **19**, 627 (1968).
4. K. Klausner, U. S. Patent 3,131,153, April, 1964.
5. P. A. Sanders, *Aerosol Age* **5**, 33 (1960).
6. P. A. Sanders, *Amer. Perfumer Cosmetics* **81**, 31 (1966).
7. P. A. Sanders, *Aerosol Age* **5**, 26 (1960).
8. G. O. Curme and F. Johnston, American Chemical Society Monograph Series, No. 114, "Glycols," Reinhold Publishing Corp, New York, N.Y., 1952.
9. Dow Chemical Company Technical Bulletin, "Dow Propylene Glycol, U.S.P."
10. Union Carbide Chemicals Company Technical Bulletin, "Carbowax" Polyethylene Glycols.
11. Celanese Chemical Company Specifications Bulletin, N–26–4, 1,3 Butylene Glycol.
12. Dow Chemical Company Technical Bulletin, "Toxicological Information on Dow Glycols."

22

AEROSOL POWDERS

An aerosol powder consists primarily of a suspension of a finely divided solid in a liquefied gas propellant. Other materials may be present, such as auxiliary solids, or perfumes and oils, but the active ingredient is a dispersed solid. When the product is sprayed the propellant evaporates rapidly leaving the dry powder. The sprays may be cool, but are not nearly as cold as alcohol based sprays.

Powder aerosols were suggested as early as 1954,[1-3] and there was a considerable amount of activity in this field because of the obvious advantages of this type of aerosol system for many materials. However, although there are a variety of powder sprays on the market at the present time, the penetration of the aerosol market has never been as large as initially predicted. The main reason for this has been the number of problems with powder sprays, and while many of these have been overcome during the past few years, powder sprays still have to be formulated with care and tested thoroughly before marketing.

Suspensions of pharmaceutical agents in the propellants, designed for inhalation therapy, were among the first aerosol powder products to be marketed.[4] Aerosol bath powders also appeared very early[5] and now constitute one of the largest groups of aerosol powders. Other powder products now available include personal deodorants, dry shampoos, foot powders, fungicides, spot removers, poison ivy sprays, and dry lubricants.

PROBLEMS WITH POWDER AEROSOLS

Some of the problems encountered during the development of powder aerosols were:

Valve Clogging

Clogging of the valve occurs when the concentration of powder is too high and is due to packing of the powder around and in the valve orifices. This can take place even if the initial particle size of the powder is quite small.

According to Beard,[1] clogging results from one or more of the following causes:

Large and/or Needle Shaped Particles. Needle shaped, fibrous, or large particles can mat or form *filter cakes* and clog a valve even at low powder concentrations.

Partially Soluble, Resinous, or Crystalline Materials. Particles that are partially soluble in the propellant tend to increase in size as a result of solution and recrystalization during storage. This can lead to the formation of aggregates that clog the valve. In addition, partially soluble compounds tend to be deposited at the expansion orifices as a result of the rapid evaporation of propellant at these locations.

Agglomerative Sedimentation. Agglomerative sedimentation is defined by Beard as the formation of hard packed cakes of powder in the bottom of the container and hard plugs in the lower end of the dip tube. Powders generally have a different density than that of the liquefied propellant in which they are suspended. If the powder particles are heavier than the liquid phase, they will settle to the bottom under the influence of gravitational forces. The close packing of the particles at the bottom of the container may lead to formation of agglomerates and hard cakes that are much larger in size than the original particles.

Agglomeration of particles can take place before the particles have settled to the bottom, and the increase in particle size of the agglomerates will increase the rate of settling. Kanig and Cohn[6] have shown that agglomeration occurs relatively soon after preparation of the aerosols and reported that the rate of agglomeration of the suspensions with which they worked was practically nil after the first week.

Leakage

Failure of the valve to close completely after operation will result in leakage. This is caused by deposits of powder on the valve seats and is more likely to occur with a hard crystalline material than with a soft product. This problem can be deceptive because the leak can be very slight and almost unnoticeable but enough so that complete loss of propellant will occur during extended storage.

Nonuniform Delivery

In products designed for inhalation therapy, a specified dosage is delivered by means of a metering valve. Agglomeration of the particles can lead to nonuniform delivery because of the change in particle size. This can occur even if the agglomeration is insufficient to cause valve clogging. In addition, since the therapeutic value of the product depends upon the particle size, agglomeration can reduce the efficacy of the product.

Caking and Wall Deposits

Powders with lower specific gravity than that of the liquid phase will rise to the top of the liquid during storage. In some cases, the powder will form deposits or cakes on the container wall above the liquid level. According to Thiel and Young,[7] this is due to a mutual repulsion between the individual particles which causes them to creep up the wall above the liquid phase where they form deposits. Mutual repulsion can arise if all of the particles have the same electric charge. This is particularly serious with a pharmaceutical product formulated to produce a specific dose with a metering valve.

Considering the problems that can occur with powder aerosols, it is apparent that any potential product must be thoroughly tested before the product is marketed. Extensive spraying tests are particularly important to be certain that valve clogging or leakage does not occur during use. Various methods, which have been reported to minimize some of these problems, are discussed in a later section.

BASIC COMPONENTS OF AEROSOL POWDER SPRAYS

Powders

Type of Powder.   The type of powder is determined by the use for which the product has been developed. Typical powders include talcs, insecticides, pharmaceuticals, antiperspirants, clays, iron oxides, and various lubricants. A number of aerosol formulations are listed in References 1 and 8–12, and these give an indication of the variety of powders that are used as active ingredients.

One of the major constitutents of many powder formulations is talc. Before World War II, the best available talcs came from France and Italy, and these talcs are still used extensively in face and bath powders. Some commercial talcs have an average particle diameter as low as 3 μ. Prussin[13] has pointed out that talcs with a platelet structure are preferred to those with needle-like particles since the platelet structure allows the particles to slide or cascade over each other and have less tendency to clog valves.

Concentration. Practically all powders will cause valve clogging or leakage if the concentration is sufficiently high, but the specific concentration at which this occurs depends upon a number of factors, including the type of powder, particle size of the powder, type of valve, and other additives that are present. Therefore, the maximum concentration of powder that can be sprayed without valve clogging has to be determined for each product.

Traditionally, for powders such as talc, 5–10% was considered to be about the highest practical concentration because of the possibility of valve clogging and leakage.[3] However, as a result of the improvements that have been made in powder systems, concentrations of 20% are commonly mentioned in the literature and some of the published formulations contain as high as 33% talc. One system uses 90% powder but this is a different type of formulation and is discussed later. The concentrations of active ingredients in aerosols designed for inhalation therapy are generally quite low and seldom exceed 3%.

Particle Size. There is some variation in the literature with respect to what is considered the maximum possible particle size for powders. 100 μ has been mentioned several times as about the upper limit but it is rather generally agreed that for most products, the particle size should be below 50 μ and preferably below 30 μ.[7–10,12] A 325 mesh screen is commonly used for preparing powders for aerosols because the maximum diameter of a spherical particle that will pass through the screen is 43 μ. Reed[3] has pointed out that since typical aerosol valves do not have orifices much smaller than 400 μ, it is evident that clogging of the valve usually is not the result of the particles being too large but more the result of an accumulation of the powder around and in the valve orifices.

For pharmaceutical products for inhalation therapy, the particles should be in the 1–10 μ range.[7,10] Particles with a diameter greater than 10 μ may not reach the desired area to be treated and particles with a smaller diameter than 1 μ are exhaled.

Solubility. The aerosol powder should be as nearly insoluble in the liquid phase as possible in order to avoid intermittent solution and recrystallization during storage.[3] This leads to an increase in particle size.

The solubility of the powder in water also determines the extent to which the powder is affected by moisture in the formulation. Kanig and Cohn,[6] determined the effect of varying quantities of moisture upon the agglomeration of three powders with differing water solubilities and found that the powder with the highest solubility in water was the least stable in the presence of moisture. For example, in suspensions stabilized with

"Arlacel" 83 (sorbitan trioleate), talc, which is insoluble in water, exhibited signs of agglomeration only at moisture levels of 1.0% or above. Prednisolone, which is slightly soluble in water, was affected by moisture concentrations above the 0.1% level, while the water soluble Isoproternol Hydrochloride was affected at the 0.005% moisture level.

Thiel and Young[7] consider 300-ppm moisture to be the limit for aerosol powder formulations where the powder is water soluble.

Propellants

A large variety of propellant blends have been suggested for use with aerosol powders. Propellant 12–Propellant 11 mixtures with ratios of Propellant 12 to Propellant 11 varying from 20/80–50/50 are commonly used,[1,3,7–9,14] as are mixtures of Propellant 12 and Propellant 114. Thiel et al., have suggested three component blends of Propellant 12, Propellant 114, and Propellant 11 as well as blends with isobutane.[7,10]

According to Reed,[3] a Propellant 12–Propellant 114 (20/80) blend gives soft, dry sprays suitable for cosmetic products while the Propellant 12–Propellant 11 (30/70) blend gives heavier sprays that would be useful for fungicidal and insecticidal products.

The concentrations of the propellants vary anywhere from about 60% in products with high powder concentrations to over 99% for pharmaceutical preparations.

Surface-Active Agents

Many formulators believe that a surface-active agent is a necessity for a stable powder suspension. One function of the surfactant is to retard agglomeration which leads to rapid sedimentation and consequent valve clogging and leakage.

Thiel and his co-workers[7,10] have been very active in the field of surfactants for powders and report that nonionic surfactants with an HLB value less than 10 and preferably in the range of 1–5 are the most effective. The surfactant should be liquid and soluble or dispersible in the liquid phase. The preferred surfactants were "Arlacel" C (sorbitan sesquioleate), "Span" 80 (sorbitan monooleate) and "Span" 85 (sorbitan trioleate). These are products of Atlas Chemical Industries, Inc., Wilmington, Delaware. A survey of the literature indicated that "Span" 85 was the most commonly used surfactant and the concentrations varied from 0.25%–1.0%.[7,9,10]

"Tween" 80 (polyoxyethylene sorbitan monooleate) and "Emcol" 14 (polyglycerol oleate) have also been reported to be helpful in retarding agglomeration.

Miscellaneous Additives

High boiling organic liquids, such as mineral oils and fatty acid esters, are often used in powder formulations. These additives must be soluble in the liquid propellant phase to be effective. Geary and West[8] have listed three ways in which additives of this type are useful in powder formulations: (1) They retard agglomeration; (2) they serve as lubricants for the particles and the valve components; and (3) they reduce dusting. The oils also increase the adherence of the powder to the skin.[3]

The concentrations of mineral oil are normally less than 1%, although some formulations with higher concentrations have been published. The concentration of the mineral oil should be kept as low as possible because of the possibility of lipoid pneumonia from inhalation of droplets of mineral oil.

One of the most commonly used esters is isopropyl myristate. According to DiGiacomo,[14] purified isopropyl myristate is an ideal suspending agent for powdered aerosols. It is reported to prevent the crystallization of certain bacteriological agents, perfume constituents and other ingredients that may crystallize out on long standing. Adhesion of the powder to the skin is also increased. Concentrations of 0.5–1.0% are usually adequate.

"Solulan" 97,* an acetylated lanolin derivative, was found to be particularly useful in retarding agglomeration of talcum powder (unpublished "Freon" Products Laboratory data). "Solulan" 97 appears to wet the powder particles and provide a physical barrier to agglomeration. It also adds lubricity to the system. 1.0% is usually sufficient.

"Cab-O-Sil,"** colloidal silica, is negatively charged and when added to a powder aerosol it gives the system a negative charge. This retards agglomeration. "Cab-O-Sil" also increases the viscosity of the system and thus decreases the rate of settling. The normal concentration in the aerosol is about 0.5%. Whether the "Cab-O-Sil" increases creeping and caking on the container wall as a result of the repulsion between the particles has not been reported.

Zinc stearate increases the adhesion of the powder to the body and also is reported to increase the *slip* of the talc. A concentration of 0.5%–1.0% should be adequate.

Valves

Beard[1] has given an excellent summary of the requirements for a powder valve. For example, the path from the valve seat to the spray orifice should be very short. This minimizes the space in which the product can

---

* "Solulan" 97 is manufactured by the American Cholesterol Company, Edison, New Jersey.
** "Cab-O-Sil" is available from Cabot Corporation, Boston, Massachusetts.

accumulate to collect moisture or fall back to cause clogging. The valve should have a sharp shut-off and a high valve seating pressure in order to minimize leakage. Beard suggests that the valve be operated wide open without any attempt to control the spray rate by varying the pressure on the actuator. Otherwise the product will build up on the valve seat.

At the present time, three companies manufacture valves specifically for powder sprays. The companies and the valves are:

Company	Valve Designation
Aerosol Research Corporation Chicago, Illinois	KN-38 and Parc-39
Risdon Manufacturing Company, Naugatuck, Connecticut	#5832
VCA (Valve Corporation of America) Bridgeport, Connecticut	B-9-F6

The Precision Valve Corporation, Yonkers, New York, has a number of valves which it recommends for powder sprays, depending upon the particular product. Several of the other valve companies either have powder valves in the blueprint-stage or are considering powder valves.

SOME MODIFIED POWDER SYSTEMS

From time to time various modifications in powder systems have been suggested as a means for minimizing problems such as agglomeration and valve clogging. Some of these modified systems are discussed briefly in the following section. Since not all the systems have a particular designation they are discussed under the name or names of the inventors.

W. C. Beard[1]

Beard reports that valve clogging due to agglomerative sedimentation can be eliminated or reduced by use of certain selected propellants or propellant mixtures. He refers to this as the selected propellant process. The system is based upon the discovery that if the propellant has a specific gravity about equal to that of the powder or if the difference between the specific gravities of the powder and the propellant is no greater than 0.4 or 0.5, the powder will remain in a soft and flocculated condition regardless of whether it rises to the top or settles to the bottom. The powder can readily be dispersed throughout the propellant by a mild shaking or swirling action. In some cases, a difference as much as 0.8 can be tolerated.

D. C. Geary and R. D. West[8]

Geary and West conceive of the volume of a powder as consisting of two parts: the volume of the solid particles and the void volume between the particles. The powder system they suggest is based upon the concept that if the void volume between the particles is either filled or enlarged by substances which prevent the particles from agglomerating, a stable suspension will result.

The two types of materials used for filling the void volumes are bulking agents and liquids, including propellants. Bulking agents are themselves very finely divided powders, and Geary and West mention that the use of bulking agents to impart flow characteristics to powders is well known. The preferred bulking agent is "Santocel" 54, a silica gel with a particle size of 0.5–3.0 μ. The amount of bulking agent required is calculated from the bulk and absolute densities of the powders.

The quantities of bulking agents required are generally quite low, being less than 2% in most cases. However, according to Geary and West, the use of such agents allows the formulation of powder aerosols containing as much as 33% talc.

P. E. Gunning and D. R. Rink[11]

The powder system proposed by Gunning and Rink is the reverse of the typical powder system in that the powder constitutes the major portion of the aerosol and is the supporting phase while the propellant is the minor portion. The system is based primarily upon the concept that some powders can absorb sufficient liquefied gas propellant so that they can be discharged while still remaining as free flowing powders. If the active powder itself does not have sufficient capacity for the propellant, a carrier which possesses absorbing power is added. Typical carriers include amorphous silica, crystalline silicates, and metals such as aluminum powder.

The system is illustrated with an insecticidal dust with a composition of 90% insecticide and 10% Propellant 12. In another example, a medicated foot powder with a composition of 87.55% talc, 1.75% zinc stearate, 0.70% dichlorophene, and 10% Propellant 12–Propellant 11 (50/50) is disclosed.

Shulton, Inc.[9]

In this system, a stable powder aerosol is obtained by gelling the propellant. This makes it difficult for the powder to move sufficiently to agglomerate. Gel formation is obtained by adding a colloidal silica with a particle size below 0.03 μ to the aerosol system. The gelled system is illustrated with examples of antiperspirant-deodorant aerosols.

C. G. Thiel, J. G. Young, I. N. Porush, and R. D. Law[7,10]

In suspensions where the specific gravity of the powder is less than that of the propellant, cake out or deposition of the powder on the wall above the liquid level may occur. Thiel et al., report that this problem may be minimized by the addition of an auxiliary solid which has a specific gravity greater than that of the liquid. Examples of solids that are suitable for this purpose include compounds such as sodium sulfate, sodium chloride, calcium chloride, and sucrose.

Thiel, Porush, and Law also report that the addition of a quaternary compound, such as cetyl pyridinium bromide, to a powder formulation not only will minimize agglomeration but also retard wall deposits above the liquid surface. The suggested concentration of the quaternaries varies from 0.05–1.0%.

FILLING OF POWDER AEROSOLS

Filling powder aerosols is not a simple operation. Powders can be difficult to handle and contamination by moisture must be kept to a minimum. In some products the quantities of active ingredients are very small. The problems in filling powder aerosols and the loading procedures are discussed by Mintzer[4] and his article should be consulted for details.

The earliest powder aerosols were pharmaceuticals designed for inhalation. The concentration of active ingredients in these products was quite low and the loading of a dry concentrate was impractical. In addition, it was usually considered advisable to combine any suspending agents with the micronized powders before loading them into the aerosol container, and it was difficult to obtain a homogeneous mixture of the dry ingredients. In order to avoid these difficulties, a slurry or suspension of the active ingredients and additives was prepared in a high boiling propellant, such as Propellant 11. The slurry was loaded into the containers and the remainder of the propellant added. This procedure has been used successfully for many years.[15]

When a sufficient quantity of an emollient, such as isopropyl myristate, was used, a concentrate was prepared by suspending the powders in the emollient. This concentrate was loaded into the containers and the propellant was either cold filled or pressure loaded.

Many of the present aerosols contain 5–15% powder. This concentration allows the powder to be filled in the dry state. This procedure has the advantage that the chances of contamination by moisture are decreased. Equipment for this operation is available and can be obtained from various manufacturers. The other additives, such as perfumes and emollients, etc., can be combined in a different operation and added as a separate stage using accurate positive displacement filling devices.

The propellant preferably should be pressure loaded in order to minimize the possibility of contamination of the product by moisture.

REFERENCES

1. W. C. Beard, Proc. 41st Ann. Meeting, CSMA, 74 (December 1954); *Soap Chem. Specialties* **31,** 139 (1955); U. S. Patent 2,959,325 (1960).
2. *Soap Chem. Specialties* **30,** 105 (1954).
3. F. T. Reed, "Freon" Aerosol Report FA–17, "Aerosol Powders."
4. H. Mintzer, *Aerosol Technicomment'* **10,** No. 3 (October 1967). Aerosol Techniques, Inc., Milford, Conn.
5. *Aerosol Age* **3,** 37 (1958).
6. J. J. Kanig and R. Cohn, *Proc. Sci. Sec. Toilet Goods Assoc.* **37** (1962).
7. C. G. Thiel and J. G. Young, U. S. Patent 3,169,095 (1965).
8. D. C. Geary and R. D. West, *Soap Chem Specialties* **37,** 79 (1961); *Aerosol Age,* **6,** 25 (1961); U. S. Patent 3,088,874 (1963).
9. Shulton, Inc., British Patent 987,301 (1962).
10. C. G. Thiel, I. N. Porush, and R. D. Law, U. S. Patent 3,014,844 (1961).
11. P. E. Gunning and D. R. Rink, U. S. Patent 3,081,223 (1963).
12. E. Huber, U. S. Patent 3,161,460.
13. S. Prussin, "Cosmetics" Fragrance and Personal Hygiene Products," in H. R. Shepherd, "Aerosols: Science and Technology," Interscience Publishers, Inc., New York, N.Y., 1961.
14. V. DiGiacomo, *Soap Chem. Specialties* **32,** 164 (1956).
15. *Aerosol Age* **9,** 42 (1964).

MISCELLANEOUS

23

FOOD AEROSOLS

J. H. Fassnacht

The concept of using a pressurized package to dispense food products is and old one.[1] Self-whipping cream was prepared by Getz and Smith who charged cream into a pressure bottle under 80 psig of nitrous oxide and obtained a whipped product having an overrun of 260%.[2] It was not until about 1947, however, that whipped cream topping in a throw-away aerosol package was available in the consumer market.

Although many new types of food aerosol products have been introduced since 1946, many have failed for one reason or another, and whipped toppings still hold a dominant position in this market. The aerosol food market is still considered by many to have great potential for other products as well as whipped toppings.

PROPELLANTS

Nitrous oxide was used almost exclusively in the early development of the aerosol food concept.[3] Although carbon dioxide, nitrogen, helium, propane, and butane have long been recognized as safe for use in foods by the Food and Drug Administration, they have never enjoyed significant use for aerosol foods.

In 1961, the Food and Drug Administration cleared "Freon" C-318 (octafluorocyclobutane) for food use and followed up with a similar clearance for "Freon" 115 (chloropentafluoroethane) in 1965. Patents have been issued claiming the use of pentafluoropropane[4] and chloroheptafluoropropane[5] as food propellants, but these have not yet received approval from the Food and Drug Administration.

In recent years much work has centered around the use of nitrous oxide, "Freon" C-318 and "Freon" 115. The propellants are classified as compressed gases (nitrous oxide) and liquefied gases (the fluorinated hydrocarbons). This designation is arbitrary since nitrous oxide is liquid at 70°F and 650 psig (saturation vapor pressure*) but the use of this material in aerosol containers (150 psig at 40°F) is always as a compressed gas. The fluorocarbon propellants have much lower saturation vapor pressures than nitrous oxide and are generally used as liquids.

To be seriously considered for food use, propellants must be very low in toxicity. Also they must be chemically stable, odorless, tasteless, and preferably nonflammable. Propellants C-318 and 115 meet these specifications, and nitrous oxide has a long history of safe use.

Table 23–1 shows some of the properties of the food propellants.

TABLE 23-1 PROPERTIES OF FOOD PROPELLANTS

Propellant	Vapor Pressure At 70°F (psig)	Solubility In Water*	Solubility In Vegetable Oil*	Density (g/cc)
C-318	25	.015	8	1.50
115	100	.05	12	1.29
Nitrous Oxide	Gas	1.3 cc/g at 0°C	—	—

* Except for the value for nitrous oxide, figures are wt% at 77°F and saturation pressure in a closed container. This would be about 29 psig for "Freon" C-318 and 110 psig for "Freon" 115.

Use of Fluorocarbon Propellants

Blodgett and Webster reported that the use of C-318 in whipped cream, either by itself or in combination with nitrous oxide, produced a whipped cream "with markedly improved stiffness and stability, with better color retention and much less seepage or drainage, as compared to similar cream whipped with nitrous oxide."[6] Subsequent work at the Du Pont "Freon" Products Laboratory has shown that 115 is just as effective as, and to be preferred to, C-318 in most blends with nitrous oxide.

The greater the percentage of fluorocarbon propellant used, the greater the foam stability of the food product. In addition, the use of a fluorocarbon as a part of the propellant system enables the fat content of the formulation to be reduced at no sacrifice in quality.

Blending Propellants

Development work at the Du Pont "Freon" Products Laboratory resulted in the design and construction of a blender which enables direct coupling

* For any given material the saturation vapor pressure is the vapor pressure exerted by the material when the gaseous and liquid states are in equilibrium.

to supply cylinders of nitrous oxide and fluorocarbon propellants.[7] The desired propellant mixture is obtained by setting the blender to specified concentrations of each component. The final propellant mix, which is continuously monitored by a thermal conductivity device in the blender, may be fed directly into the gassing line. The blender will supply an accurate propellant mix for both continuous and intermittent line operation.

Choice of Propellant Blend

Several factors determine the choice of the blend to be used; desired foam stability, propellant blend cost and liquefication point (or dew point) of the blend. Generally mixtures containing about 75/25 wt % of nitrous oxide–fluorocarbon have been found most satisfactory. As has been mentioned previously, greater foam stability will result as higher percentages of fluorocarbon are included, but a point will eventually be reached where the fluorocarbon propellant will start to condense out of the mixture. This has been referred to as the dew point. When using the nitrous oxide blender in conjunction with a conventional gasser–shaker loading line, it is essential that the propellant be in the gaseous state. Mixtures containing liquefied propellants will be discussed later.

The dew point of a mixture of gases can readily be calculated, so that the concentration of fluorocarbon in the blend can be kept below this point. To determine the dew point of a nitrous oxide–C-318 blend: Assume a gassing line pressure of 140 psig (155 psia) at 70°F. The pressure of C-318 (saturation vapor pressure) at 70°F is 25 psig (40 psia). The dew point of C-318 in a nitrous oxide–C-318 blend will be that point at which the concentration of C-318 is great enough to produce a partial vapor pressure of 40 psia at 70°F. Since the number of moles of a gas present in the mixture is proportional to its vapor pressure, then: 40/155 = .26 = number of moles of C-318 in blend at the dew point. This calculates out to 62 wt % of C-318. The theoretical maximum workable concentration of C-318 in nitrous oxide at 140 psig and 70°F is 62 wt %. At higher working pressures and/or lower temperatures this percentage would have to be decreased accordingly. These calculations assume ideal behavior from the nitrous oxide/C-318 mixture. Since this is not an ideal mixture the calculations can only serve as a reasonably accurate guideline. Actual measurements will be necessary to determine the dew point of any mixture.

Liquid Fluorocarbon Systems: The **Full-Can** Concept

The use of liquid-state fluorocarbon propellants unmixed with nitrous oxide may be advantageous in some cases. It has been determined that about 7% by weight of fluorocarbon propellant, based on the total contents of the aerosol container, is sufficient to dispense most food formulations. The use of straight fluorocarbon propellants produces very stable, but somewhat *softer* foams that as a whole have less overrun* than those produced with the nitrous oxide propellant blends. Of possible advantage is the fact that liquefied propellants require only a fraction of the space of compressed gas propellants, thus allowing the container to be filled to a substantially greater degree with food product than is possible when compressed gases are used.

The liquefied propellants afford a constant operation pressure that permits good product uniformity and adequate discharge pressures, even when the container is nearly empty. Although the full-can concept has much to recommend it from a technical and, in many cases, economical aspect, it has yet to be commercially exploited.

Additional product variations are possible with the full-can concept using liquefied fluorocarbon propellants and *topping off* with some nitrous oxide. This procedure allows backing off on the percentage of fluorocarbon to about 4–5 wt %. The foam produced usually resembles the nitrous oxide type foam with respect to appearance and overrun, but has the stability of the straight fluorocarbon-propelled foam. The products produced in this manner are generally quite impressive. A whipped cream formulation thus-formulated has sufficient foam stability to be used in baking: for example in the preparation of a whipped cream pound cake.

The disadvantage of this propellant system is that two-stage propellant gassing is required; one each for the fluorocarbon and the nitrous oxide. The individual(s) who will invent a practical, single-stage technique for gassing this type of propellant mixture will make a substantial contribution to the food aerosol industry.

Mechanism of Stabilization

The mechanism by which the fluorocarbon gases stabilize whipped toppings is not known. It is suspected to be largely a solubility phenomenon. Propellants 115 and C-318 are slightly soluble in oils but virtually insoluble in water (see Table 23–1). They are dispersible in food concentrates in aerosol containers. When dispensed, a foam structure containing a gas which is insoluble in the continuous phase of the bubble walls is formed. Since the rate of diffusion of this insoluble gas through the bubble walls is very slow, a foam of considerable durability is formed.

The fluorocarbon propellants are only contributors in part to foam stability. It has been demonstrated that the type of aerosol foam obtained depends upon the type of emulsion that is formed.**

* See sction on overrun under Formulating Food Products, Chapter 23.
** See Chapter 18.

With food products, we are dealing generally with emulsions of edible oils with water, usually of the oil-in-water type. The characteristics of the oil, the type of emulsifiers and stabilizers, and the procedure used in preparing the food concentrate are of critical importance to the properties of the end product.

A few illustrative observations with vegetable oil whipped topping formulations have been made at the Du Pont "Freon" Products Laboratory: Higher melting oils incorporated into the formulation produce stiffer foams. Lower homogenization temperatures, at a pressure of about 1000 psi, also produce stiffer whips, but at the same time are apt to cause less emulsion stability. It is best to run a series of experiments optimizing the melting point of the oil and the homogenization temperature so that the desired balance of mouth-feel, foam stiffness, and emulsion stability is obtained. It is generally important to gas the formulation with propellant as soon after homogenization as possible. Since it takes considerable time for the oils to *set* after having been heated and then cooled, it is propitious to add propellant quickly, while the oil is still partly liquid. In fact, in some cases the product has been gassed while still warm (70–80°F), and then allowed to chill slowly to 40°F in the cooler. This latter procedure has produced slightly higher overrun in both vegetable- and butter-fat toppings, presumably because the fluorocarbon propellants are better solubilized in partly liquid oils versus solid oils.

FORMULATING FOOD PRODUCTS

Overrun

Overrun may be described as the increase in volume of a whipped product over the volume of its concentrate, expressed as a percentage. A whipped cream manufacturer once said that the overrun of his product was greater than any competitive product on the market—as long as he demonstrated his product last. Then he would be able to include enough air spaces during dispensing so that he obtained a larger volume than any of the others. Indeed, many people are skillful in dispensing an aerosol whipped topping in such a way that many voids (or air spaces) are trapped in the product, producing what appears to be a much larger volume than is actually the case. The researcher will find little benefit in making measurements of this type. Overrun determinations should always be made on products which have been dispensed into a weighed container of known volume in such a way that no voids or air spaces are present in the foam.

Overrun is calculated according to the following equation:

$$\text{Overrun} = \frac{\text{Vol. of Foam } - \text{ Vol. of Concentrate}}{\text{Vol. of Concentrate}} \times 100.$$

If the density of the concentrate is nearly 1, then

$$\text{Overrun} = \frac{\text{Vol. of Foam } - \text{ Wt. of Foam}}{\text{Wt. of Foam}} \times 100.$$

This latter equation is quite accurate for whipped cream and whipped topping formulations.

Emulsifiers and Stabilizers

Emulsifiers and stabilizers are often lumped together in a single category. This author, however, prefers to consider an emulsifier in the classical sense as a chemical compound having hydrophilic and lipophilic moieties that encourage the intimate mixing of two mutually insoluble phases. Stabilizers are of two types: (1) those that aid in the preservation of an emulsion after it has been formed, and (2) those that stabilize the foam structure after the food product has been dispensed. Naturally occuring colloids (gums) make up type (1), and insoluble materials with a particle size larger than colloidal (microcrystalline cellulose) make up type (2).

The use of colloids in whipped cream and topping formulations is widespread. Their use must be judicious, however, since they will increase the viscosity of the concentrate and tend to decrease the stability of the foam. Attempts to produce whipped gelatin and similar materials with high colloid concentrations have resulted in very sloppy, unstable foams.

Microcrystalline cellulose, used at a level of only fractions of a percent, will help to maintain a stable foam structure after the product has been dispensed. The dispersed cellulose will do nothing to prevent drainage or shrinkage; the fluorocarbon propellants have this function. An aerosol foam produced with fluorocarbon propellants will sometimes tend to expand or grow on standing, which causes the surface to dry out faster and generally reduces the appeal of the whipped product. The function of microcrystalline cellulose is to prevent this type of change in the foam.

Choice of Emulsifiers

The field of choice for food emulsifiers is severely limited when compared to the massive number of emulsifiers available for general use. Emulsifiers, like propellants and other ingredients, must have a low order of toxicity and taste. This rules out fatty acid salts, or soaps, like sodium stearate, which are excellent emulsifiers. When the list is narrowed down to food-approved types we find we are able to choose only from the class of compounds known as esters. These are further restricted to naturally occurring compounds.

Naturally occurring fats are triesters of glycerol (glycerine) and long-chain carboxylic acids (fatty acids) that usually contain an even number of carbon atoms. These triglycerides* generally contain more than one kind of fatty acid residue. Only the mono- and diglycerides are useful as emulsifiers. They may be prepared from the triglycerides by partial hydrolysis, or by total hydrolysis and reesterification of the desired fatty acids with glycerin.

Polyhydric alcohols and their derivatives are also useful as emulsifiers. Sorbitol, produced by the hydrogenation of glucose, and its derivatives are most frequently used.

Fatty acids are either saturated (Table 23–2) or unsaturated (Table 23–3). Some fatty acids possess ring structures, but these are quite rare. Chaulmoogra oil contains glycerides of these fatty acids, of which one is chaulmoogric acid:

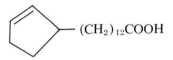

The glyceride of stearic acid is the main component of animal fats. Glycerides of unsaturated acids are major constituents of vegetable oils.

TABLE 23-2 SATURATED FATTY ACIDS

Name	Formula
Butyric acid	$CH_3(CH_2)_2COOH$
Caproic acid	$CH_3(CH_2)_4COOH$
Caprylic acid	$CH_3(CH_2)_6COOH$
Capric acid	$CH_3(CH_2)_8COOH$
Lauric acid	$CH_3(CH_2)_{10}COOH$
Myristic acid	$CH_3(CH_2)_{12}COOH$
Palmitic acid	$CH_3(CH_2)_{14}COOH$
Stearic acid	$CH_3(CH_2)_{16}COOH$

$$* \begin{array}{l} CH_2OOCR \\ | \\ CHOOCR \\ | \\ CH_2OOCR \end{array}$$

TABLE 23-3   UNSATURATED FATTY ACIDS

Name	Formula
Palmitoleic acid	$CH_3(CH_2)_5CH=CH(CH_2)_7COOH$
Oleic acid	$CH_3(CH_2)_7CH=CH(CH_2)_7COOH$
Ricinoleic acid	$CH_3(CH_2)_5CH(OH)CH_2CH=CH(CH_2)_7COOH$
Linoleic acid	$CH_3(CH_2)_3(CH_2CH=CH_2)_2(CH_2)_7COOH$

Egg yolk contains α-lecithin, a natural emulsifier which has a phosphate ester of a nitrogeneous base in place of one of the fatty acid components:

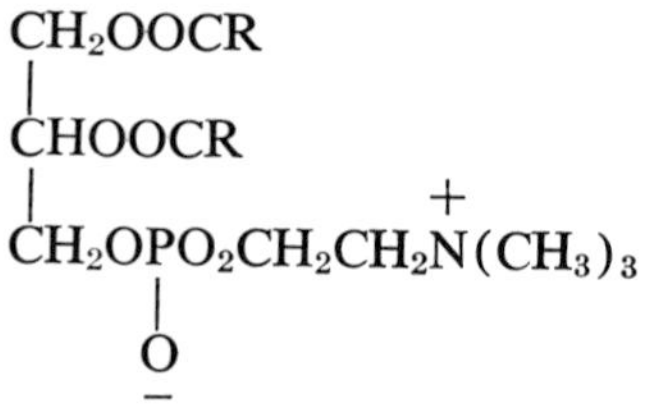

$$
\begin{array}{l}
CH_2OOCR \\
| \\
CHOOCR \\
| \qquad\quad + \\
CH_2OPO_2CH_2CH_2N(CH_3)_3 \\
| \\
O \\
-
\end{array}
$$

The most popular food emulsifiers are shown in Table 23–4. The choice of the proper emulsifier–stabilizer combination for a given food product is still more of an art than a science. There is no substitute for careful experimentation in the final choice of the most proper system. Detailed experiments have led this author to several conclusions concerning *best* combinations. The results of these researches are incorporated into the example formulations at the end of this chapter.

TABLE 23-4   FOOD EMULSIFIERS

Glyceryl monostearate
Glyceryl monooleate
Sorbitol
Sorbitan monostearate
Polyoxyethylene sorbitan monostearate
Polyoxyethylene sorbitan monooleate
α-Lecithin, and other phosphate esters
Hydroxylated α-lecithin
Lactic acid stearate (the stearic acid is esterified by the OH
group of lactic acid

Dispersion of Propellants

Soluble compressed gas propellants produce foams from aerosol food concentrates by the release of gas from solution as the concentrate is exposed to lower pressure conditions, i.e., dispensing from the container into the atmosphere. Since the fluorocarbon food propellants are essentially insoluble in all aqueous foodstuffs, foams are not naturally produced by the propellants unless they are well dispersed in the food concentrate. All attempts to produce stable emulsions of fluorocarbon food propellants in aqueous systems have been unsuccessful. Methods attempted include the use of emulsifiers and ultrasonics. Emulsions of limited stability can be produced, however, and the quality of the food foam obtained depends to a large extent on the stability of these emulsions.

The dispersion of the fluorocarbon propellants produced by pressure filling and simple hand shaking is usually sufficient to produce decent foams from properly formulated concentrates. The use of techniques that produce violent mixing of the propellant with the concentrate have produced vastly superior foams—better appearance, higher overrun and more structural stability. One such method involved the use of a standard Kartridg Pak undercap filler fed by a *saturator* that contained the concentrate and the propellant.*

The saturator is a high pressure cylinder, about the size of the 60-lb nitrous oxide cylinders still used by some food loaders. This cylinder, arranged in a vertical position, has a high pressure spray nozzle inside at the bottom and is hooked up in such a way that the product can be recirculated through the spray nozzle or fed into the filler. When a mixture of liquid fluorocarbon food propellant and food concentrate was recirculated in the saturator prior to filling, an extremely good dispersion of propellant in concentrate resulted.

The undercap filler commercially available is not satisfactory for food use, but experiments performed with this unit show that this concept of filling aerosol food products has much merit.

NEW PRODUCTS

Thus far, our discussion has avoided the mention of products except for whipped cream and whipped toppings. During our study of liquefied-fluorocarbon food propellants applications since 1962, many interesting and unusual products have been formulated. All of the rules with respect to foam

* Kartridg Pak Company, Davenport, Iowa.

stability apply to these new products; in fact, some of the concepts presented are not even practical without the stabilizing effects of the fluorocarbon propellants.

Fruit and Syrup Whips

The use of pure fruit purees and prepared sugar syrups as the base for aerosol whipped products is an intriguing idea. Early efforts to prepare fruit-flavored whips involved either the addition of fruit flavors to whipped cream toppings or the other extreme of using a fruit base with a little emulsifier or stabilizer as the concentrate. It became apparent, quite rapidly, that the food concentrate had to be an emulsion (in this case, oil-in-water) if a satisfactory aerosol whip were to be obtained. However, the addition of much oil to the fruit puree destroyed the fresh character of the fruit, and yet formulation work indicated that 10% by weight was the minimum amount of oil in the emulsion that would give satisfactory results.

A method of preparing very low fat emulsions was discovered by Fassnacht, was extended to higher fat compositions (sour cream) by Bower, and was subsequently found to be generally applicable by Fassnacht, Bower and Knipper.[8] This process consists of preparing an emulsion of oil in water that usually contains 20%–30% by weight of oil, and then mixing this emulsion with the fruit puree or syrup. The total oil content and emulsifier concentration in the finished product may be lower than when conventional methods are used to prepare the same product. This new method has provided the experimenter with great versatility in the examination of new food products. The preparation of the emulsions and their use to formulate fruit and syrup whips is described at the end of this chapter.

Cheeses

The marketing of assorted processed cheeses in piston-type aerosol cans was started several years ago by the Granny Goose Company in the San Francisco Bay area. The National Biscuit Company subsequently started marketing similar products and is now enjoying moderate success in a nationwide market.

Viscous products, such as the processed cheeses, may be packaged in piston-type cans quite satisfactorily. The shelf-life of the cheese is excellent and the package offers great convenience.

A highly satisfactory whipped cheese product has been prepared at the "Freon" Products Laboratory from spray dried cheddar cheese powder by using the emulsion technique described above (see section on formulations).

Pharmaceuticals

Foamed pharmaceuticals may also be prepared using the emulsion technique. Stable, attractive foams were prepared from most commercial cough syrups and other types of pharmaceutical syrups.[9]

Water-in-Oil Emulsions

Butter and oleomargarine are water-in-oil emulsions consisting primarily of fat (80% by weight) with about 16% by weight of water emulsified in the fat. The viscosity of both butter and oleomargarine must be reduced before they can be used as aerosol products. Butter can be reformulated by thinning with liquid vegetable oil or preferably with oil–water (80/20). After this treatment the butter will have lost its *standard of identity* and could not be referred to as *butter*.

Oleomargarine is probably a better product for aerosol packaging since it can be reformulated without loss of standard of identity. Only the melting points of the oils need be changed. However, it is a more difficult matter to prepare a formulation which will operate at refrigerator temperatures as well as summer room temperatures. Our best efforts have not produced a completely satisfactory product in this regard, but formulations which perform very well in the 50°–90°-F range have been prepared. The formulation for garlic-flavored margarine in the Formulations section is an example of a good basic margarine formulation.

When packaged in an aerosol container, margarine has a very good room temperature shelf-life—about 3–4 mo without spoilage. Although the shelf-life of most products is somewhat extended when packaged in an aerosol container, margarine probably has the most acceptable shelf-life without aseptic packaging. A discussion of aseptic packaging will follow later.

Meringue

No product has been so elusive to formulate as meringue. Almost none of the rules developed for the other food products hold for meringue. Satisfactory meringue has been obtained only from standard cookbook recipes which have been formulated at very high pressures (160–180 psig) with mixtures of "Freon" Food Propellant 115 and nitrous oxide.

The Pancake Problem

Products which are not available, or for some technical reasons not practical, always seem to be those most desired by marketers. Nearly all of the potential aerosol food marketers with whom we have had contact have expressed a desire for aerosol pancakes. Indeed, we have been able to prepare creditable pancakes, as well as cakes, biscuits, and meat spreads.

Batters, meats, and vegetables are potentially dangerous from the microbiological standpoint when packaged under anaerobic conditions. These materials are capable of supporting the growth of the Clostridium botulinum bacterium in the absence of air. Under favorable conditions, this organism produces a toxin which is usually fatal when ingested. The canned meats and vegetables being sold commercially are sterilized by heating in a retort, and thus are not contaminated with this, or other, microorganisms. Aerosol cans can not be treated in this manner, not only because the pressure buildup would be too great but because most aerosol formulations could not stand the required heating time without breaking down. The batters, of course, would be cooked with this type of heat treatment.

Experiments in the preparation of aerosol batters, meats, and vegetables are not recommended, unless strictly monitored by microbiological studies.

Sprays

Whipped products have been considered almost exclusively in this chapter since these are the products that present the real formulation challenge. There are other methods of aerosol dispensing, and in fact it is the sprays that account for the great bulk of the nonfood aerosol market.

The best sprays are formed when soluble liquefied gas propellants are formulated with the concentrate to be sprayed. Then the propellant breaks the liquid up into very fine droplets during the process of rapid evaporation as the material is dispensed from the aerosol container. When insoluble propellants are used, the valve must be entirely responsible for breaking the liquid into a spray, since in this case no help is obtained from the propellant; the propellant merely acts as a *spring* to push the contents from the container. The so-called *mechanical breakup* valves are sometimes quite effective.

With food products, the choice of solvents is limited to water, vegetable oils, ethyl alcohol, and propylene glycol. Food propellants are not soluble enough in any of these solvents to produce sprays without the use of mechanical breakup valves.

Acceptable food sprays can easily be prepared from nonviscous solvents like water and alcohol. Although vegetable oils have presented quite a problem in the past, newer valves capable of producing acceptable oil sprays have been made. The best oil sprays have been obtained experimentally by using mechanical breakup valves that have had a vapor tap of approximately one eight-thousandth of an inch drilled in the valve body.

This vapor tap will help to produce a finer spray with a propellant useage of about 10% by weight. Unfortunately, valves with such small vapor taps are not commercially available at this time.

The types of food sprays already in existence include flavorings, spices, vermouth, oil and vinegar salad dressing, and butter–flavored oil for use on popcorn. Flavorings and spices could equally well be dispensed drop-wise, or as a stream instead of a spray. Several types of cocktail mixes have been formulated at the "Freon" Products Laboratory and are best dispensed in the form of a slow stream for proper measuring.

Extreme care must be exercised when using any type of oil spray. Finely divided oil particles will lodge in the lungs and eventually cause lipoid penumonia if repeatedly inhaled.

ASEPTIC PACKAGING

Except for specialty items like fresh frozen orange juice and some dairy products, canned goods are sterilized either by retorting, hot-packing or aseptic packaging. Most food products are retorted; vegetables, fruits, soups, meats, etc. Products which are not so susceptible to microbiological degradation, such as ketchup, are hot-packed. Some other products, such as eggnog, milk shakes, diet drinks, etc., are aseptically packaged. Although the methods vary, they are each aimed at sterilization of the food products in the package.

The meaning of the word *sterile* is very often misconstrued. A sterile food product has no viable organisms in it; the product is either sterile or it is not. There is no such thing as *nearly sterile* or *almost sterile*. When speaking of food products it is most accurate to speak in terms of total microbiological counts. Generally, the lower the total count, the longer the shelf-life of a given food product. This is not to say that sterile food products will last forever. Chemical changes may occur even in sterile products which will make the food unacceptable with respect to taste. Although some canned soups and vegetables have been observed to last for two decades or more, milk products undergo a chemical change that renders them unacceptable to taste in a much shorter time.

The aseptic process involves sterilization of food product and containers in separate operations, then bringing the two together under conditions such that no contamination can occur. The food product is heated very rapidly to a temperature of about 280°F, held at that temperature for several seconds, then cooled and packaged. Any food subjected to this high temperature short-time sterilization cycle will have a more natural flavor than if sterilized by retorting.

With certain food formulations, aseptic packaging is the only method that can be used to sterilize without formulation breakdown. Emulsions (milk products, diet drinks, etc.) will separate when heated at temperatures above the metling point of the fat used in the formulation unless held under agitation while heating and then homogenizing immediately prior to cooling. The aseptic procedure keeps the food in motion at all times and allows homogenization to reset the emulsion at any stage in the heating or cooling process. Therefore, the aseptic process is particularly ideal for aerosol food products, most of which are emulsions.

Unfortunately, there is no aerosol aseptic packaging equipment commercially available at this time. We have seen designs for equipment that should be workable,* but such equipment has not been fabricated and tested as of this time.

FORMULATIONS

The formulations shown below are used as examples to point out in detail some of the items discussed earlier. Each formulation is a good, workable formulation with makeup procedures geared to laboratory size equipment.

Example 1: *Whipped Cream (30% Butterfat)*

Ingredients	Percentages
1. Heavy Cream (40% butterfat)	75.00
2. Water	14.20
3. Sugar	9.08
4. "Avicel" RC[1]	0.34
5. "Emcol" D-70-31 (10% aqueous)[2]	1.01
6. "Myverol" 18-07[3]	0.11
7. "Myverol" 18-00[3]	0.06
8. "Nu-Wip" L.P.[4]	0.20
9. Vanilla Flavor	1 g/100 g concentrate

[1] FMC Corporation, American Viscose Div., Marcus Hook, Pa.
[2] Witco Chemical Co., New York, N.Y.
[3] Distillation Products Industries, Rochester, N.Y.
[4] Germantown Manufacturing Co., Broomall, Pa.

Preparation
1. Mix all ingredients together with slow, but thorough stirring and heat at 160°F for 20 min.

* Amboy Sterile Packaging Company, Amboy, Illinois, in conjunction with Werge Engineering.

2. Cool to 100°F.
3. Add 1 g of pure vanilla extract for each 100 g of concentrate prepared.
4. Homogenize the mix at 800/200 psi, two stage.

Packaging
1. Place 250 g of product into a 16 oz lined aerosol can.
2. Cap the container with an appropriate foam valve.
3. Pressure fill 10 g (ca 4% by weight) of "Freon" Food Propellant 115–
   nitrous oxide (30/70% by weight). It is best to load the "Freon" volumetrically using a pressure burette, then pressure load the nitrous oxide
   on a gasser–shaker.
4. Cure the product for 24 hr at 40°F prior to use.

Example 2: *Whipped Cream (20% Butterfat)*

Ingredients	Percentages
1. Heavy Cream (40% butterfat)	50.00
2. Water	30.23
3. Sugar	18.15
4. "Avicel" RC[1]	0.68
5. Hydroxylated Lecithin[2]	0.20
6. "Myverol" 18-07[3]	0.22
7. "Myverol" 18-00[3]	0.12
8. "Nu-Wip" L.P.[4]	0.40
9. Vanilla	1 g/ 100 g concentrate

[1] FMC Corporation, American Viscose Div., Marcus Hook, Pa.
[2] "Centrolene" S, Central Soya, Chemway Div., Chicago, Ill.
[3] Distillation Products Industries, Rochester, N.Y.
[4] Germantown Manufacturing Co., Broomall, Pa.

Preparation
1. Mix all ingredients together with slow, but thorough stirring and heat
   at 160°F for 20 min.
2. Cool to 100°F.
3. Add 1 g of pure vanilla extract for each 100 g of concentrate prepared.
4. Homogenize the mix at 800/200 psi, two stage.

Packaging
1. Place 250 g of product into a 16 oz lined aerosol can.
2. Cap the container with an appropriate foam valve.
3. Pressure fill 10 g (ca 4% by weight) of "Freon" Food Propellant 115–

nitrous oxide (30/70% by weight). It is best to load the "Freon" volumetrically using a pressure burette, then pressure load the nitrous oxide on a gasser-shaker.

4. Cure the product for 24 hr at 40°F prior to use.

Example 3: *Whipped Vegetable Oil Topping (20% Oil)*

Ingredients	Percentages
1. Vegetable Oil, mp 92–103°F	20.0
2. Stearyllactylic Acid Emulsifier[1]	0.2
3. Polysorbate 80	0.1
4. Sodium Caseinate	0.5
5. "Avicel" RC[2]	0.2
6. Sugar	13.0
7. "Nu-Wip" L.P.[3]	0.2
8. Water	65.8
9. Vanilla Extract	1 g/100 g concentrate
10 Yellow Food Color	to suit
11. Cream Flavor	to suit

[1] Durkee Famous Foods, Chicago, Illinois. Emulsifier SGM-57
[2] FMC Corporation, American Viscose Division, Marcus Hook, Pa.
[3] Germantown Manufacturing Co., Broomall, Pa.

Preparation
1. Prepare a melt of ingredients 1, 2, 3, and 4 with stirring.
2. Prepare an aqueous dispersion of ingredients 5, 6, 7, and 8 by heating to 130°F with stirring.
3. Add the oil melt to the aqueous dispersion while stirring.
4. Heat the mix at 160°F for 20 min while stirring.
5. Cool to 100°F and add ingredients 9, 10, and 11.
6. Homogenize at 800/200 psi, two stage.

Packaging
1. Place 250 g of product into a 16-oz lined aerosol can.
2. Cap the container with an appropriate foam valve.
3. Pressure fill 10 g (ca 4% by weight) of "Freon" Food Propellant 115– nitrous oxide (30/70% by weight). It is best to load the "Freon" volumetrically using a pressure burette, then pressure load the nitrous oxide on a gasser-shaker.
4. Cure the product for 24 hr at 40°F prior to use.

Example 4: *General Purpose Emulsion*

Ingredients	Percentages
1. Water	69.0
2. "Avicel" RC[1]	2.5
3. Vegetable Oil, mp 92–103°F	23.5
4. Stearllactylic Acid Emulsifier[2]	4.5
5. Polysorbate 80	0.5

[1] FMC Corporation, American Viscose Division, Marcus Hook, Pa.
[2] Durkee Famous Foods, Chicago, Illinois. Emulsifier SGM-57.

Preparation
1. Prepare a dispersion of "Avicel" RC in water by stirring at 120°F.
2. Prepare a melt of the vegetable oil and emulsifiers.
3. Add the oil melt to the aqueous dispersion with stirring and homogenize at 600 psi at a temperature of 90–100°F.

The finished emulsion is a smooth semisolid white cream when warm. On cooling the emulsion will set to a waxy mass. Warming to about 80–90°F will soften the emulsion and make it easily workable.

Example 5: *Fruit and Syrup Whips*

Ingredients	Percentages
Fruit Puree or syrup	80
Emulsion (Example 4)	20

Preparation
1. Mix the soft emulsion thoroughly with the fruit puree or syrup by adding the syrup *slowly* to the emulsion while constantly stirring.
2. Homogenization at 500 psi is advisable.

Packaging
1. Place 250 g of product into a 16-oz lined aerosol can.
2. "Freon" Food Propellant 115–"Freon" Food Propellant C-318 (30/70% by weight) may be substituted for the "Freon" Food Propellant 115–nitrous oxide (30/70% by weight) used in the previous examples. In this case, the 16-oz container may be filled with 400 g of concentrate prior to filling with 30 g (7% by weight) of propellant.

Example 6: *Garlic Margarine Whip*

Ingredients	Percentages
1. Soft Margarine	49.00
2. Winterized Vegetable Oil	39.20
3. Water	7.15
4. Sugar	1.00
5. Salt	0.60
6. Milk Solids	0.80
7. "Atmos 300[1]	0.25
8. Garlic Powder	2.00

[1] Atlas Chemical Industries, Inc., Wilmington, Del.

Preparation
1. Place ingredients 3 through 8 in a blender (Osterizer or equivalent) and blend at high speed for 1 min.
2. Place ingredients 1 and 2 into the blender and blend for 1 min at low speed, followed by a 1-min high speed blend.

Packaging
1. Place 215 g of concentrate into a 6-oz lined, 2P-specification aerosol can.
2. Cap the can with an appropriate foam valve.
3. Pressure fill 14 g (ca 6% by weight) of "Freon" Food Propellant 115.
4. Cure the product for 24 hr at 40°F prior to use.

Example 7: *Spice Spray*

Ingredients	Percentages
1. Alcohol-soluble Spice Oil	6
2. Absolute Ethyl Alcohol	94

Preparation
1. Dissolve the spice oil in the alcohol.

Packaging

1. Place 118 g of concentrate in a 4-oz plastic-coated aerosol bottle.
2. Cap the bottle with a spray-type bottle valve.

3. Pressure fill 6.5 g (ca 5.2% by weight) of "Freon" Food Propellant 115.
4. Use a suitable actuator (mechanical breakup may be desirable for finer spray).

REFERENCES

1. U. S. Patent 1,510,975 (1926) to D. Sweeney.
2. C. A. Getz and G. F. Smith, *Trans. Ill. State Acad. Sci.* **27,** 71 (1934).
3. French Patent 820,113 (No. 4, 1937) to Food Dairies, Inc.; U. S. Patent 2,155,260 (1939) to I. M. Diller; U. S. Patent 2,212,379 (1941) to A. H. Smith, Aeration Processes, Inc.; U. S. Patent 2,250,000 (1951) to J. C. Goosman, White Dental Manufacturing Company.
4. U.S. Patent 3,369,913 (1968), S. M. Livengood and R. J. Scott, Union Carbide Company.
5. U.S. Patent 3,369,912 (1968), S. M. Livengood and R. G. Werner, Union Carbide Co.
6. *Aerosol Age* **4,** 20 (1959).
7. U.S. Patents 3,273,348 (1966); 3,330,773 (1967); Canadian Patent 717,258 (1965), J. W. DeHart, Jr., Du Pont Company.
8. U. S. Patent 3,366,494 to F. A. Bower, J. H. Fassnacht and A. J. Knipper, Du Pont Company, January 1968.
9. J. H. Fassnacht, *Product Licensing Index* **47,** 21 (1968).

24

MISCELLANEOUS AEROSOL SYSTEMS

The discussion thus far has covered homogeneous aerosol systems, where the components are mutually soluble, and emulsions and powders, which are heterogeneous in nature. The aerosol products based upon these systems are packaged in the so-called standard aerosol containers with typical aerosol valves. They account for practically all of the aerosols on the market. However, considerably different methods for dispensing products have been under investigation for many years. This group includes co-dispensing valves, aspirator or venturi systems, bag-in-can containers, and piston containers. In recent years, some of these have emerged from the development stage and become sufficiently practical and attractive so that a number of products utilizing these systems are now on the market.

These new packaging methods are particularly attractive because products can now be packaged as aerosols that previously were neither practical nor possible with the conventional systems. Products in this category included viscous materials, such as caulking compounds, heavy creams and toothpaste, as well as certain foods where it was desirable to keep the propellant from coming into contact with the food. One major advantage of keeping the propellant and concentrate separate is that the concentrate can be dispensed without alteration in form or consistency, and in the case of foods, without change in color or taste. Co-dispensing techniques allow two materials which react with each other to be packaged in the same container. When the product is discharged, the two components come into contact and react. Another attractive feature of these new products is that in many cases they increase the total aerosol market because they do not compete with older, conventional aerosols.

CO-DISPENSING TECHNOLOGY

The recent developments in co-dispensing techniques have stimulated about as much interest throughout the aerosol industry as any in its history.[1-5] Probably the most publicized product that utilizes this system is the hot shave lather, but it is generally believed that other products, such as hair colorings, may have even greater potential.

The Hot Shave Lather

The idea of an aerosol shaving lather that emerged hot from the container was conceived many years ago and has been a subject of continuing interest ever since. In recent years a number of mechanical devices have appeared on the market in which the shaving lather was either heated electrically or by hot water. In the latter system, the aerosol was equipped with a series of coils which served as heat exchangers and the product was merely held under the hot water faucet. Most of these devices were not very satisfactory because they were cumbersome or difficult to use, and in many cases the shaving lather was not very hot.

Meanwhile, research had been continuing on other methods for producing heat, including chemical reactions. As a result of this work, the first really hot aerosol shaving lather became possible and was marketed by the Gillette Company as its *Nine Flags Thermal Shaving Foam*.[1] Several other systems that produce satisfactory quantities of heat have also been developed and will be discussed in the following section.

Reactions for Producing Heat. The energy to heat the shaving lather can be obtained in several ways. The Gillette Company[6] and the DuPont Company[3] use oxidation–reduction reactions between hydrogen peroxide and a reducing agent. Another method involves the catalytic decomposition of hydrogen peroxide to water and oxygen.[7] A still different approach utilizes the heat of hydration as the source of the energy.

OXIDATION–REDUCTION REACTIONS

The Du Pont System. The Du Pont System has been described in detail by Boden.[3] In this system the hydrogen peroxide is stored in a plastic bag attached to the co-dispensing valve and is separated from the shaving lather containing the reducing agent. When the valve is actuated, predetermined proportions of peroxide and shaving lather are discharged simultaneously and come in contact in the valve. The exothermic reaction that occurs heats the shaving lather.

The reducing agents consist of sulfites, thiosulfates, or mixtures of the two. The potassium salts are preferred to the sodium salts because the latter may cause gelation. Calculations of the heat of reaction between hydrogen peroxide and potassium sulfite (Equation 24–1), followed by laboratory tests, indicated that stoichiometric concentrations of about 1.5% hydrogen peroxide (100% basis) and 7.0% potassium sulfite are necessary in order to increase the temperature of the shaving lather 50–60°F. This is equivalent to a ratio of 21 parts of a 7% hydrogen peroxide solution and 79 parts of the shaving lather with the sulfite.

$$H_2O_2 + SO_3^= \rightarrow H_2O + SO_4^=$$
$$\Delta H = 87.7 \text{ Kcal/mol} \tag{24–1}$$

When the peroxide solution and the shaving lather are mixed in the valve, the peak foam temperature is reached in approximately 30 sec. This is shown in Figure 24–1, which illustrates the variation in foam temperature with time. Accurate metering of the two components is necessary to obtain the maximum quantity of heat, and any departure from the stoichiometric proportions will reduce the amount of heat evolved. This is illustrated in Figure 24–2.

Reducing agents other than potassium sulfite can be used. Potassium thiosulfate, for example, reacts with hydrogen peroxide as shown in Equation 24–2.

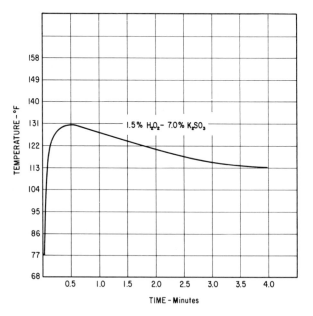

Figure 24–1   Heat production H_2O_2–K_2SO_3 oxidation reduction reaction. From Boden.[3]

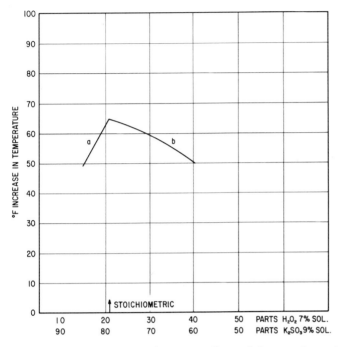

Figure 24–2   Temperature increase—effect of inaccurate metering of reacting solutions. From Boden.[3]

$$S_2O_3^= + 4\ H_2O_2 \rightarrow 2\ SO_4^= + 3\ H_2O + 2H^+$$
$$\Delta H = -\ 310.2\ Kcal/mol \qquad (24\text{–}2)$$

Potassium thiosulfate has both advantages and disadvantages compared with potassium sulfite. The heat of reaction with potassium thiosulfate is much higher than that with potassium sulfite and higher temperatures are produced. In addition, the concentration of potassium thiosulfate is much lower than that of potassium sulfite at the same hydrogen peroxide concentration. This is an advantage because high salt concentrations affect emulsion and foam stability and increase the possibility of container corrosion. One disadvantage of potassium thiosulfate is a slower rate of reaction and consequently more heat is lost to the surroundings while the reaction is taking place. As a result, theoretical peak temperatures are not realized. Also, as Equation 24–2 shows, acid is formed during the oxidation–reduction reaction so that buffering agents may be needed.

A combination of potassium sulfite and potassium thiosulfate offers considerable flexibility. Foam temperatures close to 180°F can be obtained within 30 sec with the correct mixture. Although this is too hot for appli-

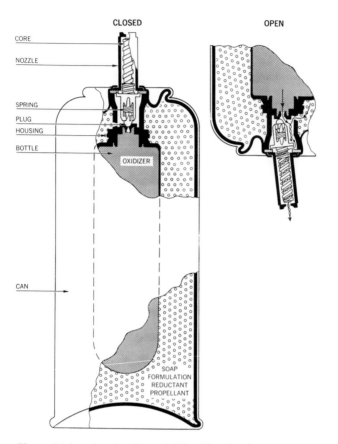

Figure 24–3   Construction of "The Hot One." (*Courtesy of the Gillette Co.*)

cation to the skin, it shows that the peak temperature can be regulated by varying the proportions of the reducing agents.

The Gillette System.   The Gillette Company has marketed two hot shaving lathers. The first was the "Nine Flags Thermal Shaving Foam,"[1] and the second was "The Hot One."[9] The Gillette and Du Pont systems are similar to the extent that both depend upon an oxidation–reduction reaction for the heat and use hydrogen peroxide for the oxidizing agent. The hydrogen peroxide is separated from the aqueous shaving lather by a plastic bag attached to the valve in both systems. The construction of "The Hot One" is illustrated in Figure 24–3.

The reducing agents disclosed in the patent issued to the Gillette Company include thiourea and substituted thiobarbituric acid derivatives.[6] The news releases concerning the "Nine Flags Thermal Shaving Foam" also indicated that the reducing agent belonged to the pyrimidine family.

HYDROGEN PEROXIDE DECOMPOSITION   The catalytic decomposition of hydrogen peroxide to form water and oxygen has also been proposed as a heat producing reaction for hot shave products.[7]

$$H_2O_2 \rightarrow H_2O + 1/2 \ O_2$$
$$\Delta H = -22.6 \ Kcal/mol \qquad (24-3)$$

Boden[3] has indicated, however, that a product based upon this system could be hazardous if there was a malfunction. The amount of hydrogen peroxide required to heat the contents of a 6-oz container could generate sufficient pressure to burst the can if all the energy were released at one time. In addition, the quantity of hydrogen peroxide required would be approximately four times that in the oxidation–reduction reaction for the same quantity of heat.

HEAT OF HYDRATION   This method is described in a patent issued to Friedenberg[8] and involves a container with a reservoir, which stores the aqueous shaving lather, and a reactant chamber, which holds a hydrophilic thermogenic agent. The term *thermogenic agent* applies to chemical compounds which produce an exothermic reaction when they hydrate or dissolve in water. The preferred agent is magnesium chloride. When the product is discharged, the shaving lather passes through the reaction chamber and is heated by the hydration energy.

Co-Dispensing Valves.   Correct metering of the hydrogen peroxide solution and the shaving lather is necessary to obtain the optimum quantity of heat from the system, as previously illustrated in Figure 24–2. Therefore the valve must ensure proper metering throughout the life of the package At the present time three companies have developed co-dispensing valves and other companies are reported to be working on this type of valve. (The Gillette Company manufactures and uses a valve of its own design.) The three commercially available valves are:

CLAYTON CORPORATION   The Clayton Corporation was one of the first companies to develop a co-dispensing valve (the *Clay-Twin*) for the hot shave lather.[3,5] The *Clay-Twin* valve, illustrated in Figure 24–4, is constructed with three 0.30-in. orifices in the body for the shaving lather and five 0.024-in. orifices in the base for the hydrogen peroxide solution. A gasket seals off the holes in the base of the valve stem when the product is not in use and prevents the hydrogen peroxide and shaving lather from reacting prematurely. The number and dimensions of the orifices are designed to deliver four parts of shave lather to one of peroxide solution.

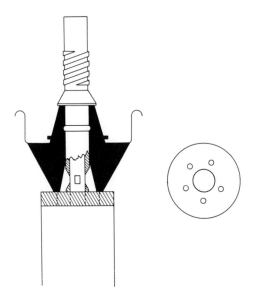

Figure 24–4   Clay-twin valve. From Boden.[3]

The tube for the peroxide solution is heat sealed to the valve stem and can be supplied in any length. The tube is constructed of either a poly-ethylene–polyester–polyethylene triple laminate or extruded polyethylene. Valves are supplied with the plastic tube attached to the valve with the bottom of the tube open for filling. In production, the plastic tube is filled with the peroxide solution, the headspace is purged, and the end of the tube is heat sealed. The valve with the loaded plastic tube is then handled as any normal aerosol valve. Equipment for filling the tubes is available.

OIL EQUIPMENT LABORATORY (OEL)   The OEL co-dispensing valve is shown in Figure 24–5.[3,10] It is constructed so that a spring controlled insert seals off the peroxide compartment when the product is not in use. When the valve is actuated, the pin is depressed and the peroxide is metered into the mixing chamber along with the shaving lather. The ratios of soap and peroxide can be varied from 2:1–5:1 by changing the inside diameter of the dip tube. Products with different viscosities can be pack-aged by selecting the correct dip tube.

The peroxide bag (or tube) is a snap-on type and the material of con-struction can be varied to suit the application. The bag is designed for easy loading on tube filling equipment and for insertion into the container. The size of the bag is limited to the 1-in. opening in the container. If greater capacity is desired, a can end valve could be used in place of the 1-in. cup and double seamed into the outer shell. This type of valve and

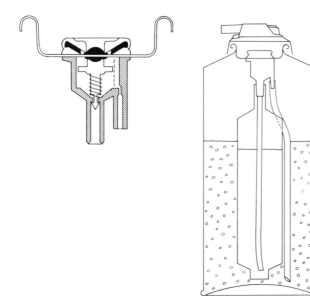

Figure 24-5   OEL co-dispensing valve. From Boden.[3]

container is currently available and has been used on commercial shaving lathers.

When the propellant is pressure loaded, the pin is depressed completely. This seals off the peroxide chamber but not the rest of the container.

VALVE CORPORATION OF AMERICA (VCA)   The VCA co-dispensing valve is illustrated in Figure 24–6.[3] It can be modified for use in either the inverted or upright position. The ratio of the peroxide and the shaving lather can be changed by varying the diameter of the dip tube in the upright model and the diameter of the tube chamber plug in the inverted model. The valve contains two distinct intake ports that operate with a single actuation. The metering is controlled by the notch in the valve stem.

Safety.   The safety of any co-dispensing system must be thoroughly checked. Laboratory tests with shaving lather systems with reducing agent and hydrogen peroxide have shown that even if the peroxide bag ruptures in the container, the resultant pressure will still be well within the limits of 2 P aerosol containers.

The stability of the hydrogen peroxide solution must also be established. Tests carried out at elevated temperatures with 8% hydrogen peroxide ("Albone" CG) showed that the solution was sufficiently stable under the test conditions. The amount of decomposition was insignificant.

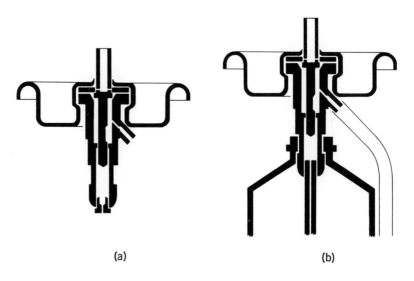

(a)   (b)

Figure 24-6   VCA co-dispensing valve. (a) Inverted use. (b) Upright use. From Boden.[3]

Other Products

The second type of co-dispensed product that reached the market was a hair conditioning aerosol for use in professional beauty salons. The product, called "Heat's On," is marketed by Helene Curtis Industries, Inc.[11]

Recently, a third product has appeared on the market. This is another hot shaving cream, "Rise Hot," marketed by Carter-Wallace, Inc.

BAG-IN-CAN SYSTEMS

The bag-in-can system consists primarily of a metal container equipped with an inner plastic bag that holds the product. The pressure that forces the product out of the plastic bag and through the valve is supplied by the propellant which is loaded through the bottom of the can and remains outside the bag. The plastic bag, therefore serves as a barrier that prevents the propellant from coming into contact with the product. The bag must be flexible enough to fold up as the product is discharged. The containers are supplied with the plastic bags already in place.

These systems have a number of advantages and should open up the field for many new products. Very viscous materials can be packaged and discharged. In addition, products can be packaged without alteration of their composition since they do not come into contact with the propellant. Other advantages include a high ratio of product to propellant. There are

Figure 24-7   The "Sepro" can.
(*Courtesy of the Continental Can Co.*)

two major bag-in-can systems, the "Sepro" Can and the "Sterigard" Dispenser.

The "Sepro" Can (Continental Can Company)

The "Sepro" Can is illustrated in Figure 24–7. Details of its construction have been discussed by Irland and Kinnavy.[12] The unit consists essentially of a standard three piece can with a plastic bag fastened to the dome, which takes the usual 1-in. valve cups. The bottom end of the can is perforated for a charging plug for the introduction of propellant. The bag is manufactured of Conoloy, a blend of nylon and polyolefins, and is reported to have a low propellant permeation. It has a bellows type construction and is flexible. It is reported that at least 95% of the product can be expelled. However, the bag is sufficiently rigid so that it can be filled without excessive deformation.

Several methods can be used to fill "Sepro" cans. Entrapment of air in the product should be avoided since air can cause *popping* or *sputter* during discharge. Standard aerosol filling equipment has been successfully employed and in addition, a spin filler, which rotates the cans while the product is being filled, has been developed. This procedure has been successful in minimizing entrapped air.

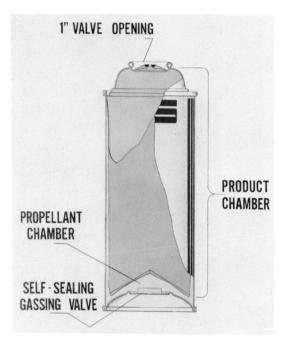

1" VALVE OPENING

PRODUCT CHAMBER

PROPELLANT CHAMBER

SELF - SEALING GASSING VALVE

Figure 24–8   The "Sterigard" dispenser. (*Courtesy of the Sterigard Co.*)

The container is designed to use fluorocarbons, hydrocarbons or compressed gases as the propellants. Some highly viscous products in "Sepro" cans are now on the market. These include a caulking compound[13] and a lithium grease.

The "Sterigard" Dispenser (Sterigard Company)

The "Sterigard" Dispenser is also a bag-in-can device[9,14] (Figure 24–8). It is constructed of standard aerosol can components and contains a bag made from a blend of low density polyethylene film. The bag occupies most of the space in the container so the product–propellant ratio is very high. The bottom of the can is fitted with a self-sealing valve for the introduction of propellant. The valve is referred to as the *Dart*. The Dart is made of laminated rubber with a soft center core.

The "Sterigard" Dispenser can be filled using standard aerosol equipment, but the propellant has to be added by means of a new piece of equipment called the Dart Propellant Injector.

The "Sterigard" Dispenser was developed originally for food aerosols but is considered to be advantageous for many cosmetic, pharmaceutical,

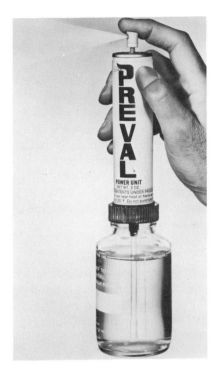

Figure 24–9 The "Preval" Spraymaker.
(*Courtesy of the Precision Valve Corp.*)

household, and industrial products. At the present time, it is reported to be under test with a number of products.

ASPIRATOR (VENTURI) SYSTEMS

Several new systems are based upon the aspirator principle. The propellant is kept apart from the product until discharge. The assembly consists of a power unit, which contains the propellant and a product container. The product container is at atmospheric pressure at all times; therefore, its construction is not limited by pressure regulations. This allows considerable flexibility in the design of the product container as well as in the materials of construction.

The "Preval" Spraymaker (Precision Valve Corporation)

The "Preval" Spraymaker is illustrated in Figure 24–9.[9] The power unit consists of a cylindrical aluminum tube containing 3 oz of liquefied gas propellant. A valve with actuator is fastened to the top of the tube. The dip tube from the valve passes through the power unit and out through the sealed bottom. A strainer is fastened to the bottom of the dip tube to screen out large particles that might cause valve clogging. The bottom of the power unit is equipped with a collar for attaching to the 6-oz product container. The dip tube extends to the bottom of the product container.

The valve contains an inlet for both the propellant and the product. When the valve is actuated, propellant is discharged past the opening leading to the product container. This creates a partial vacuum and causes the product to rise up the dip tube and out through the valve, where it mixes with the vaporizing propellant. The product container is graduated and made of glass so that the amount of product remaining in the container can be determined at any time.

One of the advantages of the "Preval" unit is that it can be used to spray any number of different liquids merely by changing the material in the product container.

The "Innovair" Aerosol System (Geigy Chemical Corporation)

The "Innovair" system illustrated in Figure 24–10, is also based upon the venturi principle.[15] It consists of a propellant unit in a nonpressurized product container. One of the ways in which it differs from the "Preval" system is that the dip tube is located outside the propellant unit. The entire propellant unit may be located either inside or outside the product container. When the three-way valve is actuated, propellant is released and passes over the opening leading to the product container. The partial

Figure 24–10   The "Innovair" system. (*Courtesy of the Geigy Chemical Corp.*)

vacuum that is created causes the product to be siphoned up the dip tube where it mixes with the propellant. At the same time, air enters the product container to replace the material that has been discharged. The system is automatically sealed as soon as the valve is released.

THE PISTON CONTAINER (American Can Company)

The aluminum "MiraFlo" piston container, manufactured by the American Can Company, is illustrated in Figure 24–11. According to Boyne,[16] it is used for packaging hand lotions, hair treatments, cheese spreads, cake icings, and caulking compounds. The unit consists of a free moving piston in a coated extruded aluminum container. The piston is constructed of a plastic, such as polyethylene, and is essentially a hollow cylinder with the upper end closed and the bottom end open. The upper end is fairly rigid but the sides are flexible in order to maintain a seal with the wall. The

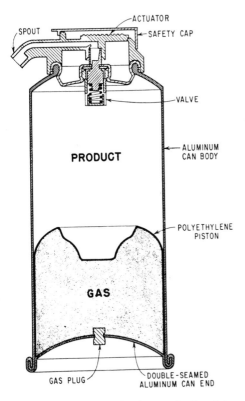

Figure 24–11   The aluminum plastic piston aerosol can. (*Courtesy of the American Can Co.*)

principle behind the piston container is that the product itself forms the seal between the piston and the wall.

The can is shipped completely assembled. The product is filled through the 1-in. opening in the piston can. In order to minimize the amount of air trapped in the container, the addition of as much product as possible is recommended, leaving only enough space for the valve cup.

The "MiraFlo" unit is pressurized through a center hole in the bottom of the container with a special gassing and plugging unit called the *Energizer*. This equipment is available as a single can gassing unit or a multihead machine. After the propellant has been loaded into the container by the Energizer, a rubber plug is pushed into the opening while the container is still under pressure. According to Hoffman and Marchak,[17] the resistance of the plug to blow out exceeds the buckling strength of the bottom of the container. The compressed gases are normally used for propellants at a pressure of about 100 psi. Nitrogen is preferred because of low product solubility. The liquefied gases can be used provided the plastic used for the piston is impervious to the liquefied gases.

Since the piston can depends upon the product to form the seal between the piston and the container wall, the product must be quite viscous. Otherwise, the propellant will leak past the seal or the product will seep down into the propellant chamber. Even if the viscosity of the product is satisfactory at room temperature, higher temperatures may decrease it sufficiently so that the seal is broken. In order to overcome these disadvantages, smaller pistons with either three or four piston rings have been proposed.[18,19] It is reported that this type of construction permits products with low viscosity to be packaged without danger of propellant bypassing the seal. In addition, the rings maintain an effective seal even if the container is dented. An improved gassing and sealing device called the *Prestoplug* closure is reported to allow both compressed and liquefied gases to be loaded

THREE-PHASE SYSTEMS

Three-phase systems are of consequence mostly from a historical standpoint, and the background of this system has been discussed in detail by Johnsen.[20] There was a period during the 1950's when interest in this system was fairly high because at that time it was about the only practical method for spraying aqueous solutions. Several three-phase products were marketed but because of the inherent disadvantage of this system, it never achieved any significant popularity and today it is a rarity.

The term *three-phase* in the areosol industry refers specifically to a system composed of two liquid layers and a vapor phase. This does not include emulsions. One of the liquid phases consists of a liquefied gas propellant and the other is the aqueous phase. The propellant can be either fluorocarbon or hydrocarbon and since it does not mix with the aqueous phase its only function is to supply the pressure that forces the aqueous phase up the dip tube and out through the valve when the product is discharged. Three-phase systems are not shaken before use since this would tend to disperse the propellant momentarily throughout the aqueous phase and give a nonuniform spray.

When the propellant is heavier than water, it forms the bottom layer with the aqueous phase on top. Fluorocarbon propellants, such as "Freon" 114, give this type of system. The length of the dip tube has to be adjusted so that it extends only to the bottom of the aqueous phase. (Figure 24–12a). If the dip tube were not shortened, propellant would be discharged initially instead of the aqueous phase. When the propellant is lighter than water, it floats on top of the aqueous phase (Figure 24–12b). This type of system is obtained with hydrocarbon propellants, such as n-butane and isobutane.

If a three-phase system is equipped with a standard valve and actuator, it will stream when discharged. This is because the solubility of propellant in water is so low that not enough propellant is present in the aqueous phase to break it up during discharge. Therefore, the aqueous phase has to be broken up mechanically, and this is accomplished with mechanical breakup actuators.

The quantity of liquefied propellant in the container at the start should be sufficient to maintain a constant pressure during the lifetime of the package. Theoretically, the last of the liquefied propellant should vaporize

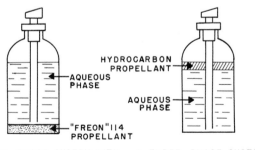

a. THREE–PHASE SYSTEM WITH   b. THREE–PHASE SYSTEM WITH
   "FREON" 114 PROPELLANT   HYDROCARBON PROPELLANT

Figure 24–12   Three-phase systems.

just as the last of the product is being discharged, but this is impossible to achieve practically. Such factors as the solubility of the propellant in the aqueous phase, possible leakage during storage, and loading variations must be considered. Johnsen[20] estimated that for a 100-cc bottle, the quantity of n-butane required would be 1.54 g, and the amount of "Freon" 114 would be 2.81 g taking into account the factors mentioned above.

The first commercial three-phase aerosol was *"Larvex,"* a moth proofer. It was pressurized with "Freon" 114 and packaged inside a protective paper tube.[21,22] Subsequently, Meissner[23] suggested the use of boiling chips to provide a more uniform discharge and Mina[24] disclosed an improved design for a mechanical breakup valve. Other improvements in containers and actuators followed and several other three-phase products appeared on the market, but the number was never very large.

There are a variety of reasons why three-phase aerosols have not been popular. In the first place, the spray is quite coarse and wet since there is no propellant present to assist in the breakup. In addition, the difficulty of cutting off the dip tube to exactly the right length for products formulated with the heavier-than-water propellants can well be imagined. Another disadvantage with three-phase systems with "Freon" propellants is that they could not be designed for use in an inverted position without dip tubes because the propellant, being on the bottom, would be discharged first.

The use of hydrocarbons as propellants eliminated the problem of adjusting the length of the dip tube, since it could be extended to the bottom of the container, and also made it possible to use the container in an inverted position without a dip tube. There was always some concern about the aerosols formulated with hydrocarbons since if an excess of propellant was present, ultimately the consumer would be spraying flammable propellant after the aqueous phase had been completely discharged.

Other disadvantages included the extraction of organic soluble ingredients from the aqueous phase by the propellant with a consequent change in the composition of the aqueous phase. Also, by this time, the consumer had become so used to shaking emulsified products, such as shave lathers, that he was inclined to shake all aerosol products, including three-phase systems. This resulted in a sputtery, nonuniform discharge and tended to discourage the user from repeating his purchase after the container had been emptied.

The three-phase system undoubtedly hastened the development of mechanical breakup actuators, and this very well may be its greatest contribution to the aerosol industry.

REFERENCES

1. *Aerosol Age* **12,** 124 (1967).
2. *Aerosol Age* **13,** 19 (1968).
3. H. Boden, *Aerosol Age* **13,** 19 (1968); "Freon" Aerosol Report A–74.
4. *Detergent Age* **5,** 122 (1968).
5. N. E. Platt, *Detergent Age* **5,** 137 (1968).
6. R. E. Moses and P. Lucas, U. S. Patent 3,341,418 (1967).
8. R. M. Friedenberg, U. S. Patent 3,240,396 (1966).
9. *Aerosol Age* **13,** 32 (1968).
10. J. M. Wittke, *Detergent Age* **5,** 153 (1968).
11. *Aerosol Age* **13,** 97 (1968).
12. L. F. Irland and J. W. Kinnavy, *Drug Cosmetic Ind.* **101,** 42 (1967).
13. *Soap Chemical Specialties* **44,** 104 (1968).
14. *Detergent Age* **5,** 142 (1968).
15. M. L. Thornton, *Detergent Age* **5,** 151 (1968).
16. R. W. Boyne, *Detergent Age* **5,** 147 (1968).
17. H. T. Hoffman and N. Marchak, *Modern Packaging* **34,** 129 (1961).
18. *Aerosol Age* **13,** 52 (1968).
19. T. C. Clark, *Aerosol Age* **11,** 28 (1966).
20. M. A. Johnsen, *Aerosol Age* **1,** 12 (1956).
21. W. C. Beard, "Valves," in H. R. Shepherd, "Aerosols: Science and Technology," Interscience Publishers, Inc., New York, N.Y., 1961.
22. S. Eaton, U. S. Patent 2,728,495 (1955).
23. H. P. Meissner, U. S. Patent 2,705,661 (1955).
24. F. A. Mina, *Aerosol Age* **1,** 28 (1956).

25

SAMPLING AND ANALYSIS OF AEROSOL PRODUCTS

T. D. Armstrong, Jr.

The components of typical aerosol products are amenable to numerous proven analytical methods—chemical, physical, and instrumental. However, in an aerosol package, the analyst is confronted by a pressurized sample, sealed in a container, and comprised of a mixture of components with widely varying vapor pressures. Vapor pressure differences are great enough to cause major components to be distributed unevenly in different phases in the sample, and loss of volatile components by fractionation is the overriding analytical concern. The aerosol valve is poorly designed for analytical sampling, and the analyst must devise a means of obtaining a representative sample from the aerosol package. The following discussion centers on sampling techniques and analysis involving either the total aerosol package or the volatile components. Details of the analyses of the components of the aerosol concentrates is left to the ample technical literature already available in the various fields encompassed by aerosol products.

Sampling Techniques

The most obvious way to sample is through the aerosol valve. The method is adequate for qualitative work, but it is difficult to effect a quantitative transfer into a suitable container. Brook and Joyner[1] have described a type of adapter that can be used to sample through some aerosol valve types into evacuated bottles. It is possible to invert the can, purge the dip tube, and draw a vapor sample into a gas syringe for air analysis. The ratio of propellant to concentrate can also be determined for some products by discharging part of the aerosol through the valve and into a container to recover the concentrate. The aerosol is weighed and the weight of recovered concentrate is determined. In general, however, it is better to find sampling methods giving better control and recovery than can be obtained with the aerosol valve.

If the contents of the aerosol container are chilled to below the boiling points of the propellants, the container may be opened and the contents transferred to a more adaptable container. Quantitative transfers are difficult, and the method has limited application for analysis involving the propellant. Jenkins and Amburgey,[2] however, reported pipetting samples for propellant analysis from chilled aerosol containers. The method is satisfactory for obtaining concentrate samples, and it is very useful for sampling foam products.

A method for sampling that offers great versatility is the use of a puncturing device. Several types are available. The easiest is the piercing valve used in the refrigerant industry for emptying disposable cans of refrigerant. These valves are available at refrigerant supply stores. The valve comes with a clip for attachment to the cap of the refrigerant can. The flanges on the clip can be straightened, and the clip can then be fastened to the side of an aerosol can with hose clamps. The piercing valve is screwed into the can until it punctures the side of the can. The piercing needle is enclosed in an elastomeric gasket that prevents leakage around the puncture. The valve is joined to ¼-in. copper tubing. Samples can then be obtained from the liquid or vapor phases depending upon the orientation of the aerosol container. Figures 25–1 and 25–2 illustrate this device.

In some instances, particularly with containers of small diameters, it is better to puncture the bottom of the aerosols. Figure 25–3 illustrates a device used for this purpose. The metal plates are securely fastened over the ends of the aerosol by means of the three screw posts. A piercing device is then screwed through an opening in the center of the bottom plate until the can is punctured.

Figure 25–1   Piercing valve, clip, and hose clamps.

Figure 25–2   Piercing valve attached to aerosol sample.

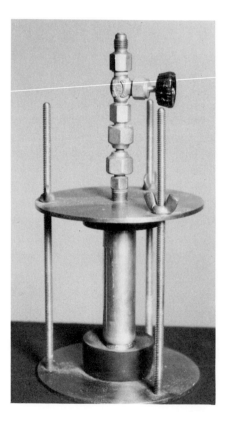

Figure 25–3   Bottom-puncturing device for aerosol samples.

Analysis

Gas Chromatography.   Gas chromatography can be used for a number of the more important analyses performed on aerosol products. Root and Maury[3,4] and Gaglione and Sciarra[5] have discussed the principles of gas chromatography and its application to aerosol analysis. Briefly, the chromatograph is an instrument that separates the components of a gas or vapor mixture. The separated components are passed through a detector that gives a response proportional to concentration. The samples may be vapor or liquid. Liquid samples are introduced into a heated injection port where they are instantly vaporized, and they are kept in the vapor state during their passage through the instrument.

The sample is carried through the instrument by a carrier gas, normally helium. Separation into individual components occurs during passage through the column, which is a length of tubing packed with a fine-mesh solid material called the support. The support is coated with a non-volatile liquid. Separation occurs because of differences in the absorbancies of the sample components in this liquid coating. The selection of the liquid is critical to obtaining separation. For propellants, di-n-butyl maleate, di-2-ethylhexyl sebacate, silicone oils, nujol, and similar partitioning liquids can be used.

The separated components of the sample then pass through a detector which gives a signal that can be amplified and recorded, usually as a peak on a strip chart. The most common detectors measure the differences in the thermal conductivity of the pure carrier gas and the carrier gas–sample component mixture. A number of other types of detectors exist. Flame ionization detectors pass the sample through a hydrogen flame and into an electrical field. The flame ionizes the sample and in the electrical field. The ions migrate to collectors. The current resulting from the neutralization of the collected ions is amplified and recorded. Electron capture detectors pass the sample through an electron beam, and some of the electrons are held by the sample component. As with the ionization detector, the charged sample components are collected, and the resultant current is amplified and recorded. A detector that may be of interest for aerosol analysis detects differences in the densities of the carrier gas and the carrier gas–sample mixture. Other detectors, which are more or less specific for compounds containing certain elements such as chlorine, phosphorous, and sulfur, have been developed.

A number of methods can be used to introduce an aerosol sample into a chromatograph. Instruments can be equipped to take vapor, liquid, or liquefied gas samples, and the method chosen depends upon the analysis desired, the constituents of the sample, the accuracy required, and the equipment available to the analyst.

Most analyses require samples from the liquid phase. Materials which are liquids at STP conditions can be injected into a chromatograph by syringe and hypodermic needle or by use of a liquid sampling valve, which injects a measured volume of liquid sample into the instrument. Problems arise with mixtures containing liquefied gases. The samples are under pressure, and vaporization tends to occur whenever any of the sample is transferred from the sample container. Syringes cannot be recommended for this type of sample, but it is possible to use liquid sampling valves. Sampling must be done in a manner insuring that the calibrated volume in the valve is liquid-full. Cannizzaro[6] described a method using a liquid sampling valve for aerosol propellant analysis.

Jenkins and Amburgey[7] applied Henry's Law to sampling aerosol containers for chromatographic analysis. Henry's Law states that the vapor pressure of a solution is proportional to the mole fraction of the solute in solution. In this method, the aerosol sample is chilled in dry ice and then punctured. A sample is pipetted from the chilled aerosol into a volumetric flask containing a chilled nonvolatile solvent. The weight of the pipetted sample is determined, and the volumetric flask is filled to volume. The partial pressures of the dissolved propellants are low enough to permit the solution to be syringed into a chromatograph without fractionation. Calibration curves plotting weight of propellants versus peak heights are prepared, and the milligrams of propellants in the solution are determined from these curves. The concentrations of the propellants in the aerosol are calculated by

$$C = \frac{A}{B} \times 100,$$

where C = percent propellant in the aerosol; A = weight of propellant in solution (determined from calibration curve); and B = weight of pipetted sample.

Another method based on Henry's Law is to sparge a vaporized sample into a relatively nonvolatile solvent. The aerosol sample is punctured with a piercing valve, and a fritted glass sparger tube is connected to the valve. Figure 25–4 is a picture of an aerosol container fitted with a piercing valve and sparger tube. 85 ml of solvent are placed in a 4-oz glass bottle. The sparger tube is placed below the surface of the solvent, the valve is opened slightly, and the sample is sparged into the solvent for 15 sec. Samples that clog the fritted glass can be run through straight tubing, but better contact with the solvent is obtained with the fritted glass, and this method is preferred. The bottle is then sealed with a serum cap. A standard mixture of approximately the same composition as the sample is prepared and sparged in the same manner as the sample. Aliquots of the solutions are

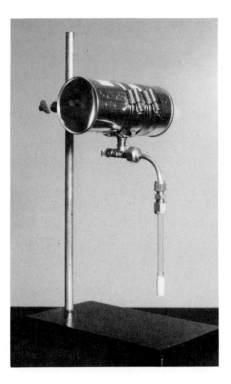

Figure 25-4 Aerosol sample with piercing valve and sparger tube attached.

then withdrawn through the serum caps into a syringe and injected into the chromatograph. The peak heights of the sample components and of the standard components are measured, and the composition of the sample is calculated by the equations:

$$X,\ Y,\ Z,\ \text{etc.} = \frac{(Ax,y,z,\ \text{etc.}) \times (Bx,y,z,\ \text{etc.})}{(Cx,y,z,\ \text{etc.})},$$

$$Px = \frac{X}{X + Y + Z + \text{etc.}} \times 100,$$

$$Py = \frac{Y}{X + Y + Z + \text{etc.}} \times 100,$$

where

$X,Y,Z,$etc. = relative concentrations of components $X,Y,Z,$ etc. in the sample;

$A_{x,y,z,\text{etc.}}$ = the peak height of $X,Y,Z,$ etc. from the chromatogram of the sample solution;

$B_{x,y,z,\text{etc.}}$ = the peak height of X,Y,Z, etc. from the chromatogram of the standard solution

$C_{x,y,z,\text{etc.}}$ = the concentration of X,Y,Z, etc. in the standard and

$P_{x,y,z,\text{etc.}}$ = percent component X,Y,Z, etc. in the sample.

Benzene, carbon tetrachloride, cyclohexane, acetone, and trifluorotrichloroethane have been used as solvents.

The success of the method depends upon treating the sample and the standard in an identical manner as to sparge rate, sparge time, and sample injection.

The methods using dilute sample solutions are particularly useful for aerosol samples such as hair sprays and paints containing nonvolatile material which would soon foul an instrument. The use of dilute solutions allows a large number of samples to be run before enough nonvolatile material accumulates to cause instrument problems.

It is possible to allow a sample to vaporize in an evacuated volume, and then to introduce an aliquot of the vapors into a chromatograph. Brook and Joyner[1] used this method in their work with aerosol propellants. An adaptor with a hypodermic needle attached was fitted on the valve of the aerosol sample. The needle was inserted through a serum cap into an evacuated gas bulb. The gas bulb was connected to a manometer and a gas sampling valve, which is a valve that injects a known volume of gas into a chromatograph. The sample was discharged into the volume, the pressure was adjusted to 10 ± 0.1 cm Hg, and the sample was introduced to the chromatograph. Numerous modifications of this procedure are possible.

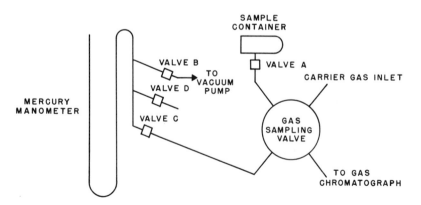

Figure 25-5   Gas chromatographic gas sampling system for aerosol products.

Figure 25–5 is a line diagram of a sampling system that can be used for sampling completely volatile samples or vapor phase samples. The sample is punctured with a piercing valve, Valve *A* in the diagram, and connected to the gas sampling valve as shown in Figure 25–6. The gas sampling valve is positioned to connect the sample container with the line from Valve *C*. Valves *B* and *C* are opened, Valves *A* and *D* remain closed, and the system is evacuated back to Valve *A*. When the system is evacuated, Valve *B* is closed, and Valve *A* is opened slightly to admit sample up to the desired pressure which is read with the manometer. When sampling from the liquid phase, Valve *A* acts as a vaporizing valve, and caution must be exercised to be sure the sample is vaporizing at the valve. When the desired sample pressure is attained, Valve *A* is closed, and the gas sampling valve is turned to inject the sample into the chromatograph. Standards are prepared and run in the same manner and concentrations are calculated in the same manner as in the sparging method.

It is occasionally necessary to determine the amount of air in the vapor space of an aerosol product. This determination can be made with the system diagramed in Figure 25–5. The aerosol is oriented to remove sample from the vapor space, and a sample is run in the manner just described. An air sample of the same size is run, and the concentration of air in the sample is calculated by

Figure 25-6   Aerosol sample with piercing valve connected to a gas sampling valve on a chromatograph.

$$C = \frac{A}{B} \times 100,$$

where:

C = vol % air in the sample;

A = area of air peak in sample (mm²); and

B = area of air peak in pure air sample (mm²).

Other noncondensibles such as CO_2 can be determined in the same way.

With most columns, the main constituents of air, oxygen and nitrogen, are not separated, but appear as one peak on the chromatogram. If it is necessary to know the oxygen concentration in the vapor space of an aerosol, it can be determined using a chromatograph with a column packed with #13X molecular sieves. The sample is run as in the air analysis just described, and an air sample of the same size is run as a standard. The air will be separated, and two peaks will show on the chromatogram. The first and smaller peak is oxygen. The concentration of oxygen in the sample is calculated by

$$A = \frac{B \times C}{D},$$

where

A = vol % oxygen in the sample;

B = area of oxygen peak in the sample chromatogram (mm²)

C = 21 (vol % oxygen in air); and

D = area of oxygen peak in the chromatogram of the air standard.

It should be noted that the concentration of noncondensibles in the vapor space will be lowered as samples are withdrawn allowing more of the liquefied propellants to vaporize into the head space.

Aerosol products that produce foams that are collapsed easily can be analyzed by the sparging method, although it may be necessary to use a straight tube rather than a fritted glass sparger. Products producing stable foams are difficult to sample by any of the methods already discussed. Stable foams usually can be collapsed by contacting them with dilute acid. The sample is connected to a volume filled with 6N hydrochloric acid. The piercing valve is opened slightly, and the foam sample is allowed to slowly displace part of the acid solution. The acid and foam are then agitated until the foam collapses. The vapor space above the acid is then sampled for propellant analysis by gas chromatography.

The retention time of a compound in a chromatograph is constant as long as the conditions of analysis are kept constant. As a result, retention times can be used as a means of identification of the components of a sample. Caution should be exercised, however, as some compounds will have identical or very similar retention times. In cases of doubt, the identification should be verified by infrared or mass spectrometer analysis. It is possible to trap the different components of a sample as they exit from a chromatograph. The trapped fractions can then be identified by their infrared or mass spectra.

Propellant to Concentrate Ratio. One of the most important determinations made in aerosol analysis is the measurement of the amount of propellant and amount of concentrate in the aerosol package.

Chromatography can be used for products that can be completely vaporized or for products in which the concentration of nonvolatile material in a volatile solvent is accurately known. It is possible to add to the aerosol a measured amount of a volatile material, termed an internal standard, which is foreign to the mixture already present. The ratio of the propellant to the internal standard is determined by chromatography. Knowing the original weight of the aerosol, the amount of internal standard added, and the ratio of the propellant to the internal standard, the concentration of the propellant in the original aerosol can be calculated.

In most instances, however, methods other than chromatography are used.

Brook and Joyner[1] passed a measured aerosol sample through a weighed calcium chloride solution, and the bath solution then was reweighed. Assuming that all of the sample except the propellant was retained in the solution, the amount of concentrate in the aerosol was determined from the weight of sample passed through the bath and the weight of sample retained in the bath.

Clapp[8] reported a procedure in which volatile and nonvolatile fractions of aerosol formulations are determined by comparing the amount of propellant expelled from an aerosol container after each of four steps with that expelled from a standard treated in the same manner.

Sciarra[9] described several gravimetric methods and a densimetric method for determining the volatile content of aerosol products. In the densimetric method, the sample is chilled to constant temperature below the boiling point of the propellant, the container is opened and the contents are transferred to a chilled cylinder. The density is determined by hydrometer, and the concentration of propellant is determined from a graph of density versus propellant concentration that has been prepared from the density determinations made at the same temperature for carefully prepared standards.

A general method that can be varied to meet the requirements of different types of samples is to transfer the contents of the aerosol to a distillation flask and distill off the propellant. The aerosol package is weighed, chilled in dry ice, and opened. The entire contents are transferred to a weighed distillation flask, and the empty aerosol container is reweighed so the sample weight can be determined. The distillation flask is fitted with a reflux condenser, and the propellant is allowed to boil away. The distillation is discontinued when temperature measurements show the propellant to be gone. The distillation flask is reweighed, and the weight of sample remaining is determined. The difference between the original sample weight and the weight of material remaining is the weight of propellant. The temperature of the distillation flask can be controlled by immersion in a thermostated bath, and loss of volatile solvent can be controlled by use of chilled water in the reflux condenser.

Moisture and Other Analyses.   Moisture in aerosol products is determined best by Karl Fischer titration. Reed and Downing[10] developed a procedure in which a measured amount of standarized Karl Fischer reagent was titrated under pressure with the aerosol sample. The weight of sample necessary to neutralize the known amount of reagent was determined, and the moisture content calculated by

$$\text{ppm } H_2O = \frac{A \times B \times 10^6}{C - D},$$

where

A = ml Karl Fischer reagent;

B = standardization factor, g H_2O/ml Karl Fischer reagent;

C = original weight of aerosol sample, grams; and

D = final weight of aerosol sample, grams.

Reed[11] also developed a method for moisture determinations in refrigerant–oil mixtures that is particularly adaptable to aerosol products. In this procedure a methanol–chloroform mixture is titrated to dryness with Karl Fischer reagent, and the titration vessel is then chilled with dry ice. The aerosol sample is punctured with a piercing valve, weighed, and connected to the titration vessel. A sample of the aerosol is let into the vessel, where it condenses. The solvent–sample mixture gradually is brought back to room temperature, the propellant escaping through a vent, and the remaining solution is titrated with Karl Fischer reagent.

It is possible to modify automatic Karl Fischer analyzers[12] for use with propellants, and the same apparatus and procedure can be used with aerosol samples. The sample is slowly passed into a titration vessel containing dried methanol or a dried methanol–chloroform solution. The sample is titrated as it enters the titration vessel. When a sufficient titer of Karl Fischer reagent has been used, the weight of the aerosol sample is determined, and the moisture content is calculated as before.

The success of any of these methods depends upon the prevention of atmospheric moisture contamination. Dry air is used to keep a positive pressure in the reaction vessel, and the vents are protected with driers.

Acidity in an aerosol sample can be determined by sparging the sample into previously neutralized isopropanol. The isopropanol is then titrated with base to the desired end point.

Many inorganic determinations can be made on water solutions obtained by sparging the sample through water. If two liquid phases result, they are separated. The organic phase is washed several times with water. The water washes are combined, and the analysis is performed by any appropriate method. Samples prepared in this way are suitable for rapid analysis by the specific ion electrodes now available for chloride, fluoride, sodium, potassium, and a number of other inorganic ions. By knowing the weight of sample sparged and the amount of water in the washes, these results can be calculated as concentrations in the original aerosol product.

REFERENCES

1. R. J. Brook and B. D. Joyner, *J. Soc. Cosmetic Chemists* **17**, 401 (1966).
2. J. M. Amburgey, *Proc. Sci. Sec. Toilet Goods Assoc.* **31**, (1959); *Aerosol Age* **4**, 35 (1959).
3. M. J. Root and M. J. Maury, Proc. 43rd Ann. Meeting, CSMA, **44**, (December 1956); *Soap Chem. Specialties,* Part I, **33**, 101 (1957); *Amer. Perf. Aromatics* **69**, 50 (1957).
4. M. J. Root and M. J. Maury, *J. Soc. Cosmetic Chemists.* **8**, 92 (1957).
5. J. J. Sciarra and O. G. Gaglione, *Paint Varnish Prod,* Part I, **54**, 63 (1964); Part II, **54**, 77 (1964).
6. R. D. Cannizzaro, *Aerosol Technicomment,* **11**, No. 1 (1968).
7. J. M. Amburgey, *Proc. Sci. Sec. Toilet Goods Assoc. No.* 31 (1959) *Aerosol Age* **4**, 35 (1959).
8. C. Clapp, Proc. 37th Mid-Year Meeting, CSMA, 21 (May 1951).
9. J. J. Sciarra, *Paint Varnish Prod* **55**, 68 (1965).
10. R. C. Downing and F. T. Reed, Proc. 38th Ann. Meeting, CSMA, 41 (1951).
11. F. T. Reed, *Refrigeration Engineering* **62** (1954).
12. "Freon" Technical Bulletin B–23, E. I. du Pont de Nemours & Co., (1956).

26

TOXICITY

J. Wesley Clayton, Jr.

INTRODUCTION

Society has now become accustomed to the convenience of pressurized packages that deliver the product to the point of use. A variety of functions are served by aerosol products, and bodily contact with aerosol products can and does occur. The manufacturers and formulators of aerosol commodities are thus brought face to face with the toxicity and hazards of their products in a way which is unique to the aerosol industry because the probability of contact is enhanced by the propellant's ability to forcefully discharge the substance from its container. It is, therefore, of the utmost importance that the propellant have a low order of toxicity and not contribute significantly to the hazard of using pressurized formulations.

Two terms are frequently used in this chapter and consequently require definition. These are *toxicity* and *hazard*. Toxicity is often defined as "too much," signifying that any agent can be toxic in sufficient quantity. Toxicity, therefore, is the *capacity* of any substance to cause derangement within a living organism. The derangement may be so severe as to eventuate in the death of the organism. The substance may produce serious and even permanent injury, or it may produce transient injury from which the organism can recover. In some instances, the organism can even adapt to the continued presence of the substance without injury but with evidence of biological adjustment having occurred. The severity of the derangement produced by a foreign substance is proportional to the amount or dose that has reached the reacting site(s) on or in the organism. It is an axiom of toxicology that any substance entering or contacting the body can cause a derangement at some degree of contact. A corollary is that the effect can be attenuated to toxicological insignificance by reducing the degree of contact. Thus no substance may be called *nontoxic*. Even water is toxic in sufficient amounts as attested by the report of Langaard and Smith.[1]

Hazard is defined as the *risk* of injury, and it is therefore a probability concept. The probability is dependent in part by the amount of the substance that can cause injury or change and in part by other factors such as vapor pressure, chemical properties, density, surface activity, viscosity, etc. These can either facilitate or limit bodily contact with a foreign material and thereby influence toxicity. For example, if a small amount of a substance can produce a toxic effect, i.e., it is high in toxicity, but if the vapor pressure is low, the risk or hazard of inhalation injury is lowered. If the viscosity of a material is so high that it is difficult to take it in by swallowing, the hazard by the oral route of intake is lowered. Conversely, the hazard is increased in those materials of high vapor pressure or low viscosity. To be sure, in these instances, if the substance is low in its *capacity* to cause toxic effects i.e., low toxicity, the resultant change on contact may be negligible. Thus the two concepts, toxicity and hazard, must be weighed in evaluating the safety of the materials of society.

Safety may be considered the inverse of hazard. In toxicological circles, safety is the practical certainty that proper use of a substance will not culminate in injury. On the other side of the coin, there is the chance that some derangement may follow upon misuse of the substance. The daily press continually reports the sorrowful consequences of abuse of an otherwise safe product. This it does often with some surprise or chagrin because the commonly accepted idea of safety includes not only the safety for intended use, but also any abuse which the careless or clever may hit upon.

The evolution of the toxicology of fluoroalkanes began in the 1920's with the advent of mechanical refrigeration. The then refrigerants were unsatisfactory. Ethylene was flammable; sulfur dioxide and ammonia were corrosive and toxic. In 1926, LeBeau and Damiens,[2] prepared tetrafluoromethane, the simplest perfluoroalkane. Then Midgley and Henne,[3] synthesized dichlorodifluoromethane specifically for refrigerant use on the grounds of its pressure–temperature relationships, and low toxicity which they inferred from its chemical stability. It is to Thomas Midgley that we owe the first recorded human toxicity experiment with fluoroalkanes. In the public eye, at the American Chemical Society Meetings of 1930, he inhaled enough dichlorodifluoromethane to extinguish a burning candle. However, the toxicity experiments conducted by the Underwriters' Laboratories to evaluate the acute inhalation toxicity of refrigerant fluoroalkanes constituted the major toxicological efforts of this era.

As yet there is no sound principle by which we can determine the toxicity of the fluoroalkanes from their chemical properties. Nevertheless, there is an important correlation between the biological activity of the fluoroalkanes and their chemistry. Fluorocarbons in general are known for their chemical stability. This is a function of the short interatomic distance between carbon and fluorine and strength of the bond joining the two elements. Furthermore, the presence of fluorine in a molecule stabilizes adjacent bonds rendering them less susceptible to rupture. For example, fluorine stabilizes neighboring C–Cl bonds in the chloroalkanes and tends to reduce the steric strain resulting from the relatively voluminous chlorine atoms. As more fluorine is added to the molecule the C–F distance is progressively shortened, and the bond energy is increased. This means that, in highly fluorinated compounds, the C–F bond requires more energy for cleavage than molecules of lower fluorine content. In biochemical systems, the degree of enzymatic dehalogenation is inversely related to bond dissociation energies (Slater,[4] Gregory,[5] Butler[6]), and in many cases, toxicity follows the same pattern. This is illustrated in Table 26–1 which shows carbon–halogen bond energies and the progression of toxicity.

TABLE 26-1   RELATIONSHIP BETWEEN BOND DISSOCIATION
ENERGY AND TOXICITY

Bond Dissociation	$Cl_3C{\rightarrow}Cl$	$Cl_2HC{\rightarrow}Cl$	$Cl_2FC{\rightarrow}Cl$	$F_3C{\rightarrow}Cl$	$F_3C{\rightarrow}F$
Energy (K cal)	68	72	74.5	83	121
Order of Toxicity	CCl_4 >	$CHCl_3$ >	CCl_3F >	$CClF_3$ >	CF_4

TOXICITY TESTING

As with most systematic testing procedures, toxicity tests have developed in response to particular needs. There are two basic needs that toxicity tests serve. One is the need to determine the effects of chemicals on the body with respect to *time* of contact. The second is the need for information about the *types* of bodily contact that can be experienced and the ensuing biological response. In this second area, toxicologists frequently utilize routes of entry into the body which are not commonly experienced except in medical procedures. Direct injection into blood vessels, body cavities, or organs are examples of experimental techniques employed for special reasons.

Focusing on the time of contact, it is customary to classify toxicity tests as acute, i.e., single contact or intake of momentary or few hours duration, or chronic, i.e., continuous or repeated contact or intake of several days or even years.

Acute Toxicity

Objectives

(a) To measure in quantitative terms the capacity of an agent to produce derangement as a result of a single contact with the agent.
(b) To discern *target* organs or systems.
(c) To gauge the range of toxicity, i.e., the relation of the response to the dose administered.
(d) To determine the nature of the toxic effect, i.e., the mode of action, the fate of the agent and microscopic changes in tissue.

Terminology and Tests

LD_{50} OR LC_{50}.

These two terms are numerical values that represent the amount of an agent which, on statistical grounds, would be lethal to 50% of a population. The amount is expressed as the dose, in the LD_{50} e.g., milligrams of substance per kilogram of body weight (mg/kg) or, in the LC_{50}, as the concentration in inhaled air, e.g., milligrams or volumes of the substance per multiple volumes of air (mg/liter, mg/m^3, or parts per million, ppm). The LC_{50} is applicable to the inhalation toxicity since it indicates the amount of agent in the air which is inhaled. It should always be accompanied by a statement of the duration of the exposure as the LC_{50} is generally inversely related to the length of the exposure period. The LD_{50} can be used in any toxicity test in which a substance contacts the body or is taken internally and thus can be related to the body weight of the organism receiving the dose.

The determination of the LD_{50} or LC_{50} is not simply a matter of empirically determining the dose or concentration that kills one-half of a particular group of animals. Rather it is derived by computation from an experiment in which a number of animals are given several, different doses of a substance. The logarithms of the doses are then plotted against the percentage of mortality on a probability scale. A straight-line is fitted to the plotted data, and the intersection of this line with the 50% point is the LD_{50} or LC_{50}. Confidence limits can also be derived. For more detained information, the reader is referred to Bliss[7] and Litchfield and Wilcoxon.[8]

ALD OR ALC

This refers to the Approximate Lethal Dose and is the lowest dose or concentration which is lethal, on an empirical basis, to a single animal or group. The units are the same as in the preceding discussion. This is a rapid way of estimating the LD_{50} or LC_{50} and is often employed to establish the appropriate range of doses or concentrations that are later used in a test to obtain the LD_{50} or LC_{50}. The determination of the ALD or ALC is given by Deichmann and Le Blanc.[9]

Chronic Toxicity
Objectives

(a) To determine cumulative effects of repeated contact or intake over a period of time which satisfies the question studied. For example a chronic experiment may be of several days duration, weeks or a lifetime, depending on the kind of information desired. Chronic experiments of short duration, say, two weeks to one or three months are often called *subacute* to distinguish them from chronic studies of longer duration, e.g., a two-year study.

(b) To extend toxicity information to several species of animals since acute toxicity work is often limited to the smaller animals, rats, mice, or guinea pigs.

(c) To ascertain the biologic response by various means, e.g., changes in appearance or behavior, body weight changes, food consumption, blood and urine studies, metabolism or detoxication of the substance, effects on reproduction and embryonic development, and microscopic examination of organs.

(d) To estimate, if possible, the dosage or concentration in air which elicits no response in the test animals by any of the criteria utilized. This so-called *no-effect* level is often then used to establish tolerance or safe levels for human exposure. In making the judgment that a particular level will be safe for man, the toxicologist usually applies a *safety factor* to the no-effect level inferred from animal data. In establishing no-effect levels of pesticides, a factor of 0.01 is applied in many cases. In the setting of Threshold Limit Values* no single factor has emerged as generally valid (Smyth[10]). The no-effect concept is subject to semantic as well as scientific difficulties in defining it. If the term is taken to mean no effect whatsoever, no amount of testing can establish it, as it is impossible for any observation to demonstrate zero. To be scientifically consistent, what can be said in defining no-effect is that certain observations of finite precision have failed to demonstrate a measurable response.

---

* The Threshold Limit Values refer to airborne concentrations of substances and represent conditions under which it is believed that nearly all workers may be repeatedly exposed, day after day, without adverse affect." (1967 TLV List)

Tests.   In general the tests for chronic toxicity are defined as the duration of the investigation, e.g., 10-day repeated test, 30-day test, 90-day test, two-year test. Often the progression in time is employed as the grand design in a detailed toxicologic investigation, and the investigator will progress to the point where he believes the data are adequate to apply to the particular situation. This is the reason why chronic toxicity studies are so diverse in design and, therefore, so difficult to compare from one investigation to another.

Dosages or concentrations in air in chronic toxicity studies are reduced several-fold from those employed in acute toxicity work; and the longer the chronic study, the lower the dosages or concentrations used. Although one dosage or concentration can provide meaningful data, a more desirable design would involve several test levels so that effects at the different levels could be evaluated. It is considered good toxicologic design if one of the test levels produces toxic effects, another, a mid level, produces milder or reversible effects, and a third or low level produces no detectable change. The last may be interpreted as the no-effect level.

The toxicologist often speaks in terms of CT, that is, the product of concentration (or dose) and time as being constant relative to a specific toxic effect, viz., $CT = K$. However, experimental work reveals that this relationship holds only over very short ranges of time. For example, rat exposures to perfluoroisobutylene, a toxic pulmonary irritant, showed a progressively lower lethal CT as the time of exposure was increased. This means that acute toxicity studies conducted at relatively high levels for brief periods cannot be divided into small increments and converted numerically into chronic toxicity by applying the appropriate time factor Chemicals administered at low levels for long periods of time may display qualitatively different biological activities than when given at higher levels in a short time period. Thus the toxicologist can offer no convenient substitute for a chronic toxicity experiment.

The second need cited above was the desirability to have information on bodily effects as related to the kind of contact experienced.

Natural Routes of Entry or Contact

1. Mouth—Oral Toxicity
2. Lungs—Inhalation Toxicity
3. Skin—Dermal Toxicity, Primary Irritation and Skin Sensitization
4. Eye—Eye Toxicity

The biologic response arising from the various routes of entry may be classified as local or systemic. Local effects are those which result from the direct contact of a substance with the tissue at the site of contact. The response to corrosive chemicals is tissue destruction; the response to less active materials is termed irritation. It is characterized by redness, varying degrees of swelling, loss of tissue layers, and cellular changes. These are reversible effects. Local effects are common consequencs of contact with eyes or skin, and the effects can be related to dose or concentration by a scoring system that lends a degree of quantitation for the evaluation of the response. Several scoring systems have been published for eye and skin contact, Draize,[11] Carpenter and Smyth,[12] and Lehman[13] et al.

In some cases small amounts of a substance may elicit local skin effects as a result of the skin having acquired an increased sensitivity through prior contact with the substance. This phenomenon is called skin sensitization and the response is reddening, with or without swelling and with or without blistering at the point of contact. These reactions may not have occurred at all or occurred only to a minimal degree on first contact. Skin sensitization reaction is typified by the reaction of human skin to poison ivy. There are no known fluoroalkanes which exhibit this property. Tests employed for the evaluation of this potential have been reviewed by Stevens.[14]

Systemic effects are those responses that occur in organs removed from the site of contact of an agent. In this case the material must be absorbed into the body and be transported by the blood or body fluids to the point(s) of attack. Systemic effects are most readily appreciated from oral intake and inhalation, however, absorption through the skin with effects on distant organs is not uncommon, and even systemic effects from eye contact are not unknown.

Unnatural Routes of Entry

These can be generally defined as any mode of entry into the body cavities that can be technically accomplished without the procedure itself producing irreversible or toxicologically significant changes. In view of the fact that these procedures involve artificial means of bodily entry, they serve special purposes and interpretation of the animal tests are applicable to humans only insofar as the same mode of entry obtains. The same holds for toxicity studies utilizing natural routes of contact or entry. Toxicity figures from one entry or contact route are not directly applicable to the toxicity resulting from other means of ingress.

It has been stressed above that there are some precautions in interpreting toxicity test results. Those emphasized were: (1) the caution about converting from acute to chronic toxicity data by improper application of the $CT = K$ principle, and (2) the danger of applying toxicity results from one mode of contact to another. However, there is a more basic warning to be sounded. No matter how detailed and varied the studies and observations, the resulting data are not absolutes but estimates and probabilities. From these the toxicologist attempts to formulate meaningful statements about the toxicity for or hazard to the human population. Without experimental data, the toxicologist often is left with inferences from the use of a chemical for many years. Who is to say, however, if the composition is the same now as years ago; if the dose, duration, and type of exposure would be the same; if exposure to other physical, chemical, and social stresses would be the same during the span of time? How can the toxicologist be assured his measurements were always capable of detecting a change? He may only be able to state that no injury was detectable, not that prolonged use actually caused no injuries. Therefore, the lack of quantitative data imposes serious strictures in the toxicological interpretations from long term use of a chemical.

The decision to market a commodity with the concomitant human exposure involves balancing the toxicological risks of use against the benefits of use in society. Many factors bear on this decision. Animal studies provide a preliminary or initial basis for the judgment. Acute and chronic toxicity studies are fundamental in establishing safety guidelines for the hazards involved in the use of a substance. Ultimately, the decision is a judgment, reflecting some personal or social system of values, and it is based on experience in actual use.

TOXICITY OF FLUOROALKANES

Aerosol propellants containing fluorine in the molecule all belong to the chemical class of compounds known as alkanes, that is, straight, branched chain or cyclic, saturated organic compounds. The ones of principal concern here are:

Empirical Formula	Propellant Number
CCl_3F	11
CCl_2F_2	12
$CHClF_2$	22
$CClF_2$-$CClF_2$	114

There are many other fluoroalkanes, however, and any discussion of the toxicity of the ones used as aerosol products must also consider other members of the series. In general, these are all nonreactive compounds of high chemical stability resulting from the strong C–F bond which, in large measure, accounts for a low degree of toxicity.

Acute Inhalation Toxicity

The early toxicity studies on fluoroalkanes were designed around their use as refrigerants and were aimed at protecting the householder or refrigeration repairman. The Underwriters' Laboratories designed an inhalation study in which groups of twelve guinea pigs were exposed to a graded series of concentrations of the test compound. At each sampling time of 5 min, 30 min, 1 hr. and 2 hr, three guinea pigs if surviving the exposure, were removed from the chamber for observation or pathological examination of vital organs within a 2–10-day period after exposure. This work also included other refrigerants in use at the time, for example, ammonia and sulfur dioxide. The comparison provided by their inclusion was invaluable in putting candidate refrigerants and propellants in perspective with existing materials.

Studies on several refrigerant compounds in the decade 1931–1941 led to a classification system that has been used to grade the safety of new compounds. In this system, Class 1 compounds are the most toxic, causing death or serious injury when inhaled continuously by guinea pigs at 0.5–1.0 vol % for 5 min, e.g., sulfur dioxide. Class 2 is composed of materials that are lethal or injurious to guinea pigs at concentrations of 0.5–1.0 vol % for 30-min exposure, e.g., ammonia and methyl bromide. Classes 3 and 4 are distinguished by an exposure that proves toxic for guinea pigs at 2.0–2.5 vol % for 1 hr and 2 hr, respectively, e.g., carbon tetrachloride and chloroform (Class 3) and dichloroethylene and methyl chloride (Class 4). Class 6 is defined as a 2-hr exposure at 20 vol % that does not produce injury to the guinea pigs, e.g., dichlorodifluoromethane, dichlorotetrafluoroethane, monobromotrifluoromethane. Classes 4–5, 5a, and 5b are not precisely defined; they accommodate compounds more toxic than those in Class 6 but less toxic than Class 4 materials, e.g., trichlorotrifluoroethane and methylene chloride (Class 4–5), trichloromonofluoromethane, chlorodifluoromethane, and CO_2 (Class 5a), ethane, propane, and butane (Class 5b).

The primary purpose of this system was to gauge the acute inhalation hazard, and it is evident that the fluoroalkanes are low in toxicity on the basis of the above groupings. Subsequent work has been confirmatory, and experience in the chemical industry attests to a low order of toxicity for man.

In view of the fact that TLV's are often used incorrectly as toxicity indices, it is important to put them in perspective. The first industrial hygienic standards for airborne contaminants proposed for some fluoroalkanes by Cook[15] ranged from 5000 ppm for dichlorofluoromethane to 100,000 ppm for dichlorodifluoromethane. The values were subsequently lowered to 1000 ppm, not from toxicity considerations but on the premise that proper engineering standards dictate that no containment vapor should be allowed to exceed 1000 ppm on an 8-hr average. Only carbon dioxide has a TLV greater than 1000 ppm, i.e., 5000 ppm. Thus, while the TLV of 1000 ppm is commensurate with the low toxicity of the fluoromethanes, its purpose is primarily for good housekeeping. Industrial experience has established its validity and utility.

Inhalation Toxicity of Aerosol Sprays

In 1958, Bergmann et al.,[16] suggested a possible relationship between hairspray inhalation and a condition of the lungs that was characterized by storage in lung tissues of particles of the resin component of the hairspray. Out of the ensuing controversy, focusing on cause and effect, Bergmann et al.,[17] Brunner et al.,[18] Schepers,[19] McLaughlin and Bidstrup,[20] there grew a body of data on the effects of inhaling fluoroalkanes, mainly, CCl_3F and CCl_2F_2, as discharged from a pressurized container. Under these conditions, it is possible to inhale droplets of liquid fluorocarbon; and consequently, it becomes important to compare these results with those stemming from the inhalation of vapor alone.

Three investigations, which were aimed at the hairspray issue, contribute to our knowledge of the biological effects of fluorocarbons 11 and 12. Calandra and Kay[21] in their evaluation of hairsprays by guinea pig exposures used a fluoroalkane propellant–ethanol formulation as a control. Exposures were for 45–90 days. Three times each day, 5 days a week, the guinea pigs were exposed in a chamber to a 15-sec spray discharge of the propellant–alcohol mixture. The chamber was then sealed and the guinea pigs retained in the chamber for 15 min. Body weights, ratiological and hematological studies, behavior, urine analyses, survival and pathology disclosed no adverse effects attributable to the inhalation of the fluoroalkane propellant and ethanol. While the authors do not report concentrations of the components of this formulation, they estimate the magnitude of these exposures to be 12×10^4 times that experienced by persons using hairspray in the conventional way. A low order of inhalation toxicity is clearly evident from this study.

Draize et al.,[22] studied the acute and chronic (45–90 days) inhalation toxicity of six types of aerosol hairsprays using rabbits. Exposures were twice a day to a 30-sec spray followed by a 15-min exposure in the closed chamber. They do not indicate what propellant system was used. However, their work disclosed no adverse toxicologic, radiologic, or histologic effects from any of the formulations tested.

Several studies on the acute inhalation toxicity of fluorocarbon propellants have been conducted at the Haskell Laboratory. In these investigations the propellant is discharged for 5–30 sec into a small exposure chamber containing four-six rats. After the timed discharge, the chamber is sealed and the rats kept in the chamber for a 15-min static exposure, i.e., no diluting air is admitted. During the 15-min residence time, the oxygen concentration gradually declines but not to a point which would adversely affect the animal. Oxygen concentration is checked at the end of each exposure. The amount of propellant discharged is determined and the chamber concentration is calculated. During the spray discharge and the static exposure which follows, the rats are observed for toxic reactions. After exposure the surviving animals are observed for about two weeks and then are sacrificed for microscopic examination. Data from one of these experiments are summarized in Table 26–2.

TABLE 26-2   ACUTE INHALATION TOXICITY OF SEVERAL
FLUOROCARBON PROPELLANTS

Fluorocarbon Propellant	Time of Discharge Seconds	Concentration		Mortality Ratio*
		PPM		
		FC-12	FC-11	
12	15	339,000		0/4
11	15		279,000	4/4
	10		183,000	4/4
	5		90,300	0/4
11/12	15	130,000	115,000	0/4

* Numerator = number dying, denominator = number exposed.

It is evident from the table that high concentrations of Fluorocarbon 11 were lethal to rats in the 10–15 sec discharge time. Exposure with Fluorocarbon 12, or 11, and 12 mixed were not lethal in the 15-sec discharge. There were no adverse changes in the lungs of rats as determined by histologic examination. Survivors gained weight and were normal in appearance and behavior following exposure. It is believed that the deaths were due to an anesthetic effect of fluorocarbon 11. Scholz[23] has shown that Fluorocarbon 11 has the highest anesthetic activity of the fluoroalkanes he studied, namely 11, 12, 113, and 114.

Giovacchini et al.,[24] investigated the chronic inhalation of hairspray discharged from containers pressurized with Fluorocarbons 11 and 12. They included a *treated control* which was composed of propellant and other hairspray ingredients excluding the resin. These authors exposed dogs twice daily to a 10-sec spray discharge; the animals then remained in the closed exposure chamber for 15 min. The exposure period lasted for two years. One-third of the dogs were sacrificed after one year of exposure and the remainder after two years for pathology. The authors report normal values for hematological measurements, liver and kidney function tests, normal chest x-rays, normal blood pressure and electrocardiograms; body weights and food consumption were maintained throughout the test. Pathology conducted at one and two years revealed no adverse effects. The authors' summary statement is: "In brief, no significant alterations attributable to long-term exposure of the experimental animals to hairspray could be found." The same conclusion is applicable to the fluorocarbon propellant control spray.

The low order of inhalation toxicity of the fluoroalkanes employed as pressurizing materials for aerosol products is evident from the preceding work. Experience gained from the manufacture and sale of fluoroalkane propellants indicates that they are safe for their intended use. When the intended use is abused and deliberate overexposure is managed, high concentrations of gaseous or liquid fluorocarbon and lack of adequate oxygen may result in injury to the respiratory tract or death.

Chronic Toxicity of Fluoroalkanes

There are limited data relating to the question, what activity do the fluoroalkanes have in repeated exposures? Tables 26–3 and 26–4, condensed from Tables 18 and 19 in Clayton[25] summarize the published experimental work on this question. The reader can readily perceive that variety is of the essence in this area. Variety in the kind, number, and sex of animals, variety in daily exposure duration, variety in the number of exposures, and variety of observation are apparent from this cursory presentation. This is no plea for uniformity of design, because the experiment should be appropriate to the need for which it is conducted.

Reasons for conducting chonic toxicity experiments are diverse. The early studies of Sayers et al.,[26] on CCl_2F_2 and Yant et al.,[27] on $CClF_2$-$CClF_2$ were directed toward safety for refrigeration engineers and repairmen with regard to these, then new, refrigerants. From this work it was possible to establish industrial hygiene standards, which have been validated by industrial and market experience. Recently, newer uses for fluoroalkanes have necessitated suitable experiments to evaluate chronic effects.

In the area of food additives an interesting evolution is apparent in the design of chronic toxicity experiments. With a background of inhalation studies for gaseous materials, the U. S. Food and Drug Administration in 1961 issued a regulation approving octafluorocyclobutane (OFCB) as a food propellant on the basis of a 90-day inhalation study (Table 7, Clayton et al.[28]). The inertness of OFCB was demonstrated by the absence of any effect detected by appearance, behavior blood, and urine studies and

TABLE 26-3 REPEATED EXPOSURES TO SEVERAL FLUOROMETHANES

Compound	Conc. PPM	Hours Day	Days* (1)	Animals	Major Observations
CCl₃F	4,000	6	28	Rats (M,F), Mice (M,F), G.Pigs, Rabbit	No mortality.
	12,000	4	10	Rats	No mortality.
	12,500	3.5	20	Dogs	No mortality.
	25,000	3.5	20	G.Pigs, Cats, Rats	No mortality.
CCl₂F₂	200,000	7–8	3.5–56	Dog, Monkeys, G.Pigs	Some mortality and CNS reactions.
	810	24	92	Rats, G.Pigs, Rabbit, Monkey, Dog	Some mortality, No Toxic signs. Liver changes in G.Pigs, Monkeys.
	840	8	30	Rats, G.Pigs, Rabbit, Monkey, Dog	Slight mortality. No toxic signs. Liver changes in G.Pigs, Monkeys.
CHClF₂	100,000	3.5	20	Rats, G.Pigs, Cats, Dogs	No mortality. No signs of toxicity.
	14,200	6	10 mo.	Rabbits, Rats, Mice	No mortality. CNS reactions.
	1,980	6	10 mo.	Rats, Mice	No signs of toxicity.
CClF₃	10,000	6	20	Rats	No mortality. No signs of toxicity.
CBrF₃	500,000	2	15	Rats, Mice, G.Pigs	Coincidental mortality. No toxic signs.
	23,000	6	18 wk	Rats, Dogs	No mortality. No signs of toxicity.
CCl₃F/	5,000	2	100	Mice, Rats, G.Pigs	No mortality. No signs of toxicity.
CCl₂F₂	15,000	2	100	Mice, Rats, G.Pigs	No mortality. No signs of toxicity.
	50,000	2	100	Mice, Rats, G.Pigs	No mortality. No signs of toxicity.

From Clayton.[25]
* (1) Except as noted.

TABLE 26-4 REPEATED EXPOSURES TO SEVERAL FLUOROETHANES

Compound	Conc. PPM	Hours Day	Days (1)	Animals	Major Observations
$CCl_2F\text{-}CCl_2F$	3,000	4	10	Rats	CNS and respiratory signs. No mortality.
	1,000	6	31	Rats, Mice, G.Pigs, Rabbit	No mortality. Slight liver change.
	1,000	18	16	Rats	No mortality. No signs of toxicity.
$CCl_2F\text{-}CClF_2$	5,000	7	30	Rats	Slight lowering of body weight gain. Pale liver. No mortality.
	2,500	7	30	Rats	No mortality. No signs of toxicity.
	25,000	3.5	20	G.Pigs, Rats	No mortality. No signs of toxicity.
	12,500	3.5	20	Cats, Dogs	No mortality. No signs of toxicity.
$CClF_2\text{-}CClF_2$	200,000	8	2–4	G.Pigs	No mortality. CNS signs. Slight liver change.
	141,600	8	21	G.Pigs	1/6 died. Slight liver change.
	200,000	8	3–4	Dogs	Death—CNS signs of toxicity.
	141,600	8	3–21	Dogs	No mortality. CNS signs.
	100,000	3.5	20	Cats, G.Pigs, Rats, Dogs	No mortality. No signs of toxicity.
$CClF_2\text{-}CF_3$	100,000	6	90	Mice, Rat, Rabbits, Dogs	No mortality. No signs of toxicity.

From Clayton.[25]

histopathology in work in which mice, rats, rabbits, and dogs were exposed 6 hrs daily for 90 exposures to 100,000 ppm (by volume). A similar problem was faced with chloropentafluoroethane (CPFE), also proposed and subsequently cleared by the FDA as a food propellant. As with OFCB, the newer food propellant was subjected to a rigorous inhalation program of similar design. Biological inertness was evident from the same array of measurements. In addition, it was shown that urinary fluoride excretion was not changed in the rats and dogs inhaling CPFE (Clayton et al.[25]). A feature of the safety evaluation of CPFE was a short-term, oral intubation experiment on rats. In this experiment CPFE, dissolved in cottonseed oil, was administered daily five times a week for two weeks. From the body weights, appearance, and behavior, as well as from histological appraisal, adverse effects were not apparent. Thus, cumulative oral toxicity was not evident. Although other oral toxicity studies are known for several fluoroalkanes, this 10-day study on CPFE was the first to be reported for a food additive use. In the future, for applications in which fluoroalkanes might contact food, chronic oral toxicity studies will receive priority and that these will probably be accompanied by an investigation of metabolism.

Toxicity—Acute Oral, Skin, and Eye

The importance of inhalation studies on fluoroalkanes is evident since the first compounds were gases at room temperature when used as refrigerant or fire extinguishing compounds, and the hazard was from inhalation. However, with the expanding uses for fluoroalkanes, it is apropos to consider the toxicity by other routes of administration. As shown in Table 26–5, the several fluoroalkanes illustrated show a low order of acute oral toxicity for the rat. Rats have been repeatedly dosed by gastric intubation with dichlorodifluoromethane (CCl_2F_2) dissolved in peanut oil. The average daily dose during ten daily treatments was 430 mg/kg. The six rats survived and showed no clinical signs of toxicity nor pathologic changes. The oral toxicity of CCl_2F_2 on the grounds of a 90-day study with rats and dogs was also shown to be of a low degree. A repeated oral experiment with the symmetrical dichlorotetrafluoroethane ($CClF_2$-$CClF_2$) was carried out with rats at a daily dosage of 1300 mg/kg; and there were no deaths, no clinical signs of toxicity, and no pathological changes. Quevauviller[29] reported that a daily oral dose of 2000 mg/kg of dichlorotetrafluoroethane was tolerated by rats for a period of 23–33 days.

Trichlorotrifluoroethane, CCl_2F-$CClF_2$, as indicated by rat studies in Table 26–5, has a low order of oral toxicity with an LD_{50} of 43,000 mg/kg. Odou[30] has administered a massive oral dose to a female dog lightly anesthetized with sodium pentobarbital. The total dose given was

92,000 mg/kg; death occurred in 1.5 hrs. Some dogs have survived oral doses of this magnitude, and recently Odou[30] reported that a human under anesthesia accidently received about one liter of cold CCl_2F-$CClF_2$ in the stomach. This produced vomiting and immediate but transient cyanosis. The individual survived and reported only severe rectal irritation and diarrhea for 3 days thereafter.

TABLE 26-5   ACUTE ORAL TOXICITY OF VARIOUS FLUOROALKANES

Compound	ALD (mg/kg)[1]
CCl_2F_2	$> 1,000$[2]
$CHCl_2CClF_2$	7,500
CCl_2FCCl_2F	$> 25,000$
$CClF_2CCl_3$	$> 25,000$
$CClF_2CCl_2F$[3]	45,000 (LD_{50} = 43,000 mg/kg)
$CClF_2CClF_2$	$> 2,250$ mg/kg[2]

[1] Rats, 10–14 day survival period.
[2] Maximum feasible dose of fluorocarbon dissolved in peanut oil.
[3] ALD, rabbits = 17,000 mg/kg.

Trichlorotrifluoroethane, CCl_2F-$CClF_2$, is also low in toxicity as judged by skin contact studies with rabbits. Application of CCl_2F-$CClF_2$ to the skin of rabbits gave an ALD of 17,000 mg/kg. There was a local irritation of the skin at the site of application, and histology disclosed alterations in the dermis and adjacent connective tissues. There were no systemic changes attributable to treatment. Scholz[23] reported that CCl_2F-$CClF_2$ exerted no harmful effects on the eyes of rabbits reciving 0.1 ml in nine applications during an eleven day period. He reported the same results for CCl_3F.

Similar results from eye studies on dichlorodifluoromethane were reported by Downing and Madinabeitia,[31] citing work conducted at Haskell Laboratory. In this work a 50% solution of CCl_2F_2 in refined mineral oil was sprayed into rabbits' eyes from a distance of about 6 in. Controls consisted of rabbits receiving mineral oil alone. The animals receiving the fluorocarbon with mineral oil in the eye showed the same eye reaction as the controls did. In both groups, slight conjunctival irritation developed, but this had disappeared in 24 hrs.

Chloropentafluoroethane, $CClF_2$-CF_3, dissolved in cottonseed oil, was administered orally to rats five times a week for two weeks. Gross and microscopic examination of major tissues revealed no change attributable to $CClF_2$-CF_3. Therefore, under the conditions of this experiment there was no evidence of cumulative toxic effects exerted by $CClF_2$-CF_3, an observation supported by repeated inhalation studies reported in the next section.

Quevauviller et al.,[32] and Quevauviller[29] have investigated the effects of various fluoroalkanes on the skin, tongue, soft palate, and auditory canal of rats; the eye of the rabbit; and the speed of healing of wounds and burns in the rat. Five materials were evaluated by these tests, namely, CCl_3F, CCl_2F_2, CCl_3F, and CCl_2F_2 mixed; CCl_3F and $CHClF_2$ mixed; and $CClF_2$-$CClF_2$. The animals were treated one or two times a day, five days a week for 5 or 6 weeks. In the case of wounds and burns, the treatment was continued until healing was complete.

The skin of the rats became irritated evincing an edema and a slight inflammatory reaction. The reactions were most marked with the CCl_3F–$CHClF_2$ mixture and $CClF_2$. Older rats were more severely affected than younger rats.

The application of the five test substances to the tongue, soft palate, and auditory canal of rats produced no significant abnormalities.

The rabbit eye responded to the five materials by a hyperemia and lacrimation. There was an inflammatory reaction of the eyelid noted in rabbits exposed to CCl_3F and $CClF_2$-$CClF_2$. There were no other histologic alterations.

The healing of wounds and burns experimentally made on the skin of rats was retarded compared to controls. There appeared to be little if any differences in this retardation among these five compounds or mixtures evaluated.

CONCLUSIONS

The fluoroalkanes as a class possess a low order of toxicity by several routes of bodily contact. There is no evidence that fluoroalkanes used as propellants for aerosol products contribute to their toxicity. It appears that the toxicity is inversely related to the number of fluorine atoms in the molecule. The high degree of stability of the C–F bond would seem to account in large measure for low toxicity. Dehalogenation of fluorocarbons in biological systems is not prominent, probably because of the stability conferred upon the molecule by fluorine. However, in spite of a low rate of metabolism, fluorocarbons in sufficient amounts are biologically active and can cause death in high concentrations. Thus, it is important to conduct toxicological studies appropriately related to the hazards in use. It is to be emphasized that metabolic studies and experiments yielding information on mechanism of action are highly desirable for the safety evaluation of fluoroalkanes. Chronic studies should be based on the projected use of the compound, i.e., the experiment should fit the need.

REFERENCES

1. H. Langaard and W. O. Smith, "Self-induced Water Intoxication Without Predisposing Illness," *New Eng. J. Med.* **266**, 378 (1962).
2. P. LeBeau and A. Damiens, "Chimie Minerale sur le Tetrafluorure de Carbone," *C. R. Acad. Sci. (Paris)* **182**, 1340 (1926).
3. T. Midgley and A. L. Henne, "Organic Fluorides as Refrigerants," *Industr. Eng. Chem.* **22**, 542 (1930).
4. T. F. Slater, "Necrogenic Action of Carbon Tetrachloride in the Rat. A speculative Mechanism Based on Activation," *Nature* **209**, 36 (1966).
5. N. L. Gregory, "Carbon Tetrachloride Toxicity and Electron Capture," *Nature* **212**, 4460 (1966).
6. T. C. Butler, "Reduction of Carbon Tetrachloride *in vivo* and Reduction of Carbon Tetrachloride and Chloroform *in vitro* by Tissues and Tissue Constituents," *J. Pharmacol. Exptl. Therap.* **134**, 311 (1961).
7. C. I. Bliss, "The Calculation of Dose-Mortality Curves," *Q. J. Pharm. Pharmacol.* **11**, 192 (1938).
8. J. T. Litchfield and F A. Wilcoxon, "Simplified Method of Evaluating Dose-Effect Experiments," *J. Pharmacol. Exptl. Therap.* **5**, 95 (1949).
9. W. B. Deichmann and T. J. LeBlanc, "Determination of the Approximate Lethal Dose with about Six Animals," *J. Industr. Hyg. Tox.* **25**, 415 (1943).
10. H. F. Smyth, Jr., "The Toxological Basis of Threshold Limit Values. I. Experience with Threshold Limit Values Based on Animal Data," *Amer. Ind. Hyg. Assoc. J.* **20**, 341 (1959).
11. J. H. Draize, G. Woodward and H. O. "Methods for the Study of Irritation and Toxicity of Substances Applied Topically to the Skin and Mucous Membranes," *J. Pharmacol. Exp. Therap,* **82**, 377 (1944).
12. C. P. Carpenter and H. J. Smyth, Jr., "Chemical Burns of the Rabbit Cornea," *Amer. J. Opthalmol.* **29**, 1363 (1946).
13. A. J. Lehman, D. W. Fassett, H. W. Gerarde, H. E. Stokinger and J. A. Zapp, Principles and Procedures for Evaluating the Toxicity of Household Substances," Publication 1138, National Academy of Sciences, National Research Council, (1964).
14. M. A. Stevens, "Use of the Albino Guinea-Pig to Detect the Skin-Sensitizing Ability of Chemicals," *Brit. J. Ind. Med.* **24**, 189 (1967).
15. W. A. Cook, "Maximum Allowable Concentrations of Industrial Atmospheric Contaminants," *Industr. Med.* **14**, 936 (1945).
16. M. Bergmann, I. J. Flance and H. T. Blumenthal, "Thesaurosis Following Inhalation of Hair Spray," *New Eng. J. Med.* 258, 471 (1958).
17. M. Bergmann et al., "Thesaurosis Due to Inhalation of Hair Spray," *New Eng. J. Med.* **266**, 750 (1962) .
18. M. J. Brunner et al., "Pulmonary Disease and Hairspray Polymers: Disputed Relationship," *JAMA* **184**, 851 (1963).
19. G. W. H. Schepers, "Thesaurosis Versus Sarcoidosis," *JAMA* **181**, 635, (1962).

20. A. I. G. McLaughlin and P. L. Bidstrup, "Effects of Hair Lacquer Sprays on Lungs," *Food Cosmet. Toxic.* **1**, 171 (1963).
21. J. Calandra and J. A. Kay, "Effects of Aerosol Hair Sprays on Experimental Animals." Proc. SCI SEC. **30**, 41 (1958).
22. J. H. Draize et al., "Inhalation Toxicity Studies of Six Types of Aerosol Hair Sprays," *Proc.* SCI. SECT. *Toilet Goods Assoc.* **31**, 28 (1959).
23. J. Scholz, "Neue Toxikologische Untersuchungen einiger als Treibgas verwendter Frigentypen," *Fortschr. Biol. Aerosolforsch* **4**, 420 (1957–1961).
24. R. P. Giovacchini et al., "Pulmonary Disease and Hairspray Polymers," *JAMA* **193**, 118 (1965).
25. J. W. Clayton, Handbuch der experimentellen Pharmakologie, Vol. (O. Eichler, A. Farah, H. Herken, and A. D. Welch, Eds.), Springer, Berlin, 1966.
26. R. R. Sayers, W. P. Yant, J. Chornyak, and H. W. Shoaf, Toxicity of Dichloro-Difluoromethane: A New Refrigerant. U. S. Bureau of Mines Report No. 3013, (May 1930).
27. W. P. Yant, H. H. Schrenk, and F. A. Patty, Toxicity of Dichlorotetrafluoro-ethane. U. S. Bureau of Mines Report of Investigations. R. I. 3185, 1–10, (1932).
28. J. W. Clayton, M. A. Delaplane, and D. B. Hood, "Toxicity Studies with Octafluorocyclobutane," *Amer. Industr. Hyg. Assoc. J.* **21**, 382 (1960).
29. A. Quevauviller, "Hygiene and Safety of Propellants in Medicated Aerosols," *Prod. Probl. Pharm.* **20**, 14 (1965).
30. B. L. Odou, Personal Communication (1963).
31. R. C. Downing and D. Madinabeitia, "The Toxicity of Fluorinated Hydrocarbon Aerosol Propellants," *Aerosol Age* **5**, 25, (1960).
32. A. Quevauviller, M. Schrenzel et Vu Ngoc Huyen, "Local Tolerance of Skin, Mucous Membranes, Sores, and Burns of Animals to Chlorofluorinated Hydro-carbons," *Therapie* **19**, 247 (1964).

27

THE CONSTRUCTION AND USE OF TRIANGULAR COORDINATE CHARTS

Triangular coordinate charts are extremely useful for illustrating graphically how the properties of a three-component mixture such as miscibility, flammability, density, and pressure vary with a change in the composition of the mixture. It is important for those associated with the aerosol industry to have an understanding of triangular coordinate charts because many of the aerosol products today have been formulated with three component propellant blends, such as Propellant A (45% Propellant 12, 45% Propellant 11, 10% isobutane) or three-component propellant–solvent combinations such as propellant–ethyl alcohol–water mixtures. For those in the technical areas of aerosols, it is almost a necessity to know how to interpret triangular coordinate charts because these charts have been used extensively throughout the technical aerosol literature to show the properties of three-component mixtures.

The discussion in the following sections covers first, the construction, step by step, of a triangular coordinate chart and second, the use of the chart in showing how the properties of a three-component system vary with a change in the composition of the system.

CONSTRUCTION OF A TRIANGULAR COORDINATE CHART

A triangular coordinate chart is constructed as follows:
1. Draw an equilateral triangle (all sides of equal length) with its base as a horizontal line (Figure 27–1).

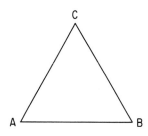

Figure 27–1

2. Specifically designate each corner of the triangle using letters *A, B, and C* (Figure 27–2).

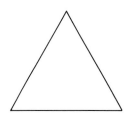

Figure 27–2

3. Draw a line *Aa* passing through Point *A*, which intersects line *CB* at right angles. The intersection of line *Aa* with line *CB* is labelled Point a (Figure 27–3).

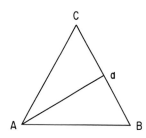

Figure 27–3

4. Similarly draw lines Bb and Cc (Figure 27–4).

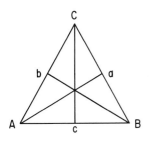

Figure 27–4

5. In discussing three-component mixtures, it is necessary to indicate the compositions of the mixtures by showing the ratios of the components in the mixture. This can be accomplished by various means, such as the weight or volume percent of each of the individual components, or by other means, such as parts by weight or parts by volume. In most applications, weight percent of each of the three components is usually selected. The combined weight percent of the three components must total 100%.

6. Assume that we have a mixture containing three components, Compound #1, Compound #2, and Compound #3. Let the letters A, B, and C located at each corner of the triangle represent each of the three components, respectively. For example, letter A = Compound #1, letter B = Compound #2, and letter C = Compound #3.

7. Let Point A represent 100% of Compound #1 and the base line CB, which is opposite Point A, represent 0% of Compound #1. Point a, which is on the base line CB, therefore, also represents a composition with 0% of Compound #1 (Figure 27–5).

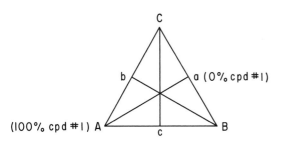

Figure 27–5

8. Similarly, Points *B* and *C* represent 100% of Compound #2 and Compound #3, respectively, and Points *b* and *c* represent 0% of Compounds #2 and #3, respectively (Figure 27–6).

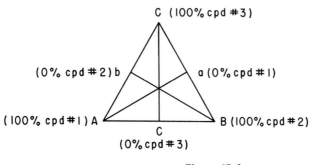

Figure 27-6

9. At this point we have constructed a diagram in which each corner (an apex) of the triangle represents 100% of a component and each base line opposite a corner or apex represents 0% of the component. Therefore, the portions of the triangle in between the apex and the base line must represent compositions which contain percentages of the component between 100% and 0%. We will now proceed to expand the triangular coordinate chart to show these compositions.

Locate points on line *Aa* which will divide line *Aa* into a number of equal segments to represent equal steps from 100%–0%. Assume 20% steps are required. These locations are represented by points $g = 80\%$, $h = 60\%$, i = 40%, and $j = 20\%$ (Figure 27–7).

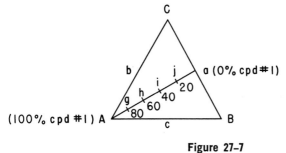

Figure 27-7

10. Draw lines parallel to line *CB* through Points *g, h, i, and j* (Figure 27–8).

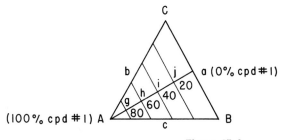

Figure 27-8

Any point on the line passing through *g* which is parallel to the base line *CB,* represents some composition which contains 80% of compound #1. Similarly, any point on the line passing through points *h,* *i,* and *j* parallel to line *CB* represents mixtures containing 60%, 40%, and 20%, respectively, of Compound #1. Remember that apex *A* represents 100% of Compound #1 and line *CB* represents mixtures containing 0% of Compound #1.

11. Similarly locate points on line *Bb* which will divide line *Bb* into a number of equal segments to represent equal steps from 100%–0% at 20% intervals. These points will be represented by points *m* = 80%, *n* = 60%, *o* = 40%, and *p* = 20% (Figure 27–9).

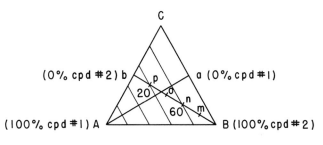

Figure 27-9

12. Draw lines parallel to line *AC* through Point *m, n, o,* and *p* (Figure 27–10).

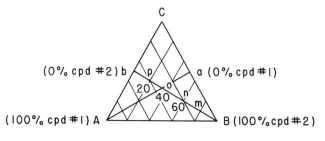

Figure 27–10

Any point on the line passing through Point *m* parallel to line *AC* represents a mixture containing 80% of Compound #2. Similarly, any point on lines passing through Points *n, o,* and *p,* parallel to line *AC* represents mixtures containing 60%, 40%, and 20%, respectively of Compound #2.

13. In like manner locate points on line *Cc* which will divide line *Cc* into a number of equal segments to represent equal steps from 100%–0% at 20% intervals as before. These points will be represented by Points $t = 80\%$, $u = 60\%$, $v = 40\%$, and $w = 20\%$ (Figure 27–11).

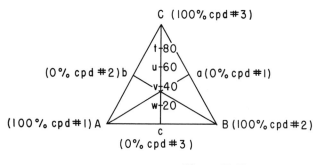

Figure 27–11

14. Draw lines parallel to line *AB* through Points *t, u, v,* and *w* (Figure 27–12).

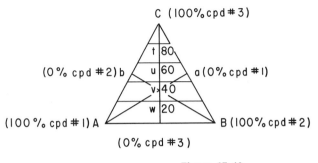

Figure 27-12

Any point on the line passing through Point t parallel to line AB represents compositions containing 80% of Compound #3. Similarly any points on lines passing through Points u, v, and w parallel to line AB represent mixtures containing 60%, 40%, and 20%, respectively, of Compound #3.

15. We are now ready to assemble a complete triangular coordinate chart by combining Figures 27–8, 27–10, and 27–12. (Figure 27–13).

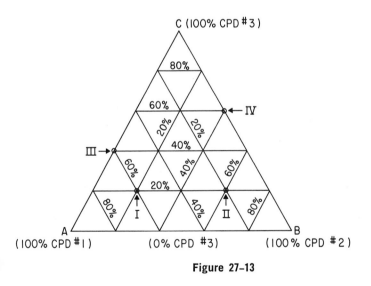

Figure 27-13

The triangular coordinate chart illustrated in Figure 27–13 shows 20% increment lines for Compound #1, Compound #2, and Compound #3. The actual commercial charts have increment lines at 1% intervals.

In Figure 27–13, line *AB* represents mixtures with Compound #1 and Compound #2 only. Line *BC* represents mixtures with Compound #2 and Compound #3. Line *AC* represents mixtures of Compound #1 and Compound #3. All points in the interior of the chart represent mixtures of all three components. In all cases the proportions of the components must add up to 100% at every point. Sample mixtures, Points I, II, III, and IV, respectively, have been circled on the chart. The compositions of mixtures corresponding to these points are:

	Weight Percent of Components in Mixtures		
Point	*Component #1*	*Component #2*	*Component #3*
I	60	20	20
II	20	60	20
III	60	0	40
IV	0	40	60

APPLICATIONS OF THE TRIANGULAR COORDINATE CHART

The major application for the triangular coordinate charts is to show how the properties of a three-component mixture vary with the composition of the mixture. By carrying out a few experiments or a few calculations and plotting the resulting data on triangular coordinate charts, it is possible to determine the properties of the three-component mixture with any composition. In many cases, it is possible to illustrate the effect of changes in composition upon a number of different properties by plotting the data on the same triangular coordinate chart. This may indicate relationships between the properties that otherwise might remain obscure.

Examples of properties of mixtures that have been illustrated on the triangular coordinate charts are: (1) flammability, (2) density, (3) vapor pressure, (4) solubility, (5) spray characteristics, and (6) cost.

In the following sections, several examples will be presented to illustrate how experimental data may be transferred to a triangular coordinate chart.

Solubility Characteristics of "Freon" 12–Propylene Glycol–Ethyl Alcohol Mixtures

The solubility characteristics of "Freon" 12–propylene glycol–ethyl alcohol mixtures will be used to provide the first example of how experimental data may be plotted on triangular coordinate charts.

Combinations of the "Freon" propellants with propylene glycol and ethyl alcohol have been used for the formulation of aerosol pharmaceutical products. Although a propellant, such as "Freon" 12, is miscible with ethyl alcohol, and propylene glycol is miscible with ethyl alcohol, "Freon" 12 is not miscible with propylene glycol. As a result, certain combinations of the "Freon" propellants with propylene glycol and ethyl alcohol are not mutually soluble. When these combinations are mixed, two liquid phases will be formed. These combinations usually are of little value in formulating aerosols. Other combinations of the three components are mutually soluble and will form a single homogeneous liquid phase. Those combinations that are not miscible and will form two liquid phases and those which are mutually soluble and form a single homogeneous phase can be determined very easily on triangular coordinate charts by carrying out a few simple experiments.

The experimental data were obtained as follows: first, the solubility of "Freon" 12 in propylene glycol was obtained by placing a known weight of propylene glycol in a glass container and adding "Freon" 12 until the mixture became cloudy and formed two phases. Knowing the initial weight of the glycol and the weight of the "Freon" 12 that was added, it was possible to calculate the composition at the point where immiscibility occurred. In the particular instance cited, it was found the "Freon" 12 was soluble to the extent of 11% in propylene glycol.

Additional data were obtained by preparing mixtures of propylene glycol and ethyl alcohol of known weight and adding "Freon" 12 until two liquid phases are formed. The weights of propylene glycol and ethyl alcohol in the initial mixture were known. By determining the weight of "Freon" 12 that was added, it was easy to calculate the composition of the "Freon" 12–propylene glycol–ethyl alcohol mixture at the point where the combination became immiscible and formed two liquid phases. By using different ratios of propylene glycol to ethyl alcohol in the initial mixture, a number of different compositions were obtained where the mixture of the three components were immiscible.

Finally, the solubility of propylene glycol in "Freon" 12 was determined by titration. This was found to be essentially zero. It should be emphasized that the solubility of propylene glycol in "Freon" 12 is entirely different from the solubility of "Freon" 12 in propylene glycol.

The data in Table 27–1 were obtained experimentally using the preceding technique.

These data will now be plotted on a triangular coordinate chart and a curve will be drawn through the points. The resulting curve divides the charts into two parts. One part will contain all compositions that are not miscible and form two liquid phases; the other side will indicate compositions that are miscible.

1. Let the corners of the triangular coordinate chart represent 100% of each of the three components (Figure 27–14).

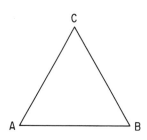

Figure 27-14

TABLE 27-1 SOLUBILITY PROPERTIES OF "FREON" 12–PROPYLENE
GLYCOL–ETHYL ALCOHOL MIXTURES

	Compositions of Mixtures at the Point Where Immiscibility Occurred (wt %)		
Mixture Number	"Freon" 12	Propylene Glycol	Ethyl Alcohol
1	11	89	0
2	24	60	16
3	35	45	20
4	55	25	20
5	70	15	15
6	100	0	0

2. Let Point A = 100% "Freon" 12; B = 100% Ethyl Alcohol; and
C = 100% Propylene Glycol (Figure 27–15).

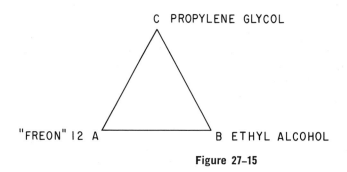

Figure 27-15

3. In the previous discussion of triangular coordinate charts, it was pointed
out that line *AB* represents compositions with 0% (Compound *C*)

propylene glycol, line *BC* represents compositions with 0% (Compound *A*) "Freon" 12, and line *AC* represents compositions with 0% (Compound *B*) ethyl alcohol. Any line that represents compositions with 0% of one component must represent compositions containing the other two components (with the exception of the corners, which represent 100% of one component) (Figure 27–16).

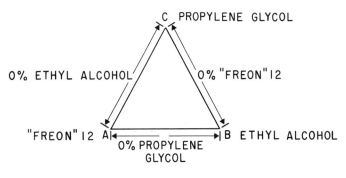

Figure 27–16

4. According to the definition of the triangular coordinate charts, any point in any line parallel to line *BC* represents a constant percentage of "Freon" 12. The particular percentage involved ranges from 100% "Freon" 12 at Point *A* (apex *A*) to 0% "Freon" 12 at the base line *BC* opposite the apex *A* (Figure 27–17).

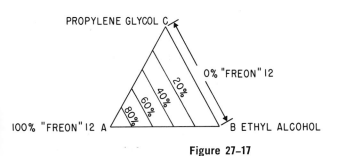

Figure 27–17

5. Similarly, the lines parallel to line *AB* represent compositions with constant percentages of propylene glycol (Figure 27–18).

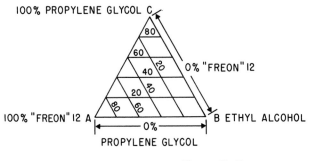

Figure 27-18

6. Mixture No. 1 contained 0% ethyl alcohol. Therefore, the point representing the composition of 11% "Freon" 12 and 89% propylene glycol will be located on line *AC* which represents compositions containing 0% ethyl alcohol. To locate this point, find the line parallel to line *BC* which represents compositions containing 11% of "Freon" 12 (Figure 27-19).

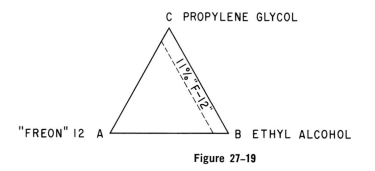

Figure 27-19

Follow this line until it intersects line *AC,* which contains compositions with 0% ethyl alcohol. Note now that at the particular place where the 11% "Freon" 12 line intersects line *AC,* the 89% line for propylene glycol also intersects line *AC* (Figure 27-20). This is because any point on the triangular coordinate chart has to total 100% for the components involved.

C PROPYLENE GLYCOL

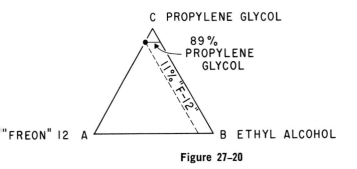

89 %
PROPYLENE
GLYCOL

"FREON" 12 A

B ETHYL ALCOHOL

Figure 27-20

The intersection of the 11% "Freon" 12 line and the 89% propylene glycol line on line *AC*, therefore, locates the composition of Mixture No. 1. This point has been encircled.

7. Mixtures 2, 3, 4, 5, and 6 contain combinations of all three components. Referring to the previous definition of a triangular coordinate chart, lines parallel to line *AC* represent mixtures containing constant percentages of ethyl alcohol. Combining this with Figures 27–17 and 27–18 we now have our complete triangular chart again (Figure 27–21).

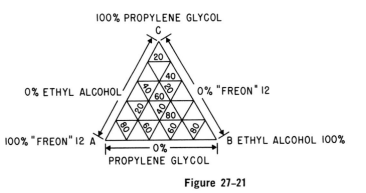

100% PROPYLENE GLYCOL

C

0% ETHYL ALCOHOL

0% "FREON" 12

100% "FREON" 12 A

B ETHYL ALCOHOL 100%

0%
PROPYLENE GLYCOL

Figure 27-21

8. In locating points corresponding to mixtures of three components, it is only necessary to locate the point using the percentages of two of the components. The percentage of the third component will be found at the same place since the total of the three components has to equal 100%. If it doesn't, an error has been made.

Mixture No. 2 has a composition of 24% "Freon" 12, 60% propylene glycol, and 16% ethyl alcohol. The point corresponding to this composition may be found by starting with any one of the three compo-nents. However, as an example, find the line parallel to line *BC* which represents compositions containing 24% "Freon" 12 (Figure 27–22).

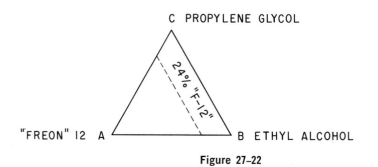

Figure 27–22

Next, locate the line parallel to line *AB* which represents compositions containing 60% of propylene glycol, and follow this line until it inter-sects the 24% "Freon" 12 line (Figure 27–23).

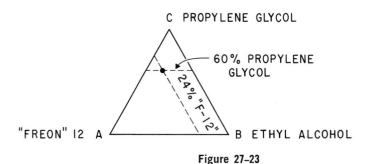

Figure 27–23

The circled intersection of the 24% "Freon" 12 line and the 60% propylene glycol line automatically falls on the 16% ethyl alcohol line since the composition of any point has to equal 100%.

The circled intersection represents the composition with 24% "Freon" 12, 60% propylene glycol, and 16% ethyl alcohol. The point could also have been located by starting either with the 60% propylene glycol line or the 16% ethyl alcohol line and following either of these lines until it intersects one of the other lines.

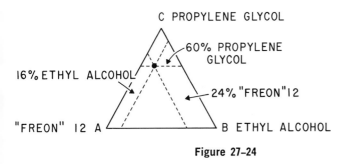

Figure 27-24

The location on the triangular coordinate chart corresponding to the compositions of Mixtures 3, 4, and 5 are located in a similar way. The composition of Mixture 6 is located at apex *A* since the mixture is essentially 100% "Freon" 12. The chart then appears as follows:

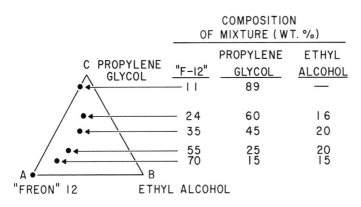

	COMPOSITION OF MIXTURE (WT. %)		
	"F-12"	PROPYLENE GLYCOL	ETHYL ALCOHOL
	11	89	—
	24	60	16
	35	45	20
	55	25	20
	70	15	15

Figure 27-25

A curve through all the points is now drawn (Figure 27-26).

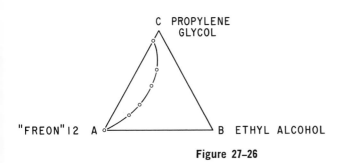

Figure 27-26

Since the points corresponding to the immiscible mixtures were obtained by adding "Freon" 12 to homogeneous mixtures of propylene glycol and ethyl alcohol until two liquid phases were formed, this means that the areas to the left of the curve indicate compositions that are immiscible and the areas to the right of the curve indicate compositions that are miscible (Figure 27–27).

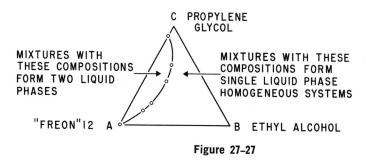

Figure 27-27

As previously mentioned, the data on the solubility of the "Freon" 12–propylene glycol–ethyl alcohol mixtures was obtained by adding "Freon" 12 to solutions of propylene glycol and ethyl alcohol until the mixtures became immiscible and formed two phases. Mixtures of propylene glycol and ethyl alcohol with ratios of 75/25, 50/50, and 25/75 could be used, for example. During the addition of the "Freon" 12, the ratio of propylene glycol to ethyl alcohol remains unchanged but the concentration of the propylene glycol and ethyl alcohol in the entire mixture decreases as more and more "Freon" 12 is added. The actual change in composition of the mixture as the "Freon" 12 is added occurs along a line drawn from the initial glycol–alcohol composition on the base line through the "Freon" 12 apex (Figure 27–28).

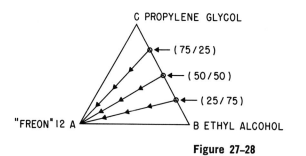

Figure 27-28

The change in composition with increasing additions of "Freon" 12 is indicated by the arrows on the lines above. The fact that the lines have to be drawn through point *A* or the "Freon" 12 apex can be understood if it is realized that as more and more "Freon" 12 is added, the mixture approaches a composition having essentially 100% "Freon" 12.

Vapor Pressures of "Freon" 12–Propylene Glycol–Ethyl Alcohol Compositions

The solubility chart illustrated in Figure 27–27 can be of considerable help to aerosol formulators since it shows which compositions are of no value because of immiscibility. In addition, vapor pressure measurements can be carried out on a variety of "Freon" 12–propylene glycol–ethyl alcohol mixtures and all the different compositions that give the same pressures; i.e., 60 psig, 50 psig, 40 psig, and 25 psig can be determined. Different compositions giving the same pressures can then be plotted on Figure 27–27 as follows (Figure 27–29).

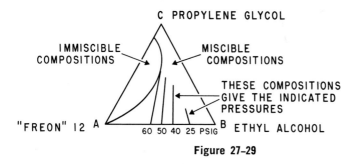

Figure 27-29

Now the data provided by the triangular coordinate chart is of greater significance to the aerosol formulator. Not only does he know which compositions to avoid because of immiscibility, but now he is able to formulate and achieve any desired pressure and thus choose a composition that can be packaged in containers having different pressure specifications.

Spray Characteristics of "Freon" 12–Propylene Glycol–Ethyl Alcohol Mixtures

Additional information can be obtained and plotted on Figure 27–29. For example, a series of formulations could be prepared with different compositions, and those compositions that gave a stream instead of a spray could be noted. The boundary that divides the stream from coarse spray-yielding compositions could be designated as the limiting spray boundary and this boundary could be plotted on Figure 27–29 as follows (Figure 27–30):

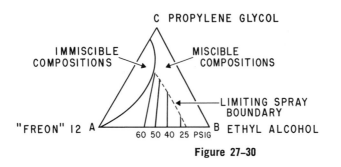

Figure 27–30

From the data on this chart,* the aerosol formulator can select a homogeneous composition which will provide him with the pressure and the spray characteristics he desires. The chart, therefore, has eliminated the necessity for carrying out a multitude of trial and error experiments.

Flammability Properties of Propellant Blends

Another example of the use of the triangular coordinate charts in presenting data in a compact form is illustrated by flammability data obtained with various "Freon" propellant–vinyl chloride blends. Because of the lower cost of vinyl chloride, these blends have aroused considerable interest. However, vinyl chloride is flammable, and it was necessary to determine what particular blends of the "Freon" propellants with vinyl chloride did not form explosive mixtures with air. After extensive tests, it was found that the following mixtures of the "Freon" propellant/vinyl chloride blends did not explode in air. However, if the concentration of vinyl chloride was increased, the mixtures became flammable. Therefore, the blends given below indicate what is termed the flammability limit of vinyl chloride in "Freon" 12/"Freon" 11 mixtures.

"Freon" Propellant–Vinyl Chloride Mixture	Nonexplosive Combination (wt % Ratio)
"Freon" 12/Vinyl Chloride	75/25
"Freon" 12–"Freon" 11 (50/50)– Vinyl Chloride	77/23
"Freon" 11–Vinyl Chloride	79/21

These data can be plotted on a triangular coordinate chart as follows (Figure 27–31).

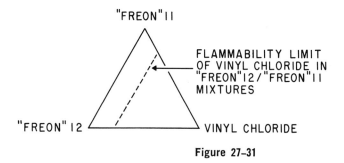

Figure 27-31

Compositions to the left of the flammability limit curve do not form explosive mixtures with air when completely vaporized, and those to the right of the curve do form explosive mixtures with air.

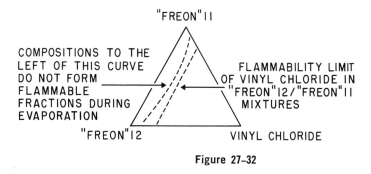

Figure 27-32

It was necessary to determine what compositions of "Freon" propellant–vinyl chloride gave flammable fractions during evaporation and what compositions gave nonflammable fractions. Again, the data could be plotted on Figure 27–31 as an addition to the flammability limits already determined.

Thus, from the data obtained on a relatively small number of "Freon" propellant–vinyl chloride blends, it was possible to determine the flammability characteristics of all the different "Freon" propellant–vinyl chloride compositions by plotting the data on triangular coordinate charts.*

---

* Reprinted from "Freon" Aerosol Report, A–70, "The Use and Function of Triangular Coordinate Charts," with permission of the copyright owner, E. I. du Pont de Nemours and Company.

INDEX